T0326615

The Embedding Method
for Electronic Structure

The Embedding Method for Electronic Structure

John E Inglesfield

Cardiff University, Wales, UK

IOP Publishing, Bristol, UK

ISBN   978-0-7503-1042-0 (ebook)
ISBN   978-0-7503-1043-7 (print)
ISBN   978-0-7503-1117-5 (mobi)

DOI   10.1088/978-0-7503-1042-0

Version: 20151201

IOP Expanding Physics
ISSN 2053-2563 (online)
ISSN 2054-7315 (print)

British Library Cataloguing-in-Publication Data: A catalogue record for this book is available from the British Library.

Published by IOP Publishing, wholly owned by The Institute of Physics, London

IOP Publishing, Temple Circus, Temple Way, Bristol, BS1 6HG, UK

US Office: IOP Publishing, Inc., 190 North Independence Mall West, Suite 601, Philadelphia, PA 19106, USA

*This book is dedicated to the memory of my mother, Winifred Inglesfield.
She wrote her book when she was 86!*

Contents

Preface

This book is about the embedding method, a technique I have been involved with for most of my research life as a theoretical and computational condensed matter physicist. As the name implies, this method enables us to 'embed' a calculation of electronic structure for a particular region of space into a neighbouring (or surrounding) region. The need to do this occurs very frequently in electronic structure calculations—for example, at surfaces where we would like to solve the Schrödinger equation for the atoms at the surface—but we must somehow take account of the fact that the surface atoms are bonded to the rest of the material: a semi-infinite substrate. In this situation the embedding method allows us to work entirely in the region of space we are interested in (the surface, say), the substrate being replaced by the *embedding potential*, which is localized at the boundary with the substrate. It was, in fact, for surface problems that I developed the embedding method, and it has indeed proved useful for accurate calculations of surface electronic structure. But as I shall show in this book, it has many other applications, and it has some remarkable links with other branches of theoretical physics.

This book is mainly intended for researchers working in the fields of electronic structure and photonics who will, I hope, find something useful or at least interesting here. I expect that some readers are using an embedding potential without knowing it; the self-energy in molecular transport studies is an embedding potential under another name. I also hope that the book will show that there is plenty of life in embedding, and that there are many new problems in condensed matter physics to which embedding may contribute. Embedding itself needs further development, especially in ways of incorporating many-body effects. This is a further motivation for writing the book—to try to bring together different aspects of embedding, so that further developments can be made. As much for my sake as for the readers'!

Some of the ideas behind embedding, as well as the Green function techniques I used in its derivation, were borrowed from the 'matching Green function method'. This was invented by Federico García Moliner and Juan Rubio as a way of constructing the Green function for a combined system from Green functions for the components; a philosophy akin to embedding. It was my former PhD supervisor, Volker Heine, who brought the Spanish work to my attention and encouraged me to work on Green function matching. This helped greatly with the development of the embedding method a few years later. Green functions will play a large part in this book, at the mathematical level of chapters 2 and 3 of *Classical Electrodynamics* by J D Jackson.

Over the years, I have benefitted from collaborating with many colleagues in Daresbury Laboratory, Nijmegen, Cardiff and elsewhere. John Pendry provided the congenial climate at Daresbury in which I developed my embedding method, and during this period visits to Tom Grimley at Liverpool, the pioneer of embedding, helped me to formulate my ideas. Greg Benesh (Baylor University, Waco) and I have had a long-standing collaboration, and together we developed the first surface embedding program. Other friends with whom I have worked closely are Hiroshi

Ishida (Nihon University, Tokyo), who has developed the best surface programs, and Simon Crampin (University of Bath), who has made many contributions to embedding and its applications. I should like to thank Jos Thijssen (Delft University of Technology) and Mario Trioni (Universitá di Milano-Bicocca) for their help with parts of this book, as well as many years of collaboration. It was always enjoyable to work with my research students, who helped me as much (if not more) than I helped them. Some of this book was written in the congenial atmosphere of the Donostia International Physics Center (San Sebastián), and I thank Pedro Echenique for his hospitality.

John E Inglesfield
September 2015

Author biography

John E Inglesfield

The English Lake District is where I come from, and where I moved back to after I retired a few years ago.

I was educated at St John's College, Cambridge University, where I read Natural Sciences. I stayed on in Cambridge for my PhD, supervised by Volker Heine in what was then the Solid State Theory group in the Cavendish. This led to a fellowship at St John's, after which I moved to the Theory Group at Daresbury Laboratory in Cheshire. It was during my time at Daresbury that I developed the embedding method, and of course this is what I am still working on thirty years later. Eventually I became head of the Theory and Computational Science Division at Daresbury, and from there I moved to the University of Nijmegen where I was professor of Electronic Structure of Materials. My next move was to a chair in physics at Cardiff, where I stayed till retirement. During my working life I have been on several editorial boards, and I continue as a deputy editor of *Journal of Physics: Condensed Matter*.

My research has always been in theoretical and computational physics, much of my work being on the electronic structure and properties of metal surfaces. My interest in surfaces was in fact the motivation for developing embedding, though my earlier work on surfaces used the matching Green function method. I enjoy not just developing the methodology of embedding in different situations, but also programming it up. In the course of writing this book I have come across many links between embedding and other techniques, and my next project will be to explore some of these.

As a Cumbrian, mountain walking is one of favourite activities, and I also love gardening. Now that I am retired I spend a lot of time at my holiday home in the south of France, in the Languedoc, where in fact much of this book was written (working in the mornings, swimming in the afternoons—a perfect summer routine!). This is an ideal base for walking and skiing in the Pyrenees, as well as being within striking distance of the Mediterranean. I try to improve my French, not only by meeting my fellow villagers, but also by reading French detective stories. My other interests include railways, botany, and atmospheric phenomena like rainbows, haloes and glories.

IOP Publishing

The Embedding Method for Electronic Structure

John E Inglesfield

Chapter 1

Introduction

The embedding method is a way of solving the Schrödinger equation for electrons in a region of space (region I in figure 1.1) joined on to a substrate (region II) [1]. Region I is in some sense the region of space we are really interested in—it may be the region where we make measurements of local properties of electrons such as their charge density or their energy density. At the surface of a metal, for example, region I may consist of the top layer or two of atoms, the region where tools like the scanning tunnelling microscope or photoemission probe the electronic structure [2]. However, the wave-functions in region I extend into the substrate, and from elementary quantum mechanics we have to match the wave-functions across the boundary S (figure 1.1) in both amplitude and derivative. Unfortunately, in anything more than simple textbook examples, like the square well, explicit wave-function matching becomes rather tedious, though this is the way some of the first realistic calculations of surface electronic structure were carried out [3]. This is where we can use embedding: in the embedding method we solve the Schrödinger equation *only* in region I, with what we call the *embedding potential* added to the Hamiltonian for this region. The embedding potential, which is in fact a generalised logarithmic

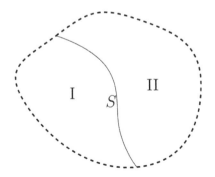

Figure 1.1. Region I is embedded on to region II over surface S.

derivative, ensures that the solution of the Schrödinger equation in region I matches the wave-function in region II in both amplitude and derivative. The embedding method allows the electronic structure at a surface to be calculated in a much simpler way than by wave-function matching, by adding an embedding potential to the Hamiltonian for the surface atoms. The method is flexible, and as well as surface electronic structure, it has been used to study interfaces, conductance through molecules, confined electrons—all this and much more will be described in this book.

1.1 A brief history of embedding

Methods of embedding in one form or another, even the terms *embedding* and *embedding potential*, have been around for many years. The idea of embedding the region of interest into the rest of the system has a long history, preceding my 1981 paper [1] by at least 30 years.

For a long time chemists have used cluster calculations to find the electronic structure of impurities in solids, or adsorbates on surfaces [4, 5], but there are well-known problems with this approach, such as the appearance of spurious electronic states (dangling bonds) on atoms on the surface of covalently bonded clusters. In many cases, the problems can be limited by the simple expedient of taking a bigger cluster, and with modern computers, depending on the computational approach—Hartree–Fock, density functional theory (DFT), correlated wave-functions—large clusters can be treated. But this is a relatively recent development, and it was the necessity of using small clusters, especially when computer power was limited, which originally led to the development of embedding methods.

In a cluster description of an impurity atom in an ionic solid [6], or an adsorbate on the surface of such a solid [5], it is often enough to embed the cluster Hamiltonian in the electrostatic potential due to the surrounding crystal. This is justified, when a localized basis set is used, because the ions have closed shells. This approach is used in the quantum mechanics/molecular mechanics method [7]: the cluster itself is treated quantum mechanically, whereas the surrounding lattice can be treated classically, with pair interactions between the atoms. This is an example of one embedding philosophy—to treat the embedded region at a greater level of sophistication and accuracy than the region into which it is embedded.

In a covalent or metallic system, embedding must describe the way that the outermost atoms of the cluster are bonded to the rest of the system, going beyond electrostatic embedding. Carter and her colleagues have recently developed a way of doing this based on DFT, in which the charge density is the fundamental variable [8, 9]. The atomic cluster is placed in an energy-independent, local embedding potential chosen so that the ground-state charge density, calculated in DFT with one-electron wave-functions, is the same as the density of the cluster + the atomic environment. The electrons in the cluster, with the embedding potential simulating the bonding to the rest of the system, can then be treated in a more sophisticated way, using configuration interaction methods to describe the electron–electron interaction, for example. At the same time, atoms within the cluster can be varied and the changes in bonding studied, always assuming that the embedding potential is a local property. The range of problems tackled by this method is impressive, including the excited states of adsorbed molecules [8] and the Kondo state of a Co adsorbate on noble metal surfaces [10].

What is missing in this, and similar approaches using a local, energy-independent embedding potential [11], is a proper treatment of the cluster wave-functions, so that they match the wave-functions in the substrate. This means that electronic excitations which extend beyond the cluster into the substrate cannot be properly described. There are two general approaches in which the wave-functions are properly embedded, classified by Grimley and Pisani [12], and Fisher [13], as 'perturbed-crystal' and 'perturbed-cluster' methods. In perturbed-crystal methods (I quote from Fisher) 'the solution for the perfect solid is modified to account for the presence of the defect', whereas in perturbed-cluster methods 'the solution for a small region around the defect is modified to account for the rest of the crystal' [13]. The first approach corresponds more or less to using the Green function of the perfect crystal with Dyson's equation to find the Green function for the defective crystal, with the change in potential due to the impurity or defect treated as a perturbation. In the second—embedding—approach, an embedding potential is added to the cluster Hamiltonian, to account for the rest of the system; in a localized orbital basis, the embedding potential is usually called a 'self-energy' [14]. Of course, the two approaches must give the same answer. If both are calculated to the same degree of approximation, and in a localized orbital basis it can be shown that they are identical [13, 15].

The impurity calculations of Dederichs, Zeller and co-workers [16] are excellent examples of the perturbed crystal approach. In these calculations, the Green function for a perfect crystalline solid is calculated within the Korringa–Kohn–Rostocker (KKR) scheme, and then the Green function for the solid containing the impurity is calculated using Dyson's equation, treating the change in potential due to the impurity, and in the neighbouring atoms, as the perturbation. As well as impurities in metals and semiconductors, these authors have also studied the interaction of adsorbate atoms on surfaces. An example of the impurity problem treated by the perturbed cluster approach—and one of my first applications of embedding—was a calculation of the electronic structure of interstitial H in Cu: a cluster of four Cu atoms surrounding the H impurity was embedded into the surrounding Cu [17].

One of the earliest examples of embedding is the Anderson impurity model [18], developed in 1961 to model transition metal impurities in free-electron metals, where an embedding potential (not called as such) appears in the expression for the impurity density of states. This model was subsequently used by Grimley [19, 20] and Newns [21] in the late 1960s in their studies of adsorbates on metal surfaces; in the context of chemisorption, the embedding potential was called the 'chemisorption function'. This was further developed by Grimley and Pisani in their 1974 paper [12] into a method for embedding an adsorbate cluster into the surface, within Hartree–Fock and using a localized basis set. This paper explicitly refers to 'embedding' the cluster, and the embedding potential is here referred to as the 'chemisorption matrix'. Subsequently, Pisani [22] and his co-workers extended this quantum chemical type of approach into

EMBED, a part of the CRYSTAL suite of computer programs [23], which has been used to calculate the binding energies of impurities and adsorbates. Unfortunately, the further development of EMBED has been largely discontinued in recent years.

In 1971, Caroli *et al* [24] introduced the concept of self-energy into the theory of electron transport through a metal–insulator–metal tunnelling junction, as a way of including the coupling of the insulator to the electrodes (embedding the insulator into the electrodes, in other words). The electrode self-energy appears in the expression, published in 1992 by Meir and Wingreen [25], for the transmission of electrons in ballistic transport, through a molecule for example. As molecular transport remains one of the most active areas of nanoscience, both experimentally and theoretically, this is probably the most productive use of the self-energy. An important aspect of this work is including electron–phonon scattering as well as the electron–electron interaction in the molecule—in other words, how to embed an interacting system into a 'non-interacting' environment, in this case the metal leads [14].

The embedding method which I developed in 1981 [1], the focus of this book, differs from this tight-binding self-energy approach and from the work of Grimley and Pisani in that it divides real space (configuration space), rather than the space of orbitals, into regions I and II (figure 1.1). Another novel feature was the development of the variational principle for the one-electron energy in terms of the trial wave-function only in region I. This was based on real-space Green function techniques, which did not come out of the blue—I had been using them for many years in my work on the 'matching Green function method' [26]. This method, invented by García-Moliner and Rubio in 1969 [27], shows how the real-space Green function for a composite system can be built up from the Green functions for the component parts, by matching over the boundary surfaces. This contains the essence of embedding.

The embedding potential provides a boundary condition (a generalised logarithmic derivative) on the wave-function in region I. This idea is related to R-matrix theory introduced by Wigner and Eisenbud in 1947 [28] to describe nuclear reactions, and subsequently applied with great success to electron scattering by atoms and molecules [29]. The R-matrix, defined on a sphere surrounding the atom or molecule, determines the scattering of electrons outside, by providing a generalized logarithmic derivative, similar to the embedding potential. The formal links were first pointed out by Zou [30], who showed how R-matrix theory provides a simple expression for the embedding potential to replace a finite region II (such as an atom or molecule).

In the more mathematical literature, the embedding potential (derived, it seems, quite independently) appears as the *Dirichlet-to-Neumann map* [31, 32]. The embedding potential gives the normal derivative on S (the Neumann boundary condition) of the solution of the Schrödinger equation in region II in terms of the amplitude (the Dirichlet boundary condition), an operation which mathematicians call Dirichlet-to-Neumann mapping. The Dirichlet-to-Neumann map is precisely 'my' embedding potential, to within a factor of -2. The Dirichlet-to-Neumann work has been used in the context of the Helmholtz equation [31], an equation which occurs in acoustic problems, elasticity theory as well as electromagnetism [33]. This equation is often solved using finite element methods, and the Dirichlet-to-Neumann map can be used in exactly the same way that the embedding potential is used to eliminate the substrate in surface problems: the Dirichlet-to-Neumann map defined over a boundary provides the correct boundary conditions for the solution in region I (as we call it) to match the solution in the outer domain [34].

Despite other approaches to embedding, when I talk about the 'embedding method' in this book (and elsewhere), it is always the method derived in my 1981 paper which I am referring to. There have been many improvements to this method, particularly in applications to surfaces and interfaces, where the work of Ishida has led to very accurate and stable computer programs [35]. Unfortunately, the method has *not* proved the answer to all the applications I originally envisaged, and as a glance at any issue of Physical Review B will confirm, most surface calculations are still carried out using slab geometry. This approach works well, particularly for surface energy calculations and structural optimization, as these are local properties. Nevertheless, there are many examples of surface electronic structure where it is important to include the coupling of the surface atomic layers to a semi-infinite substrate, to distinguish between surface states and surface resonances, for example. This is where embedding comes into its own, and as a recent example, a relativistic version of embedding has proved very powerful in calculations of spin–orbit induced Rashba surface states [36, 37]. (It's worth noting as well that FLEUR[1], developed at Forschungszentrum Jülich as one of the most accurate electronic structure program packages, incorporates the embedding method as an option for surfaces and interfaces.)

I end this brief history on an optimistic note: the embedding method in one form or another is alive and well, as the rest of the book will show.

1.2 Overview

Our embedding method is based on a variational principle, in which a trial wavefunction in region I is extended into region II by an exact solution of the Schrödinger equation at a trial energy [1]. This is then eliminated by Green function techniques, to be replaced by the embedding potential on the boundary S between regions I and II (figure 1.1). This is the basis for all the work described in this book, and we shall discuss the variational underpinning of embedding in chapter 2. Throughout the book, and starting in chapter 2, we use simple model calculations to illustrate the embedding method—here to find bound states, the local density of states in the energy continuum and resonances in open systems.

Probably the most successful application of embedding has been in calculations of surface electronic structure, particularly the properties of surface states and surface resonances. These calculations will be the subject of chapters 3 and 4. In chapter 3 we shall describe different ways of calculating the embedding potential for surface calculations, as well as the very accurate linearized augmented plane-wave method (LAPW) method for solving the Schrödinger equation in the surface region [38]. This will bring out some general points about basis functions in the embedding method.

---

[1] http://www.flapw.de

Self-consistency and the electron–electron interaction are of course fundamental to electronic structure calculations: embedding is a one-electron scheme, but the electron–electron interaction can be taken into account by the self-consistently determined Hartree and exchange-correlation potentials [39]. The important point is that screening is very effective (at least in metallic systems), so we can go to self-consistency in region I alone and use the same embedding potential throughout. If this were not the case, embedding calculations would become much more difficult. In chapter 4 we shall present various applications of surface embedding [40]. There is a lot of interest at the moment in surface states for which relativistic effects are important: these can be calculated very accurately with a relativistic version of embedding, involving the Dirac equation rather than the Schrödinger equation [41], and this is the subject of chapter 8.

A surprising aspect of embedding is that it provides a remarkably efficient way of solving the Schrödinger equation for electrons confined by an essentially infinite potential barrier of arbitrary shape [42]. This technique, which has been applied to finding the energy distribution of electrons confined by nanostructures, will be described in chapter 5. It is an example of something unusual—a local, energy-independent embedding potential.

The embedding method as described in the first few chapters is in coordinate space, with an embedding potential which is a function of points r_S, r'_S on the boundary S of region I. This real-space formulation has the advantage that any convenient basis set can be used to solve the embedded Schrödinger equation within region I. However, the whole method can be reformulated in a tight-binding representation without any reference to the coordinate space embedding potential [15]. In chapter 6 we shall discuss embedding in an orbital representation, showing that the embedding potentials in coordinate space and tight-binding are actually identical. This orbital representation of the embedding potential is the same as the self-energy in calculations of transport through molecules [14]; the self-energies replace the metallic leads at the ends of the molecule. Referring to the self-energy reminds us of another context in which the self-energy is used—many-body Green function theory [43] (we shall see in the course of the book that there are several conceptual links). Transport is the subject of chapter 7, and as well as the local orbital approach, we shall discuss molecular transport using real-space embedding potentials to replace the contacts [44].

In recent years, various techniques for calculating electronic structure in solids have been applied to finding the electromagnetic modes in dielectric structures—the science of photonics and plasmonics [45, 46]. The embedding method can also be generalized to the vector electromagnetic fields of Maxwell's equations [47], and we shall discuss this in chapter 9. There are two problems associated with photonics calculations which embedding can solve. The first is the discontinuity in fields across the dielectric boundary, which means that plane-wave expansions of the fields converge relatively slowly. Our solution is to embed away the dielectric object, replacing it by an embedding potential over the surface. The second problem is how to handle waves propagating away from the structure into the vacuum, sometimes resolved by using approximate absorbing boundary conditions. This problem is identical to the substrate problem at surfaces and can be tackled in the same way, by using an embedding potential to replace the vacuum.

In chapter 10 we shall look at the development of a time-dependent embedding method [48]. The advent of ultra-short laser pulses in pulse-probe experiments makes it possible to access time-dependent electronic processes in atoms, molecules and at solid surfaces [49]. Simulating these experiments means solving the time-dependent Schrödinger equation, but electrons excited by an external field inevitably reach the edge of the computational region. Time-dependent embedding can come to the rescue, adding an embedding potential at the boundary to provide the correct boundary conditions for the excited electron. These calculations are still in the preliminary stages, but further motivation to develop time-dependent embedding comes from time-dependent density functional theory (TDDFT) as a way of including exchange-correlation in time-dependent processes [50].

The embedding method has some surprising links with other techniques in theoretical physics, and two of these form the subject of chapter 11, entitled (appropriately enough) *Connections*. The first link is with *R*-matrix theory, widely used in electron scattering calculations in atomic physics [29]. The *R*-matrix is a generalised logarithmic derivative for the scattering electron wave-function, defined over a sphere containing the atomic electrons. It is in a very real sense a many-body embedding potential, as the external electron can excite the atomic electrons in the process of scattering. The second part of this chapter explores the links between the embedding method and resonances in open systems, developing the work on resonances which we introduce in chapter 2. This is relevant, because of interest in calculating the wave-functions of resonances for use as a basis set in treating perturbed systems [51].

1.3 A note on Green functions

As the embedding method is based on Green functions, it is important to fix the sign convention—whether we have ± 1 (the identity operator) on the right-hand side of the inhomogeneous Schrödinger equation [52, 53]. In my original embedding paper [1], the Green function was given by

$$G(E) = (H - E)^{-1}, \tag{1.1}$$

where H is the Hamiltonian and E is the energy. However, the standard usage in most papers and textbooks nowadays is to define G by

$$G(E) = (E - H)^{-1}, \tag{1.2}$$

and this is the convention I shall use in this book. Likewise, in chapter 10 I shall define the time-dependent Green function by

$$\left(i\frac{\partial}{\partial t} - H \right) G(t) = 1\delta(t). \tag{1.3}$$

This is in line with general usage [54], but there is a sign-change compared with my papers on time-dependent embedding [48].

1.4 Units

In this book we use atomic units (a.u.) with $e = \hbar = m_e = 1$. In these units the speed of light $c = 137.036$. From time to time we also use the convenient energy unit of electron volt (eV), with 1 a.u. of energy ≈ 27.2 eV.

References

[1] Inglesfield J E 1981 A method of embedding *J. Phys. C: Solid St. Phys.* **14** 3795–806

[2] Horn K and Scheffler M (ed) 2000 *Handbook of Surface Science volume 2, Electronic Structure* (Amsterdam: Elsevier)

[3] Appelbaum J A and Hamann D R 1972 Self-consistent electronic structure of solid surfaces *Phys. Rev.* B **6** 2166–77

[4] Illas F, Rubio J and Ricart J M 1993 The cluster model configuration interaction approach to the study of chemisorption on metal and semiconductor surfaces *J. Mol. Struct. THEOCHEM.* **287** 167–78

[5] Pacchioni G, Ricart J M and Illas F 1994 Ab initio cluster model calculations on the chemisorption of CO_2 and SO_2 on MgO and CaO (100) surfaces. A theoretical measure of oxide basicity *J. Am. Chem. Soc.* **116** 10152–8

[6] Colbourn E A and Kendrick J 1982 Ab initio calculation for defects in the solid state In ed C R A Catlow and W C Mackrodt *Computer Simulation of Solids (Lecture Notes in Physics 116)* (New York: Springer-Verlag) pp 67–93

[7] Sherwood P *et al* 2003 QUASI: a general purpose implementation of the QM/MM approach and its application to problems in catalysis *J. Mol. Struct. THEOCHEM.* **632** 1–28

[8] Huang C, Pavone M and Carter E A 2011 Quantum mechanical embedding theory based on a unique embedding potential *J. Chem. Phys.* **134** 154110

[9] Libisch F, Huang C and Carter E A 2014 Embedded correlated wavefunction schemes: theory and applications *Acc. Chem. Res.* **47** 2768–75

[10] Sharifzadeh S, Huang P and Carter E A 2009 Origin of tunneling lineshape trends for Kondo states of Co adatoms on coinage metal surfaces *J. Phys.: Condens. Matter* **21**(35) 355501

[11] Abarenkov I V, Boyko M A and Sushko P V 2010 Embedding and atomic orbitals hybridization *Int. J. Quantum Chem.* **111** 2602–19

[12] Grimley T B and Pisani C 1974 Chemisorption theory in the Hartree-Fock approximation *J. Phys. C: Solid St. Phys.* **7** 2831–48

[13] Fisher A J 1988 Methods of embedding for defect and surface problems *J. Phys. C: Solid St. Phys.* **21** 3229–49

[14] Datta S 2005 *Quantum Transport: Atom to Transistor* (Cambridge: Cambridge University Press)

[15] Baraff G A and Schlüter M 1986 The LCAO approach to the embedding problem *J. Phys. C: Solid St. Phys.* **19** 4383–91

[16] Dederichs P H, Lounis S and Zeller R 2006 The Korringa-Kohn-Rostocker (KKR) Green function method II. Impurities and clusters in the bulk and on surfaces In ed J Grotendorst, S Blügel and D Marx *Computational Nanoscience: Do It Yourself!* (Jülich: John von Neumann Institute) pp 279–98

[17] Inglesfield J E 1981 Electronic structure of interstitial H in Cu *J. Phys. F: Metal Phys.* **11** L287–91

[18] Anderson P W 1961 Localized magnetic states in metals *Phys. Rev.* **124** 41–53

[19] Grimley T B 1967 The indirect interaction between atoms or molecules adsorbed on metals *Proc. Phys. Soc.* **90** 751–64

[20] Grimley T B 1967 The electron density in a metal near a chemisorbed atom or molecule *Proc. Phys. Soc.* **92** 776–82

[21] Newns D M 1969 Self-consistent model of hydrogen chemisorption *Phys. Rev.* **178** 1123–35

[22] Pisani C 1978 Approach to the embedding problem in chemisorption in a self-consistent-field-molecular-orbital formalism *Phys. Rev.* B **17** 3143–53

[23] Pisani C, Dovesi R, Roetti C, Causà M, Orlando R, Casassa S and Saunders V R 2000 CRYSTAL and EMBED, two computational tools for the ab initio study of electronic properties of crystals *Int. J. Quantum Chem.* **77** 1032–48

[24] Caroli C, Combescot R, Nozières P and Saint-James D 1971 Direct calculation of the tunneling current *J. Phys. C: Solid St. Phys.* **4** 916–29

[25] Meir Y and Wingreen N S 1992 Landauer formula for the current through an interacting electron region *Phys. Rev. Lett.* **68** 2512–5

[26] Inglesfield J E 1970 Green functions, surfaces, and impurities *J. Phys. C: Solid St. Phys.* **4** 14–8

[27] García-Moliner F and Rubio J 1969 A new method in the quantum theory of surface states *J. Phys. C: Solid St. Phys.* **2** 1789–96

[28] Wigner E P and Eisenbud L 1947 Higher angular momenta and long range interaction in resonance reactions *Phys. Rev.* **72** 29–41

[29] Burke P G 2011 *R-Matrix Theory of Atomic Collisions* (Berlin: Springer-Verlag)

[30] Zou P F 1992 A regional embedding method *Int. J. Quantum Chem.* **44** 997–1013

[31] Colton D and Kress R 1998 *Inverse Acoustic and Electromagnetic Scattering Theory* 2nd edn (Berlin: Springer-Verlag)

[32] Szmytkowski R and Bielski S 2004 Dirichlet-to-Neumann and Neumann-to-Dirichlet embedding methods for bound states of the Schrödinger equation *Phys. Rev.* A **70** 042103

[33] Zangwill A 2013 *Modern Electrodynamics* (Cambridge: Cambridge University Press)

[34] Keller J B and Givoli D 1989 Exact non-reflecting boundary conditions *J. Comput. Phys.* **82** 172–92

[35] Ishida H 2001 Surface-embedded Green-function method: A formulation using a linearized-augmented-plane-wave basis set *Phys. Rev.* B **63** 165409

[36] James M and Crampin S 2010 Relativistic embedding method: The transfer matrix, complex band structures, transport, and surface calculations *Phys. Rev.* B **81** 155439

[37] Ishida H 2014 Rashba spin splitting of Shockley surface states on semi-infinite crystals *Phys. Rev.* B **90** 235422

[38] Inglesfield J E and Benesh G A 1988 Surface electronic structure: Embedded self-consistent calculations *Phys. Rev.* B **37** 6682–700

[39] Kohn W and Sham L J 1965 Self-consistent equations including exchange and correlation effects *Phys. Rev.* **140** 1133–8

[40] Inglesfield J E 2001 Embedding at surfaces *Comput. Phys. Commun.* **137** 89–107

[41] Crampin S 2004 An embedding scheme for the Dirac equation *J. Phys.: Condens. Matter* **16**(49) 8875–89

[42] Crampin S, Nekovee M and Inglesfield J E 1995 Embedding method for confined quantum systems *Phys. Rev.* B **51** 7318–20

[43] Abrikosov A A, Gorkov L P and Dzyaloshinski I E 1963 *Methods of Quantum Field Theory in Statistical Physics* (New York: Dover)

[44] Wortmann D, Ishida H and Blügel S 2002 Embedded Green-function approach to the ballistic electron transport through an interface *Phys. Rev.* B **66** 075113

[45] Joannopoulos J D, Johnson S G, Winn J N and Meade R D 2008 *Photonic Crystals* 2nd edn (Princeton: Princeton University Press)

[46] Maradudin A A, Sambles J R and Barnes W L (ed) 2014 *Handbook of Surface Science volume 4, Modern Plasmonics* (Amsterdam: Elsevier)

[47] Inglesfield J E 1998 The embedding method for electromagnetics *J. Phys. A: Math. Gen.* **31** 8495–510

[48] Inglesfield J E 2008 Time-dependent embedding *J. Phys.: Condens. Matter* **20** 095215

[49] Neppl S, Ernstorfer R, Bothschafter E M, Cavalieri A L, Menzel D, Barth J V, Krausz F, Kienberger R and Feulner P 2012 Attosecond time-resolved photoemission from core and valence states of magnesium *Phys. Rev. Lett.* **109** 087401

[50] Andrade X *et al* 2012 Time-dependent density-functional theory in massively parallel computer architectures: the OCTOPUS project *J. Phys.: Condens. Matter* **24** 233202

[51] Muljarov E A, Langbein W and Zimmermann R 2010 Brillouin-Wigner perturbation theory in open electromagnetic systems *Europhys. Lett.* **92** 50050

[52] Barton G 1989 *Elements of Green's Functions and Propagation* (Oxford: Oxford University Press)

[53] Merzbacher E 1998 *Quantum Mechanics* 3rd edn (New York: John Wiley)

[54] Newton R G 1982 *Scattering Theory of Waves and Particles* 2nd edn (New York: Springer-Verlag)

Chapter 2

The variational embedding method

This chapter describes the basis of the embedding method, how it is derived and how it can be applied [1, 2]. I shall start with the original derivation using the variational principle: a trial wave-function in region I, the region we are interested in, is extended into the substrate region II by an exact solution of the Schrödinger equation. This form of trial wave-function is substituted into the variational expression for the energy, and the terms involving the wave-function in region II are eliminated by Green function methods, replaced by the embedding potential added to the Hamiltonian of region I. The embedding potential, a non-local, energy-dependent potential on the boundary of region I with II, can replace a finite or extended region II, and in this book we shall see applications of both.

In this chapter we shall use the method to calculate both discrete states and the continuum electronic structure of a simple one-dimensional potential, with extended regions on each side replaced by an embedding potential. This model brings out features of embedding which we use later in the book in more realistic situations. We shall finish this chapter with a new discussion of the analytic properties of the embedding potential, and show how resonances and their wave-functions—so-called 'resonant states' [3]—can be calculated using the embedding method.

2.1 The variational principle

The derivation of the embedding method is based on a variational approach for solving the Schrödinger equation [1],

$$H\Psi = E\Psi,$$

with
$$H = -\frac{1}{2}\nabla^2 + V(\mathbf{r}). \tag{2.1}$$

H is the one-electron Hamiltonian containing the potential $V(\mathbf{r})$; we approximate the wave-function $\Psi(\mathbf{r})$, energy E, by a trial function $\Phi(\mathbf{r})$, which we now construct in the two regions shown in figure 1.1.

It is in region I where we want full variational freedom in our choice of trial wavefunction—we call it $\phi(\mathbf{r})$ in this region—but this must, in principle, be extended into the rest of space, the substrate region II. On the outer surface, indicated by the dashed line in figure 1.1, there is a fixed homogeneous boundary condition, such as $\Psi = 0$; in many cases this boundary is pushed away to infinity. We assume that we can solve the Schrödinger equation exactly in region II, and for the trial function in this region we take the exact solution $\psi(\mathbf{r})$ at a trial energy E_0 (actually a variational parameter), which matches ϕ in amplitude over boundary S,

$$\left(-\frac{1}{2}\nabla^2 + V(\mathbf{r})\right)\psi(\mathbf{r}) = E_0\,\psi(\mathbf{r}), \quad \mathbf{r} \in \text{region II}, \tag{2.2}$$

$$\psi(\mathbf{r}_S) = \phi(\mathbf{r}_S), \tag{2.3}$$

where $\mathbf{r}_S$ indicates a point on S. This uniquely prescribes ψ. We have, then a trial function Φ given by

$$\Phi(\mathbf{r}) = \begin{cases} \phi(\mathbf{r}), & \mathbf{r} \in \text{region I} \\ \psi(\mathbf{r}), & \mathbf{r} \in \text{region II}, \end{cases} \tag{2.4}$$

which is continuous in amplitude (equation (2.3)), but unless we are very lucky, not matched in derivative. It is eliminating the discontinuity in derivative which is the key to the embedding method.

With this trial function, the expectation value of the Hamiltonian—the energy E—is given by

$$E = \frac{\int_I \mathrm{d}\mathbf{r}\, \phi^*(\mathbf{r})H\phi(\mathbf{r}) + E_0\int_{II} \mathrm{d}\mathbf{r}\, \psi^*(\mathbf{r})\psi(\mathbf{r}) + \frac{1}{2}\int_S \mathrm{d}\mathbf{r}_S\, \phi^*(\mathbf{r}_S)\left(\frac{\partial\phi(\mathbf{r}_S)}{\partial n_S} - \frac{\partial\psi(\mathbf{r}_S)}{\partial n_S}\right)}{\int_I \mathrm{d}\mathbf{r}\, \phi^*(\mathbf{r})\phi(\mathbf{r}) + \int_{II} \mathrm{d}\mathbf{r}\, \psi^*(\mathbf{r})\psi(\mathbf{r})}. \tag{2.5}$$

In addition to the integrals through regions I and II, there is a surface integral over the boundary S, containing the discontinuity in derivative of the basis function (n_S is the surface normal, which we always take *outwards* from I into II). This surface term arises from the kinetic energy operator: the kink in the trial function Φ costs kinetic energy. Let us take the surface normal at some point on S as the local z-direction, with $z = 0$ on the boundary; then the boundary contribution to the kinetic energy at this point is given by

$$T_S = -\frac{1}{2}\int_{-\eta}^{+\eta} \mathrm{d}z\, \Phi^*\frac{\partial^2\Phi}{\partial z^2}, \tag{2.6}$$

where η is a small positive number. Integrating (2.6) by parts gives

$$T_S = -\frac{1}{2}\left(\Phi^*\frac{\partial\Phi}{\partial z}\bigg|_{-\eta}^{+\eta} - \int_{-\eta}^{+\eta} \mathrm{d}z\, \left|\frac{\partial\Phi}{\partial z}\right|^2\right). \tag{2.7}$$

The integrand in the second term is finite, so in the limit of $\eta \to 0+$, and using equation (2.3) we obtain

$$T_S = \frac{1}{2}\phi^*\left(\frac{\partial\phi}{\partial z} - \frac{\partial\psi}{\partial z}\right). \tag{2.8}$$

Integrating over the boundary, this is the extra surface term in equation (2.5).

The next step in our derivation is to eliminate explicit reference in (2.5) to ψ, the trial function in region II. To do this we introduce a Green function for this region satisfying the inhomogeneous Schrödinger equation,

$$\left(-\frac{1}{2}\nabla^2 + V(\mathbf{r}) - E_0\right)G_0(\mathbf{r}, \mathbf{r}'; E_0) = -\delta(\mathbf{r} - \mathbf{r}'), \quad \mathbf{r}, \mathbf{r}' \in \text{region II}. \tag{2.9}$$

Note the $-\delta$ on the right-hand side, in line with our convention (see equation (1.2)). On the outer surface G_0 satisfies the same homogeneous boundary condition as the wave-functions, but on S it satisfies a zero-derivative boundary condition,

$$\frac{\partial G_0}{\partial n_S}(\mathbf{r}_S, \mathbf{r}'; E_0) = 0. \tag{2.10}$$

We now multiply equation (2.9) by ψ, equation (2.2) by G_0, subtract, and integrate through region II to obtain

$$\psi(\mathbf{r}) = \frac{1}{2}\int_{\text{II}} d\mathbf{r}'\left[\psi(\mathbf{r}')\nabla^2_{\mathbf{r}'}G_0(\mathbf{r}, \mathbf{r}') - G_0(\mathbf{r}, \mathbf{r}')\nabla^2_{\mathbf{r}'}\psi(\mathbf{r}')\right], \tag{2.11}$$

and then from Green's theorem [4, 5],

$$\psi(\mathbf{r}) = \frac{1}{2}\int_S d\mathbf{r}_S\, G_0(\mathbf{r}, \mathbf{r}_S)\frac{\partial\psi(\mathbf{r}_S)}{\partial n_S} \tag{2.12}$$

(remember that $\partial/\partial n_S$ is directed into region II). Putting $\mathbf{r}$ on S gives an integral equation relating the surface amplitude of the solution in region II to the surface derivative,

$$\psi(\mathbf{r}_S) = \frac{1}{2}\int_S d\mathbf{r}'_S\, G_0(\mathbf{r}_S, \mathbf{r}'_S)\frac{\partial\psi(\mathbf{r}'_S)}{\partial n_S}, \tag{2.13}$$

and the inverse of this equation gives us what we require—the surface derivative in terms of the *known* surface amplitude $\phi(\mathbf{r}_S)$,

$$\frac{\partial\psi(\mathbf{r}_S)}{\partial n_S} = 2\int_S d\mathbf{r}'_S\, G_0^{-1}(\mathbf{r}_S, \mathbf{r}'_S)\psi(\mathbf{r}'_S) \tag{2.14}$$

$$= 2 \int_S dr'_S G_0^{-1}(\mathbf{r}_S, \mathbf{r}'_S) \phi(\mathbf{r}'_S). \tag{2.15}$$

We rewrite equation (2.15) as

$$\frac{\partial \psi(\mathbf{r}_S)}{\partial n_S} = -2 \int_S dr'_S \, \Sigma(\mathbf{r}_S, \mathbf{r}'_S; E_0) \phi(\mathbf{r}'_S), \tag{2.16}$$

where Σ is the *embedding potential*, given by

$$\Sigma(\mathbf{r}_S, \mathbf{r}'_S; E_0) := -G_0^{-1}(\mathbf{r}_S, \mathbf{r}'_S; E_0). \tag{2.17}$$

Equation (2.16) is the equation which really defines the embedding potential as a generalized logarithmic derivative [6]; the factor of -2 looks arbitrary, but it simplifies subsequent formulae. We note that equation (2.14), which gives the normal derivative on S (the Neumann boundary condition) in terms of the amplitude (the Dirichlet boundary condition), is called Dirichlet-to-Neumann mapping, and G_0^{-1} (or Σ) is the Dirichlet-to-Neumann map [7, 8].

The volume integral of $|\psi|^2$ over region II, which appears in the numerator and denominator of equation (2.5), can also be replaced by a surface integral involving $\phi(\mathbf{r}_S)$ and the embedding potential. Differentiating equation (2.2) with respect to E_0, but maintaining the boundary condition that $\psi(\mathbf{r}_S) = \phi(\mathbf{r}_S)$ over S, we have

$$H \frac{\partial \psi}{\partial E_0} = \psi + E_0 \frac{\partial \psi}{\partial E_0}. \tag{2.18}$$

We multiply this equation by ψ^*, the complex conjugate of equation (2.2) by $\partial \psi / \partial E_0$, and subtract to obtain

$$|\psi(\mathbf{r})|^2 = \frac{1}{2} \left(\frac{\partial \psi(\mathbf{r})}{\partial E_0} \nabla^2 \psi^*(\mathbf{r}) - \psi^*(\mathbf{r}) \nabla^2 \frac{\partial \psi(\mathbf{r})}{\partial E_0} \right). \tag{2.19}$$

Then, integrating through region II, once again with the help of Green's theorem, gives

$$\int_{\text{II}} d\mathbf{r} \, |\psi(\mathbf{r})|^2 = \frac{1}{2} \int_S dr_S \left(\psi^*(\mathbf{r}_S) \frac{\partial^2 \psi(\mathbf{r}_S)}{\partial E_0 \, \partial n_S} - \frac{\partial \psi(\mathbf{r}_S)}{\partial E_0} \frac{\partial \psi^*(\mathbf{r}_S)}{\partial n_S} \right)$$

$$= \frac{1}{2} \int_S dr_S \, \phi^*(\mathbf{r}_S) \frac{\partial^2 \psi(\mathbf{r}_S)}{\partial E_0 \, \partial n_S}. \tag{2.20}$$

The final step is to use equation (2.16) to evaluate the normal derivative, giving us

$$\int_{\text{II}} d\mathbf{r} \, |\psi(\mathbf{r})|^2 = -\int_S dr_S \int_S dr'_S \, \phi^*(\mathbf{r}_S) \frac{\partial \Sigma(\mathbf{r}_S, \mathbf{r}'_S; E_0)}{\partial E_0} \phi(\mathbf{r}'_S). \tag{2.21}$$

Substituting equations (2.19) and (2.21) into equation (2.5) gives us an expression for E in terms of the trial function ϕ defined only in region I and on S,

$$
E = \left(\int_I d\mathbf{r}\, \phi^*(\mathbf{r}) H \phi(\mathbf{r}) + \frac{1}{2} \int_S d\mathbf{r}_S\, \phi^*(\mathbf{r}_S) \frac{\partial \phi(\mathbf{r}_S)}{\partial n_S} + \int_S d\mathbf{r}_S \int_S d\mathbf{r}'_S\, \phi^*(\mathbf{r}_S) \right.
$$

$$
\left. \times \Sigma(\mathbf{r}_S, \mathbf{r}'_S; E_0) \phi(\mathbf{r}'_S) - E_0 \int_S d\mathbf{r}_S \int_S d\mathbf{r}'_S\, \phi^*(\mathbf{r}_S) \frac{\partial \Sigma(\mathbf{r}_S, \mathbf{r}'_S; E_0)}{\partial E_0} \phi(\mathbf{r}'_S) \right) \Bigg/
$$

$$
\left(\int_I d\mathbf{r}\, \phi^*(\mathbf{r}) \phi(\mathbf{r}) - \int_S d\mathbf{r}_S \int_S d\mathbf{r}'_S\, \phi^*(\mathbf{r}_S) \frac{\partial \Sigma(\mathbf{r}_S, \mathbf{r}'_S; E_0)}{\partial E_0} \phi(\mathbf{r}'_S) \right). \quad (2.22)
$$

This is a variational expression [9], and by finding the stationary values of E with respect to variations in ϕ, and the trial energy E_0 at which the embedding potential Σ is evaluated, we obtain the solutions of the Schrödinger equation in region I which match the solution in region II in amplitude and derivative. Interestingly, Szmytkowski and Bielski [8] have derived a variational principle involving the Neumann-to-Dirichlet operator G_0 (equation (2.13)) instead of the Dirichlet-to-Neumann operator G_0^{-1} (or Σ).

2.2 The embedded Schrödinger equation

We can derive the embedded Schrödinger equation from equation (2.22) by varying the trial function ϕ. As E is stationary with respect to small changes $\delta\phi(\mathbf{r})$, ϕ satisfies the equation

$$
-\frac{1}{2}\nabla^2 \phi(\mathbf{r}) + \frac{1}{2}\delta(n - n_S)\frac{\partial \phi(\mathbf{r}_S)}{\partial n_S} + V(\mathbf{r})\phi(\mathbf{r}) + \delta(n - n_S)\int_S d\mathbf{r}'_S
$$

$$
\times \left(\Sigma(\mathbf{r}_S, \mathbf{r}'_S; E_0) + (E - E_0)\frac{\partial \Sigma(\mathbf{r}_S, \mathbf{r}'_S; E_0)}{\partial E_0} \right) \phi(\mathbf{r}'_S) = E\phi(\mathbf{r}), \quad \mathbf{r} \in I, \quad (2.23)
$$

where n is the component of $\mathbf{r}$ perpendicular to the embedding surface S, and $\delta(n - n_S)$ gives the surface S. We see that $\Sigma(\mathbf{r}_S, \mathbf{r}'_S)$ is a surface potential acting on ϕ, and that $(E - E_0)\partial\Sigma/\partial E_0$—which started life as the normalization of the trial function in region II—provides a first-order correction to give this embedding potential at the actual energy eigenvalue E. This is a remarkable aspect of the variational embedding method.

To solve this Schrödinger equation, we can substitute an expansion of the trial wave-function in terms of a set of basis functions [9],

$$
\phi(\mathbf{r}) = \sum_n a_n \chi_n(\mathbf{r}). \quad (2.24)
$$

Then, multiplying the equation by $\chi_m^*(\mathbf{r})$ and integrating through region I and over S gives the matrix equation

$$
\sum_n H_{mn} a_n = E \sum_n O_{mn} a_n, \quad (2.25)
$$

where the matrix elements of the embedded Hamiltonian are given by

$$H_{mn} = -\frac{1}{2} \int_I d\mathbf{r}\, \chi_m^*(\mathbf{r}) \nabla^2 \chi_n(\mathbf{r}) + \frac{1}{2} \int_S d\mathbf{r}_S\, \chi_m^*(\mathbf{r}_S) \frac{\partial \chi_n(\mathbf{r}_S)}{\partial n_S} + \int_I d\mathbf{r}\, \chi_m^*(\mathbf{r}) V(\mathbf{r}) \chi_n(\mathbf{r})$$

$$+ \int_S d\mathbf{r}_S \int_S d\mathbf{r}_S'\, \chi_m^*(\mathbf{r}_S) \left(\Sigma(\mathbf{r}_S, \mathbf{r}_S'; E_0) - E_0 \frac{\partial \Sigma(\mathbf{r}_S, \mathbf{r}_S'; E_0)}{\partial E_0} \right) \chi_n(\mathbf{r}_S'), \quad (2.26)$$

and O_{mn} is the overlap matrix

$$O_{mn} = \int_I d\mathbf{r}\, \chi_m^*(\mathbf{r}) \chi_n(\mathbf{r}) - \int_S d\mathbf{r}_S \int_S d\mathbf{r}_S'\, \chi_m^*(\mathbf{r}_S) \frac{\partial \Sigma(\mathbf{r}_S, \mathbf{r}_S'; E_0)}{\partial E_0} \chi_n(\mathbf{r}_S'). \quad (2.27)$$

Note that by including $\int_S d\mathbf{r}_S \int_S d\mathbf{r}_S'\, \chi_m^* \frac{\partial \Sigma}{\partial E_0} \chi_n$ in the overlap rather than in the Hamiltonian, where it appears in equation (2.23), our matrix equation (2.25) becomes a generalized eigenvalue equation, which can be solved by standard routines from program libraries like LAPACK[1]. We can also obtain the matrix equation by substituting the basis expansion (2.24) into the variational expression (2.22), and then the requirement that E is stationary with respect to variations in the expansion coefficients a_n gives equation (2.25) directly.

An important point is that the integral through region I of the kinetic energy term in the Hamiltonian (the first term on the right-hand side of equation (2.26)), combined with the surface derivative term (the second term), together give an hermitian matrix; in atomic physics, the surface derivative term is called the Bloch operator [10, 11]. The hermiticity can be brought out explicitly by combining the two terms,

$$-\frac{1}{2} \int_I d\mathbf{r}\, \chi_m^*(\mathbf{r}) \nabla^2 \chi_n(\mathbf{r}) + \frac{1}{2} \int_S d\mathbf{r}_S\, \chi_m^*(\mathbf{r}_S) \frac{\partial \chi_n(\mathbf{r}_S)}{\partial n_S} = \frac{1}{2} \int_I d\mathbf{r}\, \nabla \chi_m^*(\mathbf{r}) \cdot \nabla \chi_n(\mathbf{r}). \quad (2.28)$$

If the system I+II has discrete states, the embedding potential is real at real E_0, and the whole Hamiltonian matrix H_{mn} is hermitian; this is not the case with 'open' systems, where Σ is complex (section 2.4).

2.3 A first application

As an initial example we shall consider the one-dimensional potential shown in figure 2.1, which we can think of as a rather simplified (!) model of a surface with an overlayer of atoms (the 'atoms' constitute the potential well between $-d/2$ and $+d/2$, the 'solid' extends to the left, and the 'vacuum' to the right). The first step is to construct the embedding potentials to represent the two regions of space which comprise region II: in the course of the book we shall learn about the different ways to find Σ, but in this one-dimensional case we can use the definition (2.16), that Σ is the logarithmic derivative of the solution in region II multiplied by $-\frac{1}{2}$. From the

---

[1] Available on the web from www.netlib.org/lapack

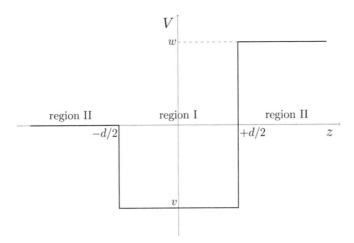

Figure 2.1. One-dimensional potential representing a layer of atoms on a surface. Region I is the atomic layer, $-d/2 < z < +d/2$; region II for $z < -d/2$ is the crystal substrate, and region II for $z > +d/2$ is the vacuum.

solution of the free-electron Schrödinger equation with $z < -d/2$, decaying or travelling to the left, we obtain the embedding potential at $z = -d/2$,

$$\Sigma(-d/2) = \begin{cases} \gamma/2, & \gamma = \sqrt{-2E_0}, & E_0 < 0 \\ -ik/2, & k = \sqrt{2E_0}, & E_0 > 0. \end{cases} \tag{2.29}$$

The embedding potential at $z = +d/2$ is the same, with the energy shift $E_0 \rightarrow E_0 - w$. At this stage we apply equation (2.25) to the bound states of the system, so we are only concerned with $E_0 < 0$, but in the next section we shall consider positive energies, and generalize equation (2.29) to complex energies.

A convenient basis set to expand the trial function in region I is

$$\chi_n(z) = \begin{cases} \cos(n\pi z/D), & n = 0, 2, 4... \\ \sin(n\pi z/D), & n = 1, 3, 5..., \end{cases} \tag{2.30}$$

where D, which defines the basis, is somewhat larger than d—we can be vague about the choice of D because as long as $D > d$ it doesn't really matter. We then use the method described in the last section to find the eigenvalues E, the variational estimates of the bound state energies.

Figure 2.2 shows E as a function of E_0, for the surface potential with $v = -0.45$ a.u., $w = 0.5$ a.u., $d = 8$ a.u., and a basis set of 10 functions with $D = 12$ a.u. With these well parameters there are three bound states, and we see that for each bound state E takes a minimum value when $E_0 = E$. This means that we can find the bound state energies by simple iteration, starting off with a guessed E_0, setting E_0 equal to the output E for the next iteration, and so on, until the output E equals the input E_0 to whatever accuracy we require. Only two or three iterations are required, even for the highest bound state where $E(E_0)$ has a rather narrow minimum. This useful behaviour comes from the form of equation (2.23), with the first-order energy correction to $\Sigma(E_0)$ coming from the $(E - E_0)\partial\Sigma/\partial E_0$ terms.

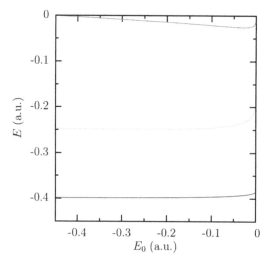

Figure 2.2. Variational estimate of the energy E of bound states of the model potential (figure 2.1, $v = -0.45$ a.u., $w = 0.5$ a.u., $d = 8$ a.u.) as a function of E_0, the energy at which the embedding potential is evaluated. Ten basis functions are used, with $D = 12$ a.u.

Table 2.1. Bound state energies (in a.u.) of the model potential (figure 2.1, $v = -0.45$ a.u., $w = 0.5$ a.u., $d = 8$ a.u.), for different numbers of basis functions, defined by D. Also shown are the exact energies, calculated by wave-function matching.

	Number of basis functions		
D (a.u.)	10	15	20
	-0.395053	-0.396561	-0.397023
8	-0.239842	-0.242573	-0.244538
	-0.018986	-0.021750	-0.022789
	-0.398674	-0.398694	-0.398695
10	-0.248899	-0.248937	-0.248940
	-0.025974	-0.026125	-0.026126
	-0.398694	-0.398695	-0.398695
12	-0.248939	-0.248940	-0.248940
	-0.026125	-0.026127	-0.026127
	-0.398695		
Exact	-0.248940		
	-0.026127		

The bound state energies obtained in this way are shown in table 2.1 for different values of D and different basis set sizes, compared with exact energies calculated by simple wave-function matching. We see that for $D = 10$ a.u. and $D = 12$ a.u. there is rapid convergence, with $D = 12$ a.u. somewhat better. The reason why we should choose $D > d$ is that this ensures a range of logarithmic derivatives of the basis functions at the embedding boundaries, so that the trial wave-function ϕ can match with the embedding potentials at $\pm d/2$. A possible problem with basis sets like this is that linear dependency can occur in the expansion through region I, particularly if $D \gg d$ and the basis set is large. The problem is reduced by computing in double precision, and can be eliminated by a transformation of basis functions (we shall see how to do so in section 5.6). We *can* take $D = d$, but the error in derivative at the embedding boundaries results in poorer convergence: we see from table 2.1 that the bound state energies with $D = 8$ a.u. converge rather slowly towards the correct values—though they *do* converge.

As well as finding the eigenvalues of equation (2.25), we can find the corresponding eigenvectors and wave-functions. So that the full wave-function $\Phi(\mathbf{r})$ (equation (2.4)) is normalized in regions I and II together, we use the embedding potential result for the integral through region II (equation (2.21)) to obtain the *renormalized* $\phi(\mathbf{r})$,

$$
\phi(\mathbf{r}) = \frac{\hat{\phi}(\mathbf{r})}{\sqrt{1 - \int_S d\mathbf{r}_S \int_S d\mathbf{r}'_S \, \hat{\phi}^*(\mathbf{r}_S) \dfrac{\partial \Sigma(\mathbf{r}_S, \mathbf{r}'_S; E_0)}{\partial E_0} \hat{\phi}(\mathbf{r}'_S)}}, \quad \mathbf{r} \in \text{region I}, \quad (2.31)
$$

where $\hat{\phi}(\mathbf{r})$ is normalized in region I. Identical renormalization, involving the energy-derivative of the self-energy, occurs in the weight of quasiparticle poles in many-body theory [12]; and in pseudopotential band-structure calculations, the energy-derivative of the pseudopotential appears in the renormalization of the wave-function [9].

The bound state wave-functions for the model surface potential are shown by the continuous lines in figure 2.3, calculated using the converged basis set with $D = 12$ a.u.

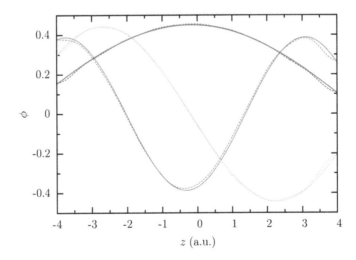

Figure 2.3. Bound state wave-functions $\phi(z)$ in region I of the model potential (figure 2.1, $v = -0.45$ a.u., $w = 0.5$ a.u., $d = 8$ a.u.) calculated with 15 basis functions, $D = 12$ a.u. (solid lines) and $D = 8$ a.u. (dashed lines). Normalization is over the whole of space. The order of the states is red (most bound), green, blue (least bound), with energies given in table 2.1. Note that the wave-functions with $D = 8$ a.u. have zero derivative at the embedding boundaries, $z = \pm d/2$.

and 15 basis functions, and renormalized using equation (2.31). The dashed lines show the results with $D = 8$ a.u., also with 15 basis functions, and we can clearly see the drawback of taking $D = d$, namely the zero derivative of the wave-functions at the embedding boundaries at $z = \pm d/2$. Results with $D = 10$ a.u. are indistinguishable from the $D = 12$ a.u. results.

2.4 The embedded Green function

With an unbounded external region II, there is generally a continuum of states over some ranges of energy, where the wave-functions extend to infinity. Rather than calculating individual continuum states—which in any case cannot be normalized over space—it is convenient to work with the Green function $G(\mathbf{r}, \mathbf{r}'; E)$ satisfying

$$\left(-\frac{1}{2}\nabla^2 + V(\mathbf{r}) - E\right)G(\mathbf{r}, \mathbf{r}'; E) = -\delta(\mathbf{r} - \mathbf{r}'), \quad \mathbf{r}, \mathbf{r}' \in \text{regions I+II}. \quad (2.32)$$

From the Green function we can calculate all the useful static properties like the charge density and the energy spectrum of the electrons; the Green function plays a fundamental role in transport theory (chapter 7), as well as forming the basis of many-body calculations.

We first consider a large but finite system, so that all the states are discrete and G can be written as a sum over the eigenstates Ψ_i and eigenvalues E_i of equation (2.1) [4],

$$G(\mathbf{r}, \mathbf{r}'; E) = \sum_i \frac{\Psi_i(\mathbf{r})\Psi_i^*(\mathbf{r}')}{E - E_i}. \quad (2.33)$$

To make contact with static quantities like the electron density and density of states we consider the *local density of states* [13], defined for real E as

$$n(\mathbf{r}, E) = \sum_i |\Psi_i(\mathbf{r})|^2 \delta(E_i - E), \quad (2.34)$$

the probability density of electrons with energy E. This is directly related to the Green function: if we put the energy in equation (2.33) just above the real axis at $E + i\eta$, with η a small positive number, and take the imaginary part of G we obtain

$$n(\mathbf{r}, E) = -\frac{1}{\pi}\Im G(\mathbf{r}, \mathbf{r}; E + i\eta), \quad E \text{ real}, \eta \to 0+. \quad (2.35)$$

We note here that the Green function evaluated at $E + i0+$ is the *outgoing* or *retarded* Green function, corresponding in the case of an infinite system to outgoing or decaying waves [6]; we shall discuss this further in section 2.6.

From the local density of states we can find the usual density of states by integrating over all space, and—most usefully—the electron density, by integrating over E up to the Fermi energy. In the limit of the extended system the sums over states in equations (2.33) and (2.34) are replaced by integrals over a continuous variable like energy, with energy-normalization of the wave-functions [6]. But equation (2.35) remains true, for continuum as well as discrete states.

The embedding method allows us to find the Green function in region I, matching it automatically to region II so that for $\mathbf{r}$, $\mathbf{r}'$ in region I it is identical to the Green function satisfying equation (2.32) throughout the whole system. The embedded Green function satisfies the inhomogeneous version of equation (2.23),

$$-\frac{1}{2}\nabla_{\mathbf{r}}^2 G(\mathbf{r}, \mathbf{r}'; E) + \frac{1}{2}\delta(n - n_S)\frac{\partial}{\partial n_S}G(\mathbf{r}_S, \mathbf{r}'; E) + V(\mathbf{r})G(\mathbf{r}, \mathbf{r}'; E) + \delta(n - n_S)$$

$$\times \int_S d\mathbf{r}_S'' \Sigma(\mathbf{r}_S, \mathbf{r}_S''; E)G(\mathbf{r}_S'', \mathbf{r}'; E) - EG(\mathbf{r}, \mathbf{r}'; E) = -\delta(\mathbf{r} - \mathbf{r}'), \quad \mathbf{r}, \mathbf{r}' \in I. \quad (2.36)$$

As we *know* the energy E at which to evaluate the embedding potential, the term in equation (2.23) involving $\partial\Sigma/\partial E_0$ does not appear in the Green function version of the embedded Schrödinger equation. The solution of this equation is the Green function in region I, which matches in amplitude and derivative with the Green function in region II. In other words, it is the Green function for the combined system I + II, when $\mathbf{r}$ and $\mathbf{r}'$ are both in region I.

As before, we solve equation (2.36) by expanding G in a basis set,

$$G(\mathbf{r}, \mathbf{r}'; E) = \sum_{mn} G_{mn}(E)\chi_m(\mathbf{r})\chi_n^*(\mathbf{r}'). \quad (2.37)$$

Substituting into equation (2.36) gives

$$\sum_{mn} G_{mn}\left(-\frac{1}{2}\nabla^2\chi_m(\mathbf{r}) + \frac{1}{2}\delta(n - n_S)\frac{\partial\chi_m(\mathbf{r}_S)}{\partial n_S} + V(\mathbf{r})\chi_m(\mathbf{r})\right.$$

$$\left. + \delta(n - n_S)\int_S d\mathbf{r}_S'' \Sigma(\mathbf{r}_S, \mathbf{r}_S''; E)\chi_m(\mathbf{r}_S'') - E\chi_m(\mathbf{r}'')\right)\chi_n^*(\mathbf{r}') = -\delta(\mathbf{r} - \mathbf{r}'), \quad (2.38)$$

and multiplying by $\chi_k^*(\mathbf{r})\chi_l(\mathbf{r}')$ and integrating over $\mathbf{r}$, $\mathbf{r}'$ through region I we obtain

$$\sum_{mn}[H_{km}(E) - EO_{km}]G_{mn}O_{nl} = -O_{kl}. \quad (2.39)$$

H has the same form as equation (2.26) without the $\partial\Sigma/\partial E_0$ term,

$$H_{mn}(E) = \frac{1}{2}\int_I d\mathbf{r}\, \nabla\chi_m^*(\mathbf{r}) \cdot \nabla\chi_n(\mathbf{r}) + \int_I d\mathbf{r}\, \chi_m^*(\mathbf{r})V(\mathbf{r})\chi_n(\mathbf{r})$$

$$+ \int_S d\mathbf{r}_S \int_S d\mathbf{r}_S'\, \chi_m^*(\mathbf{r}_S)\Sigma(\mathbf{r}_S, \mathbf{r}_S'; E)\chi_n(\mathbf{r}_S'). \quad (2.40)$$

From equation (2.39) we see that

$$\sum_m [H_{km}(E) - EO_{km}]G_{mn} = -\delta_{kn}, \quad (2.41)$$

so the matrix elements of the embedded Green function are given by

$$G_{mn}(E) = [EO - H(E)]_{mn}^{-1}, \quad (2.42)$$

perhaps the form which we could have guessed all the time!

2.5 Application to continuum states

We now apply the embedded Green function method of the last section to study the electronic states in the continuum of our simple surface model (figure 2.1). Actually there are two continua here, one with $E > 0$ on the left, the 'crystal' side; and the second with $E > w$ on the right, the vacuum side of the system. The most useful quantity to calculate here is the local density of states integrated through region I, $n_I(E)$, which from equations (2.35) and (2.37) is given by

$$n_I(E) = \int_I d\mathbf{r}\, n(\mathbf{r}, E) = -\frac{1}{\pi} \Im \sum_{mn} G_{mn}(E + i\eta) O_{nm}, \quad E \text{ real}, \eta \to 0+. \quad (2.43)$$

As we want the density of states to contain bound states as well as the continuum, we usually take η in equation (2.43) to be small but finite—this broadens the δ functions of the discrete states into narrow Lorentzians.

As in equation (2.29), we find the embedding potentials at complex energy $E + i\eta$ from the solution of the Schrödinger equation in region II, using equation (2.16). The required solution to the left of $-d/2$ is

$$\psi = \exp(-ikz), \quad \text{with } k = \sqrt{2(E + i\eta)}, \quad (2.44)$$

and we choose the root in the upper right quadrant of the complex k-plane so that ψ is travelling away and decaying into region II. This gives for the embedding potentials,

$$\Sigma(-d/2) = -i\sqrt{\frac{E + i\eta}{2}}, \quad \text{and similarly, } \Sigma(+d/2) = -i\sqrt{\frac{E - w + i\eta}{2}}. \quad (2.45)$$

The choice of root is, conveniently, the definition of the intrinsic Fortran function sqrt when $\eta \geqslant 0$ [14].

It is now straightforward to calculate the matrix G, hence $n_I(E)$, using the basis set (equation (2.30)) which we used to find the bound states and eigenvalues. The convergence properties are just the same as for the bound states discussed in section 2.3, and we take 15 basis functions with $D = 12$ a.u. The results are shown in figure 2.4, with an energy broadening of $\eta = 10^{-5}$ a.u. For $E < 0$ the bound states calculated via the Green function are at precisely the right energies, while for $E > 0$ we see a structured density of states. Immediately above the continuum edge at $E = 0$, the density of states varies like $n_I(E) \sim E^{\frac{1}{2}}$, characteristic of a one-dimensional band edge at a surface [15]; this is to be compared with the $E^{-\frac{1}{2}}$ behaviour in the bulk. There is a similar square-root singularity above the vacuum continuum edge at $E = 0.5$ a.u., though this is much weaker.

There is actually some physics we can extract from this simple model! Figure 2.5 shows $n_I(E)$ close to the continuum edge for different depths of the surface potential well, varying between -0.40 a.u. and -0.44 a.u. We see that at -0.40 a.u. (the red curve) there is a large peak just above the continuum edge, with a tiny feature just below the edge which is a very weakly bound state. As the well deepens, the bound state is pulled further off the edge, and at the same time the peak above the edge gets

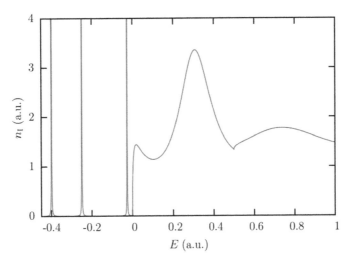

Figure 2.4. $n_I(E)$, the local density of states integrated though region I for the model surface potential (figure 2.1, $v = -0.45$ a.u., $d = 8$ a.u., $w = 0.5$ a.u.). n_I is calculated with 15 basis functions, $D = 12$ a.u., and the imaginary part of the energy is $\eta = 10^{-5}$ a.u.

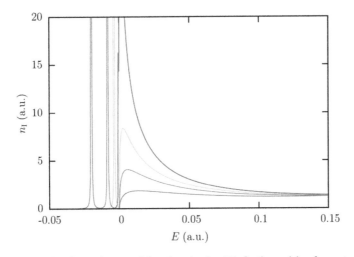

Figure 2.5. The local density of states integrated though region I, $n_I(E)$, for the model surface potential (figure 2.1) with $v = -0.40$ a.u. (red curve), -0.41 a.u. (green curve), -0.42 a.u. (blue curve), and -0.44 a.u. (violet curve). Other parameters in the calculation are the same as in figure 2.4.

smaller: the bound state is pulled out of the peak at the bottom of the continuum. We have, in fact, some sort of conservation of the local density of states, behaviour which is seen in several 'real' contexts, including electronic surface states pulled off the band edge [16], and bound plasmons dropping below the light-line in plasmonic structures [17]. There is another significant feature in figure 2.4, a resonance at $E \approx 0.3$ a.u. As the well deepens, this resonance moves down in energy, ultimately becoming a new peak at the bottom of the continuum.

2.6 Resonances and complex eigenvalues

Narrow peaks in the density of states, like the peak at $E = 0.3$ a.u. in figure 2.4, are usually associated with resonances—states which are almost localized but can leak out into the continuum. If we imagine putting an electron in the square well (figure 2.1) with energy 0.3 a.u., this will leak out in a time Δt given by

$$\Delta t \sim \frac{\hbar}{\Delta E}, \qquad (2.46)$$

where ΔE is the width of the peak [18]—this is the (so-called—see [19]) energy–time uncertainty principle. To explore the resonances in the surface model more thoroughly, in figure 2.6 we plot the well density of states over a wider range of energies, and by inspection we would say that there are resonances at about $E = 0.3$, 0.7, 1.4 a.u., with a slight wiggle at $E = 2.2$ a.u. We don't include the weak peak just above $E = 0$ in the resonances, as the bound state at $E = -0.026$ a.u. has already been pulled from the bottom of the band.

To understand the resonances in more detail we consider the analytic properties of the embedding potential $\Sigma(E)$ as a function of *complex* energy E.[2] When region II is bounded, $\Sigma(E)$ has discrete poles on the real energy axis. We can see this from equation (2.16): $\Sigma(E)$ blows up at energies at which the solution $\psi(\mathbf{r})$ of the Schrödinger equation in region II satisfies the homogeneous boundary condition $\psi(\mathbf{r}_S) = 0$. When region II is bounded these correspond to discrete, real, energy eigenvalues E_i. The variational derivation of the embedding method which we gave

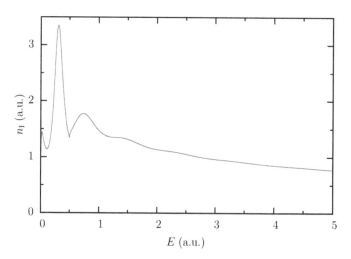

Figure 2.6. The local density of states integrated through region I, $n_I(E)$, for the model surface potential (figure 2.1). Parameters are the same as in figure 2.4.

[2] To avoid too many symbols, we are using E for the complex energy variable, even though we use it elsewhere for real energy. This should always be clear from the context.

earlier in this chapter really corresponds to the case of finite region II, as we assumed a normalizable trial function in this region. As region II gets bigger, the poles of $\Sigma(E)$ get closer together on the real axis, ultimately merging into a branch cut. With $V = 0$ in region II, the embedding potential (in one dimension) is given by equation (2.45), which we write in terms of complex E as

$$\Sigma(E) = -ik/2, \quad \text{with } k = \sqrt{2E}, \tag{2.47}$$

and the square root gives the branch cut starting at $E = 0$ (figure 2.7). When E is in the upper half-plane, both $\Re k$ and $\Im k$ are positive, corresponding to outgoing and decaying waves (as in section 2.5). However, we sometimes need an embedding potential to describe *incoming* waves, in particular when we apply the embedding potential to transport (section 7.1.1) [20]. This corresponds to evaluating Σ with E in the lower half-plane, so that $\Re k$ is negative while $\Im k$ remains positive— incoming waves but still decaying in the outward direction. By analogy with Green functions, when E is in the upper half-plane we have the *retarded* embedding potential, while E in the lower half-plane gives the *advanced* embedding potential [6].

If we stick to the retarded embedding potential, we can rotate the branch cut to extend down the negative imaginary axis (figure 2.8). In the lower right quadrant, $\Re k$ is positive and $\Im k$ negative, giving unphysical outgoing waves which increase into region II. The unphysical Σ in this quadrant is the analytic continuation of the

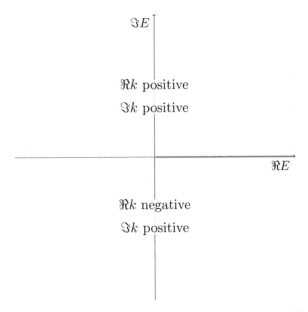

Figure 2.7. Complex E-plane: when region II is an extended one-dimensional system with $V = 0$, $\Sigma(E)$ has the branch cut shown by the red line, extending from $E = 0$ along the positive real axis. The upper half-plane (shaded green) corresponds to outgoing ($\Re k$ positive) and decaying ($\Im k$ positive) waves, while the lower half-plane (shaded yellow) corresponds to incoming and decaying waves.

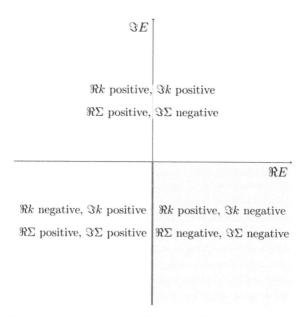

Figure 2.8. Complex E-plane: the branch cut of $\Sigma(E)$ is rotated to lie along the negative imaginary axis. Green shading corresponds to retarded waves, yellow shading to advanced waves and pink shading to the analytic continuation of the retarded waves on to the unphysical sheet, where the waves are outgoing but increasing. Resonances correspond to poles of the Green function in the region shaded pink.

physical Σ in the upper half-plane on to the *unphysical sheet* [21]. Resonances in the local density of states then correspond to poles in $G(E)$, the Green function in region I, at values of E in this lower right quadrant [18]. We can see how such poles can only occur in this quadrant by writing the Green function in region I as

$$G(E) = \frac{1}{E - H - \Sigma(E)}. \tag{2.48}$$

Here, H is the Hamiltonian for region I by itself, representing the first line of equation (2.36). The conditions for a pole are then

$$\Re E = H + \Re\Sigma(E), \quad \Im E = \Im\Sigma(E). \tag{2.49}$$

From figure 2.8 we see that if $\Im E \neq 0$, the second condition can only be satisfied in the lower right quadrant. Let us suppose that these conditions are satisfied at complex energy E_i. The contribution of this pole to the Green function is then given by

$$G(E) = \frac{a_i}{E - E_i}, \tag{2.50}$$

where a_i is the residue of the pole (let us assume for the moment that this is real). So the local density of states in region I—a function of *real* energy—is given by the Breit–Wigner resonance formula [21]

Table 2.2. Complex eigenvalues E_i for the model surface potential with parameters given in figure 2.4. $\Re E_i$ gives the centre of the resonance and $|\Im E_i|$ the half-width at half-maximum.

$\Re E_i$	$\Im E_i$
0.29904	−0.09714
0.68095	−0.30108
1.34963	−0.53503
2.17220	−0.77563
3.15098	−1.02814
4.28585	−1.29251

$$n_l(E) = -\frac{1}{\pi}\Im G(E), \quad E \text{ real},$$

$$= -\frac{a_i}{\pi}\frac{\Im E_i}{(E - \Re E_i)^2 + (\Im E_i)^2}, \tag{2.51}$$

which is a Lorentzian centred on $\Re E_i$ with half-width at half-maximum given by $|\Im E_i|$.

We can find the 'unphysical' poles corresponding to resonances in the density of states by solving the eigenvalue equation (2.25), but instead of looking for real eigenvalues we look for complex eigenvalues on the unphysical sheet (the region shaded pink in figure 2.8) [3]. As in section 2.2, we take a trial energy E_0 (complex in this case), at which we evaluate $\Sigma(E_0)$ and $\partial\Sigma/\partial E_0$, and find the eigenvalues by iteration. Having $\partial\Sigma/\partial E_0$ in the matrix equation ensures very rapid convergence. Results for the first few complex eigenvalues for the model surface potential, with the same parameters as in figures 2.4 and 2.6, are shown in table 2.2. We see that the first four are in good agreement with the 'estimates' we made by inspecting figure 2.4 (I actually made these estimates *before* calculating the eigenvalues!). As a realistic application of this approach, in section 9.3.4 we use complex eigenvalues to study frequencies and linewidths of electromagnetic eigenmodes in a lattice of metal cylinders.

2.6.1 Resonances in the spherical square well

Using the complex poles to calculate the physical properties of a system—particularly resonance scattering in nuclear physics—has a very long history [21], and here we ask how well can we describe $n_l(E)$, using a generalization of equation (2.51). We shall see in this section that at least in some systems we can find the local density of states *exactly*, as long as we include a correction coming from the branch cut in the Green function[3]. I discussed complex poles and their relation to $n_l(E)$ in my first embedding paper [1], and we shall consider the same system here: the spherically symmetric square well.

---

[3] Dr Egor Muljarov helped me greatly in understanding this topic, and showed me how to include the branch cut contribution.

We take a spherically symmetric square well, depth v, radius r_s, calculating the s-wave Green function and the corresponding density of states. This one-dimensional problem can be solved analytically, but to make contact with embedding in general we shall expand the Green function inside the well, region I, in terms of the basis set

$$\chi_n(r) = \frac{\sin(n\pi r/D)}{r}, \tag{2.52}$$

where as usual the parameter D is larger than r_s. In region II outside the well, the s-wave solutions of the free-space Schrödinger equation are given by

$$\psi(r) = \frac{\exp(\mathrm{i}kr)}{r}, \tag{2.53}$$

so from equation (2.16) the embedding potential at the surface of the spherical well is given by

$$\Sigma(E) = \frac{\left(1 - \mathrm{i}kr_s\right)}{8\pi r_s^3}, \quad \text{with } k = \sqrt{2E}. \tag{2.54}$$

The Fortran function `sqrt` automatically gives the required root when E is in the unphysical region, that is, $\Re k$ positive and $\Im k$ negative (figure 2.8) [14].

To show my first attempts many years ago, figure 2.9 gives $n_{\mathrm{I}}(E)$ for the spherical square well with $v = 1$ a.u., $r_s = 2$ a.u. calculated with a very small basis set [1]. In this paper I calculated the complex eigenvalues of the first two resonances, giving $(1.3, -1.2)$ a.u. and $(6.2, -3.2)$ a.u. The lower part of figure 2.9 shows the corresponding Lorentzians (equation 2.51), corresponding not too badly to the shape of $n_{\mathrm{I}}(E)$.

The obvious problem with describing $n_{\mathrm{I}}(E)$ in terms of Lorentzians by themselves is that this cannot describe the singularity at the bottom of the band accurately—here, the singularity has the form $\sim E^{1/2}$. To do better we have to consider the full analytic structure of the the Green function in region I as a function of complex energy E,

$$G(E) = (EO - H - \Sigma(E))^{-1}. \tag{2.55}$$

This is the same as equation (2.42), except that we separate $\Sigma(E)$ explicitly from H, to show how complex E enters G. If we evaluate $\Sigma(E)$ with the branch cut shown in figure 2.8, we obtain the analytic structure of $G(E)$ shown schematically in figure 2.10, with a branch cut extending from $E = 0$ to $-\mathrm{i}\infty$, and simple poles at the bound states and resonances.

We can now use the Mittag-Leffler theorem [22, 23] to find $G(E)$. Following from Cauchy's theorem[4], this tells us that $G(E)$ inside contour C (in blue, figure 2.10) is given by

$$G(E) = \sum_i \frac{R(E_i)}{E - E_i} + \frac{1}{2\pi\mathrm{i}} \oint_C \mathrm{d}E' \frac{G(E')}{E' - E}, \tag{2.56}$$

---

[4] The epigraph of chapter 12 (*Contour Integration and Bromwich's Integral*) in Jeffreys and Jeffreys [22] is: "Go round about, Peer Gynt!" (in reference to the Henrik Ibesen play).

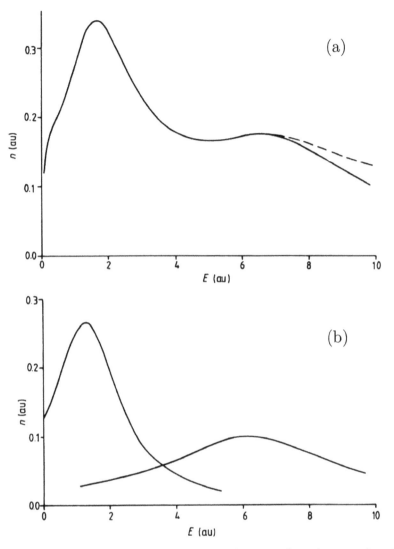

Figure 2.9. Continuum calculations for the spherically symmetric square well, $v = 1$ a.u., $r_s = 2$ a.u. (a) s-wave local density of states integrated through region I, $n_I(E)$: full curve, four basis functions; dashed curve, eight basis functions. (b) Lorentzians corresponding to resonances with complex eigenvalues, at $\omega_i = (1.3, -1.2)$ a.u. and $(6.2, -3.2)$ a.u. (from [1]).

where the sum is over the poles E_i of G inside C, with $R(E_i)$ the residue at the pole, and the integral is round the contour C in the anti-clockwise direction. We now let the radius of C tend to infinity, and then, if $G(E)$ decreases quickly enough, the only contribution to the integral in equation (2.56) comes from the branch cut, and we are left with

$$G(E) = \sum_i \frac{R(E_i)}{E - E_i} + \frac{1}{2\pi i} \int_0^{-i\infty} dE' \frac{\Delta G(E')}{E' - E}, \qquad (2.57)$$

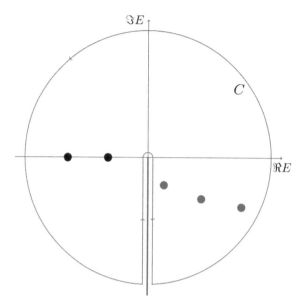

Figure 2.10. Analytic structure of $G(E)$ in the complex E-plane: black circles indicate bound states with negative real E and red circles resonances at which G has simple poles. The red line indicates the branch cut in G, which (following our choice of branch cut in Σ) extends down the negative imaginary E-axis. The blue contour C is used to evaluate G with contour integration.

where $\Delta G(\omega')$ in the integrand is the difference between G on the right- and the left-hand sides of the branch cut. The branch cut integral is the correction to the sum over the Lorentzians which we are looking for.

Not only can we find the poles themselves from the embedded eigenvalue equation, but also the residues. From the definition of the Green function (equation (2.33)), the residue at bound states is the product of the eigenstates, normalized by equation (2.31). We use the analytic continuation of this normalization to obtain

$$R(E_i) = \frac{\phi_i(\mathbf{r})\phi_i(\mathbf{r}')}{\displaystyle\int_I d\mathbf{r}\,\phi_i(\mathbf{r})^2 - \int_S d\mathbf{r}_S \int_S d\mathbf{r}'_S\,\phi_i(\mathbf{r}_S)\left.\frac{\partial\Sigma}{\partial E}\right|_{E=E_i}\phi_i(\mathbf{r}'_S)}, \tag{2.58}$$

where ϕ_i is the unnormalized eigenvector with eigenvalue E_i. Note that we must take care not to take the complex conjugate of ϕ_i in equation (2.58) as this is not an analytic operation. The $\phi_i(\mathbf{r})$ are *resonant states*, the wave-functions of the resonances [24, 25].

We now apply the Mittag-Leffler theorem to calculate the s-wave Green function of the spherical square well, with the same parameters as in figure 2.9, $v = 1$ a.u., $r_s = 2$ a.u. We calculate the local density of states integrated through a sphere of radius ρ inside the square well, given by

$$n_\rho(E) = -\frac{1}{\pi}\Im g_\rho(E)(E\text{ real}), \quad \text{with } g_\rho(E) = 4\pi\int_0^\rho dr\, r^2 G(r, r; E). \tag{2.59}$$

We cannot take $\rho = r_s$ because $g_{r_s}(E)$ does not decrease fast enough for the circular part of the contour integral in equation (2.56) to go to zero as $|E| \to \infty$; however, for $\rho < r_s$ we can apply equation (2.57) to $g_\rho(E)$. (This was pointed out to me by Egor Muljarov, and it is something which needs further study.) Here, we take $\rho = 1.5$ a.u.

Table 2.3. Poles and residues of the s-wave Green function for the spherically symmetric square well, $v = 1$ a.u., $r_s = 2$ a.u. The first two columns give real and imaginary parts of the poles of $g_\rho(E)$ (equation (2.59)), and the last two columns real and imaginary parts of the residues. Calculated with the embedded eigenvalue equation using 30 basis functions defined with $D = 2.5$ a.u.

E_i		$R(E_i)$	
−0.3772	0.0000	0.5051	0.0000
1.2917	−1.1838	0.6787	0.0562
5.9107	−3.1961	1.1172	0.0963
13.0707	−5.4960	1.0294	−0.4760
22.7412	−8.0130	0.3041	−0.3808
34.9079	−10.7008	0.3796	0.4461
49.5629	−13.5293	1.2800	0.3792
66.7011	−16.4771	1.2074	−0.5921
86.3193	−19.5285	0.1603	−0.5184
108.4151	−22.6715	0.2282	0.5961
132.9867	−25.8965	1.4012	0.5320

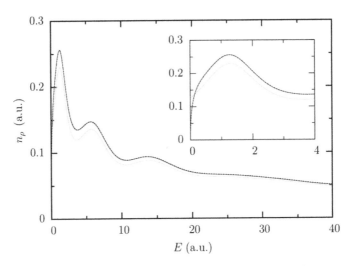

Figure 2.11. Local density of states $n_\rho(E)$ integrated through a sphere of radius $\rho = 1.5$ a.u., for the spherically symmetric square well, $v = 1$ a.u., $r_s = 2$ a.u. Red curve, calculated from the poles + branch cut contribution, using Mittag-Leffler (equation (2.57)); green curve, calculated from the poles alone; dashed black line, exact result. The inset shows the results at small E near the bottom of the continuum.

As in section 2.6, we calculate the poles and residues of $g_\rho(E)$ using the embedded eigenvalue equation, iterating on the eigenvalue to find $\Sigma(E_i)$. Table 2.3 gives our results for the first 11 poles, corresponding to the one bound state of the system and the 10 resonances with the lowest energy. We calculate the discontinuity across the branch cut, $\Delta g_\rho(E)$, from embedded Green function calculations using Σ evaluated on each side of the cut; $\Delta g_\rho(E)$ goes rapidly to zero as E moves along the negative imaginary axis, making the branch cut integral easy to evaluate. Putting the poles and the branch cut integral together in the Mittag-Leffler equation (2.57), we can then evaluate $g_\rho(E)$ and $n_\rho(E)$. Figure 2.11 shows $n_\rho(E)$ calculated in this way, compared with the contribution from the poles by themselves (the first term in equation (2.57)) and the exact value. We see that the Mittag-Leffler results agree beautifully with the exact results for $n_\rho(E)$—to within the linewidth of the plot. The inset of figure 2.11, which shows $n_\rho(E)$ close to the band edge, demonstrates how the contribution from the branch cut integral gives the correct singularity in the Green function at the bottom of the band. We shall come back to this model in section 11.2, where we shall discuss resonant states and their relation with embedding in more detail.

References

[1] Inglesfield J E 1981 A method of embedding *J. Phys. C: Solid St. Phys.* **14** 3795–806
[2] Inglesfield J E 2001 Embedding at surfaces *Comput. Phys. Commun.* **137** 89–107
[3] More R M and Gerjuoy E 1973 Properties of resonance wave-functions *Phys. Rev.* A **7** 1288–303
[4] Barton G 1989 *Elements of Green's Functions and Propagation* (Oxford: Oxford University Press)
[5] Jackson J D 1998 *Classical Electrodynamics* 3rd edn (New York: John Wiley)
[6] Merzbacher E 1998 *Quantum Mechanics* 3rd edn (New York: John Wiley)
[7] Colton D and Kress R 1998 *Inverse Acoustic and Electromagnetic Scattering Theory* 2nd edn (Berlin: Springer-Verlag)
[8] Szmytkowski R and Bielski S 2004 Dirichlet-to-Neumann and Neumann-to-Dirichlet embedding methods for bound states of the Schrödinger equation *Phys. Rev.* A **70** 042103
[9] Thijssen J M 2007 *Computational Physics* 2nd edn (Cambridge: Cambridge University Press)
[10] Bloch C 1957 Une formulation unifiée de la théorie des réactions nucléaires *Nucl. Phys.* **4** 503–28
[11] Burke P G 2011 *R-Matrix Theory of Atomic Collisions* (Berlin: Springer-Verlag)
[12] Negele J W and Orland H 1987 *Quantum Many-Particle Systems* (California: Addison-Wesley)
[13] Datta S 2005 *Quantum Transport: Atom to Transistor* (Cambridge: Cambridge University Press)
[14] Metcalf M, Reid J and Cohen M 2004 *Fortran 95/2003 Explained* (Oxford: Oxford University Press)
[15] Inglesfield J E 1978 The electronic structure of surfaces with the matching Green function method I. *Surf. Sci.* **76** 355–78
[16] van Hoof J B A N, Crampin S and Inglesfield J E 1992 The surface state-surface resonance transition on Ta(011) *J. Phys.: Condens. Matter* **4** 8477–88

[17] Giannakis N A, Inglesfield J E, Jastrzebski A K and Young P R 2013 Photonic modes of a chain of nanocylinders by the embedding method *J. Opt. Soc. Am.* B **30** 1755–64

[18] Moiseyev N 1998 Quantum theory of resonances: calculating energies, widths and cross-sections by complex scaling *Phys. Rep.* **302** 211–93

[19] Landau L D and Lifschitz E M 1977 *Quantum Mechanics (Non-relativistic Theory)* 3rd edn (Oxford: Elsevier Butterworth-Heinemann)

[20] Inglesfield J E, Crampin S and Ishida H 2005 Embedding potential definition of channel functions *Phys. Rev.* B **71** 155120

[21] Newton R G 1982 *Scattering Theory of Waves and Particles* 2nd edn (New York: Springer-Verlag)

[22] Jeffreys S H and Swirles B (Lady Jeffreys) 1972 *Methods of Mathematical Physics* 3rd edn (Cambridge: Cambridge University Press)

[23] Zeidler E 2004 *Oxford Users' Guide to Mathematics* (Oxford: Oxford University Press)

[24] Siegert A J F 1939 On the derivation of the dispersion formula for nuclear reactions *Phys. Rev.* **56** 750–2

[25] Lind P 1993 Completeness relations and resonant state expansions *Phys. Rev.* C **47** 1903–20

IOP Publishing

The Embedding Method for Electronic Structure

John E Inglesfield

Chapter 3

Embedding at surfaces

Surface electronic structure calculations have been the most successful application of the embedding method [1], and very accurate embedding codes have been developed for solving the Schrödinger equation for surfaces and interfaces [2]. Embedding has the advantage of economy—the surface electronic structure is given very accurately with only the top two or three atomic layers constituting region I—but more importantly, the accurate treatment of the semi-infinite substrate enables true surface states to be distinguished from surface resonances [3].

The starting point for electronic structure calculations in condensed matter physics is the one-electron Schrödinger equation [4]

$$\left(-\frac{1}{2}\nabla^2 + V(\mathbf{r})\right)\Psi_i(\mathbf{r}) = E_i\Psi_i(\mathbf{r}), \quad \text{where } V(\mathbf{r}) = \underbrace{V_{\text{ion}}(\mathbf{r}) + V_{\text{H}}(\mathbf{r})}_{V_{\text{es}}(\mathbf{r})} + V_{\text{xc}}(\mathbf{r}). \quad (3.1)$$

$V(\mathbf{r})$ is the effective potential felt by an electron, shown in figure 3.1 for a typical metal surface, Au(111). V_{ion} is the electrostatic potential due to the ionic or nuclear charge and V_{H} is the Hartree potential of the electronic charge, which we group together as V_{es}; V_{xc} is the exchange-correlation potential, which takes into account the exchange interaction and the correlated motion of the electrons, calculated within the framework of DFT [5]. V_{H} and V_{xc} depend self-consistently on the electron density [5].

In the bulk crystal, $V(\mathbf{r})$ has three-dimensional periodicity, and the wave-functions can be labelled by the three-dimensional Bloch wave-vector $\mathbf{k}$; it is then only necessary to solve equation (3.1) in one unit cell to find the band structure, hence the electronic structure of the whole solid. However, $V(\mathbf{r})$ ar the surface has only two-dimensional periodicity, as we can see in figure 3.1, and the periodicity in the perpendicular direction is lost. The wave-functions can still be labelled by a two-dimensional wave-vector $\mathbf{k}_\parallel$ parallel to the surface [6], but in the perpendicular direction equation (3.1) has to be integrated from the vacuum (the easy bit), through

doi:10.1088/978-0-7503-1042-0ch3

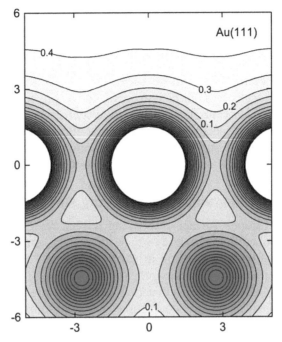

Figure 3.1. Potential $V(\mathbf{r})$ at the Au (111) surface calculated with embedding, plotted on a $(11\bar{2})$ plane intersecting nearest-neighbour surface atoms. The deeper the colour, the deeper the potential, except at the centres of the atoms, where the potential is too deep to plot. The zero of energy is the average interstitial potential, contours are plotted at intervals of 0.1 a.u., and the asymptotic vacuum potential is at 0.517 a.u. (Figure courtesy of H Ishida.)

the surface, and all the way into the bulk to find $\Psi_{\mathbf{k}_{\parallel},i}(\mathbf{r})$. But this is where embedding comes into its own, as the region we are generally interested in—the region which determines surface properties—is the top few layers of atoms, extending into the vacuum. This will be embedded on to the semi-infinite bulk crystal, allowing us to treat the continuum of states associated with bulk energy bands, as well as the discrete, localized surface states with energies in bulk band gaps.

In this chapter we shall derive the embedding potential for embedding the surface on to the bulk (section 3.1), and describe how the embedded Schrödinger equation can be solved (section 3.2). Other ways of calculating the bulk embedding potential will be discussed in sections 3.4 and 3.6. One important aspect of embedding, which we explore throughout this chapter, is the way in which the embedding surface can be shifted to a geometrically convenient surface, something which is very important for simplifying the calculations. Of course, it's the physics which counts, and we shall look at charge density and work-function results in section 3.3 and other properties of electrons at surfaces in the next chapter. Although most surface applications of embedding are to systems with two-dimensional periodicity, embedding has been applied to the isolated adsorbate problem [7], and we shall discuss this in section 3.7. This is also an example of a total energy calculation and energy minimization within the framework of embedding.

Before we discuss surface embedding, we ought to say that this method has only been used for a small fraction of surface electronic structure calculations. Most surface calculations treat a slab typically ~10 atomic layers thick: if this is repeated in the perpendicular direction, we recover full three-dimensional periodicity, and bulk electronic structure codes can be used. This works fine for properties like electron density, surface energy, surface structure and so on, because the effect of the surface on the other side of the slab is screened away (a very recent example is [8]). However, particularly for spectroscopic studies it is often important to calculate the electronic structure of the surface of a semi-infinite solid, something for which embedding is ideally suited (a recent embedding example is [9]).

3.1 Surface embedding and the embedding surface

Not too confusing I hope—surface embedding is what we are trying to do in this chapter, and the embedding surface is where we put the embedding potential! In implementing the embedding method at surfaces, region I comprises the atomic layers in which the potential differs significantly from the bulk, extending into the vacuum. In metals, where there is efficient screening of the potential, this means that we typically consider the top two or three layers of atoms, which we embed on to the perfect semi-infinite crystal (figure 3.2). The 'natural' embedding interface between the surface and the semi-infinite bulk crystal, S_c in figure 3.2, weaves its way between the atoms, but this can be moved to a more convenient planar boundary. On the vacuum side there are two approaches: region I can be terminated by a second embedding surface S_{vac}, with an embedding potential replacing the asymptotic vacuum potential (figure 3.2) [2, 10], or it can extend indefinitely into the vacuum region, with the basis functions providing the correct asymptotic behaviour [11].

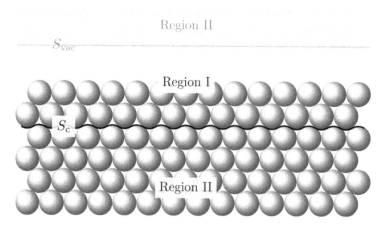

Figure 3.2. Surface region I embedded on to the semi-infinite bulk crystal, region II. The 'natural' embedding surface is S_c, weaving its way between the atoms. Region I can either extend into the vacuum, beyond the tails of the wave-functions, or be terminated by the vacuum embedding surface S_{vac}, over which a vacuum embedding potential replaces the vacuum region II (pale blue).

3.1.1 Embedding and the reflection matrix

The crystal embedding potential can be found very easily from the reflection properties of the semi-infinite bulk crystal, and this was used in the original surface embedding calculations [11, 12]. The potential in the bulk crystal is assumed to have muffin-tin form—a spherically symmetric potential within a 'muffin-tin' at each atom, and a constant potential in the interstitial region between the atoms (this form of potential is often used in scattering methods for calculating electronic structure [4]). This does not restrict the form of potential within the surface region I, where a full-potential method like the linearized augmented plane-wave (LAPW) method can be used [13, 14]. In the LAPW method, $V(\mathbf{r})$ is again partitioned into muffin-tins and an interstitial region—a natural partitioning, as we can see from figure 3.1—but non-spherical terms within the atoms and a varying potential between the atoms are added.

We take a semi-infinite bulk crystal, continuing the constant interstitial potential (taken as $V = 0$) into the positive z-direction. With an incident electron wave as shown in figure 3.3, the total wave-function above the top layer of muffin-tins can be written as

$$\Psi(\mathbf{r}) = \exp\left[i(\mathbf{k}_\parallel + \mathbf{g}) \cdot \mathbf{r}_\parallel\right]\exp(-ik_g z) + \sum_{\mathbf{g}'} R_{\mathbf{gg}'}\exp\left[i(\mathbf{k}_\parallel + \mathbf{g}') \cdot \mathbf{r}_\parallel\right]\exp(ik_{g'}z), \quad (3.2)$$

where the first term is the incident wave, and the summation is over the reflected waves, with $R_{\mathbf{gg}'}$ the reflection matrix. Here, $\mathbf{r}_\parallel$ represents the coordinates (x,y) parallel to the surface and z is the surface normal; the $\mathbf{g}$'s are two-dimensional surface reciprocal lattice vectors, and we have

$$\left|\mathbf{k}_\parallel + \mathbf{g}'\right|^2 + k_{g'}^2 = \left|\mathbf{k}_\parallel + \mathbf{g}\right|^2 + k_g^2 = 2E, \quad (3.3)$$

with E the energy of the incident wave. We can now substitute equation (3.2) in equation (2.16) to find the embedding potential over the planar surface S; at first sight this seems contradictory, as equation (3.2) is the wave-function *outside* the

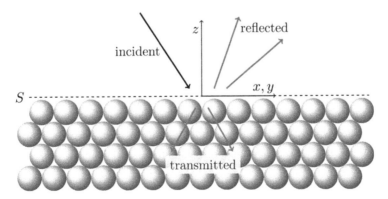

Figure 3.3. An electron wave is incident on the semi-infinite bulk crystal, with the constant interstitial potential continued in the positive z-direction. The embedding potential can be found on the planar surface S from the reflection amplitudes.

semi-infinite crystal—region II—whereas equation (2.16) was derived for the wave-function *within* region II! But $\Psi(\mathbf{r})$ given by equation (3.2) is part of the wave-function for the entire system, and asymptotically this becomes the transmitted wave travelling or decaying away from S. On the embedding plane S we expand the embedding potential Σ as a Fourier series,

$$\Sigma\big(\mathbf{r}_\|, \mathbf{r}_\|';\, \mathbf{k}_\|\big) = \frac{1}{A} \exp\Big[i\mathbf{k}_\| \cdot \big(\mathbf{r}_\| - \mathbf{r}_\|'\big)\Big] \sum_{\mathbf{gg}'} \Sigma\big(\mathbf{k}_\|\big)_{\mathbf{gg}'} \exp\Big[i\big(\mathbf{g} \cdot \mathbf{r}_\| - \mathbf{g}' \cdot \mathbf{r}_\|'\big)\Big], \qquad (3.4)$$

where A is the area of the two-dimensional surface unit cell and $(\Sigma_{\mathbf{k}_\|})_{\mathbf{gg}'}$ are the embedding potential matrix elements. Then, from equation (2.16) the matrix elements are given by

$$\Sigma\big(\mathbf{k}_\|\big)_{\mathbf{gg}'} = -\frac{ik_{\mathbf{g}}}{2}\big[(I - R)(I + R)^{-1}\big]_{\mathbf{gg}'}, \qquad (3.5)$$

where I is the unit matrix.

The reflection matrix R can be found very efficiently using the layer-doubling method, used in low-energy electron diffraction (LEED) calculations [15]. The crystal is divided into layers, and multiple scattering techniques used to determine the reflection and transmission matrices of a single layer. These can then be used to find the reflection and transmission of a pair of layers, and the process repeated for four, eight,… layers. With a small imaginary part in the energy, the process converges rapidly to give the reflection matrix for the semi-infinite crystal. R can in fact be found relative to a shifted plane surface, even one overlapping the muffin-tins; R describes the scattering between plane-waves by the semi-infinite bulk, and there is freedom in choosing the position of S where this is evaluated.

3.1.2 Shifting the embedding surface

Embedding the surface region I over the flat embedding plane S is equivalent to embedding over the natural embedding surface S_c [12]. (It is because of the muffin-tin form of potential that we talk about S_c in the first place.) Assume that S lies on the bulk side of S_c, so that it does not intersect any of the muffin-tins in region I; we then extend region I through S_c up to S, and in this buffer region we take $V = 0$, the same as in the substrate interstitial region (figure 3.4). The trial wave-function, which satisfies the Schrödinger equation in extended region I, satisfies the free-electron equation in the buffer region. But the scattering of *any* free-electron wave by the bulk crystal is described by the reflection matrix, hence by Σ. Uniqueness ensures that this is *the* solution of the Schrödinger equation matching on to the bulk, the same as if we used an embedding potential over S_c. This is convenient because it is much easier to evaluate the matrix elements of the Hamiltonian in region I when it is bounded by a flat embedding surface.

There is in fact more freedom in the choice of embedding plane than we have implied so far: S can lie between the top atomic layer of region II and the bottom atomic layer of region I, intersecting the muffin-tins in *both* regions [12, 16].

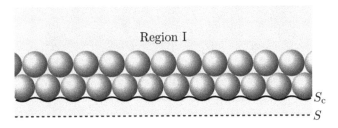

Figure 3.4. Region I (shaded grey) extends to the planar embedding surface S and includes a buffer region between the natural embedding surface S_c and S. Above S_c the full surface potential is used in the Hamiltonian; between S_c and S, $V = 0$ (the same as the bulk interstitial potential).

To understand this we use the variational expression for the energy (equation (2.22)), with the trial wave-function ϕ taken as a linear combination of LAPWs—an LAPW consists of a plane-wave in the interstitial region outside the muffin-tins, joined on to a linear combination of atomic functions inside each muffin-tin (we shall give their precise form in section 3.2.1). Writing it out term by term (this is how we actually evaluate the integrals), the embedded variational expression (2.22) becomes

$$
E = \left(\int_I d\mathbf{r}\, \phi_{pw}^* H_0 \phi_{pw} + \int_{int} d\mathbf{r}\, \phi_{pw}^* V \phi_{pw} - \int_{MT} d\mathbf{r}\, \phi_{pw}^* H_0 \phi_{pw}^* + \int_{MT} d\mathbf{r}\, \phi_{at}^* H \phi_{at} \right.
$$
$$
\left. + \frac{1}{2} \int_S d\mathbf{r}_S\, \phi_{pw}^* \frac{\partial \phi_{pw}}{\partial n_S} + \int_S d\mathbf{r}_S \int_S d\mathbf{r}_{S'}\, \phi_{pw}^* \left\{ \Sigma - \epsilon \frac{\partial \Sigma}{\partial \epsilon} \right\} \phi_{pw} \right) \Bigg/
$$
$$
\left(\int_I d\mathbf{r}\, \phi_{pw}^* \phi_{pw} - \int_{MT} d\mathbf{r}\, \phi_{pw}^* \phi_{pw} + \int_{MT} d\mathbf{r}\, \phi_{at}^* \phi_{at} - \int_S d\mathbf{r}_S \int_S d\mathbf{r}_S'\, \phi_{pw}^* \frac{\partial \Sigma}{\partial \epsilon} \phi_{pw} \right),
$$

$$(3.6)$$

where ϕ_{pw} is the plane-wave part of ϕ (in fact a linear combination of plane-waves), and ϕ_{at} is the atomic-like part of ϕ inside the muffin-tins (a linear combination of atomic solutions).

We now go through the integrals in equation (3.6) one by one. The first integral in the numerator is over the whole of region I up to the planar surface S—chosen as in figure 3.4—with ϕ_{pw} continued through the muffin-tins; H_0 is the kinetic energy operator, the Hamiltonian with $V = 0$. The second integral is over the interstitial region: through region I, excluding the muffin-tins, up to the curvy surface S_c, and with V the full interstitial potential. The third and fourth integrals are over the muffin-tins, the third integral removing the muffin-tins from the first integral, and the fourth integral containing ϕ_{at} with the full Hamiltonian H. When equation (3.6) is stationary, ϕ is embedded correctly on to the bulk crystal over S_c. We can now transfer part of the first integral in both numerator and denominator, between S and S', to region II (upper part of figure 3.5), so that region I in these integrals is bounded by S' (the grey shaded area in the lower part of figure 3.5). This is compensated by evaluating an embedding potential on S' with $V = 0$

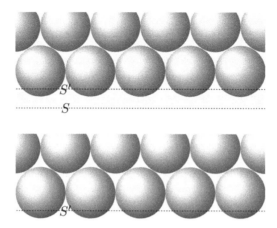

Figure 3.5. Grey shading in the top figure shows region I extending beyond the muffin-tins to embedding surface S. The free-electron Hamiltonian between S and S' can be transferred from region I to region II, shifting the embedding plane from S to S' (lower figure).

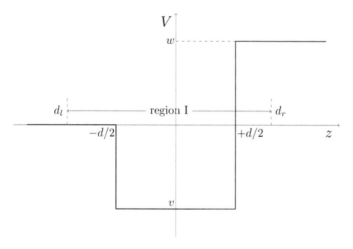

Figure 3.6. One-dimensional square-well potential at a surface, embedding at d_l and d_r.

between S and S'—this is how it is calculated anyway in equation (3.5). Embedding on to S' is then equivalent to embedding on to S, which in turn is equivalent to embedding on to S_c.

3.1.3 Testing the shifted embedding plane

To show that an embedding surface overlapping the muffin-tins really works, and to understand *how* it works, we now study the one-dimensional model of a surface with a potential well which we introduced in chapter 2 (figure 2.1), but with embedding planes at d_l and d_r (figure 3.6). We use the same sinusoidal basis functions that we used earlier (equation (2.30)), except that they are now 'augmented' inside the square well by exact solutions at 'pivot' energies ϵ_1 and ϵ_2,

$$\chi_n(z) = \begin{cases} \cos(n\pi z/D), & |z| > d/2 \\ A_{e,n}\cos(\kappa_1 z) + B_{e,n}\cos(\kappa_2 z), & |z| < d/2 \end{cases} \quad (n \text{ even})$$

$$\chi_n(z) = \begin{cases} \sin(n\pi z/D), & |z| > d/2 \\ A_{o,n}\sin(\kappa_1 z) + B_{o,n}\sin(\kappa_2 z), & |z| < d/2 \end{cases} \quad (n \text{ odd}),$$

(3.7)

with $\kappa_1 = \sqrt{2(\epsilon_1 - v)}$, $\kappa_2 = \sqrt{2(\epsilon_2 - v)}$. The coefficients $A_{e,n}$ and $B_{e,n}$ are chosen so that the linear combination of $\cos(\kappa_1 z)$ and $\cos(\kappa_2 z)$ inside the well matches $\cos(n\pi z/D)$ in amplitude and derivative at the boundary of the well, and similarly for the odd functions. We calculate the Green function for the system, evaluating the matrix elements H_{mn} and O_{mn} à la equation (3.6). So, writing $\cos(n\pi z/D)$ and $\sin(n\pi z/D)$ as $\phi_n^{pw}(z)$, and the linear combination of well solutions as $\phi_n^{well}(z)$ we have

$$H_{mn} = \frac{1}{2}\int_{d_l}^{d_r} dz \frac{\partial \phi_m^{pw}}{\partial z}\frac{\partial \phi_n^{pw}}{\partial z} - \frac{1}{2}\int_{-d/2}^{+d/2} dz \frac{\partial \phi_m^{pw}}{\partial z}\frac{\partial \phi_n^{pw}}{\partial z} + \frac{1}{2}\int_{-d/2}^{+d/2} dz \frac{\partial \phi_m^{well}}{\partial z}\frac{\partial \phi_n^{well}}{\partial z}$$

$$+ w\int_{d/2}^{d_r} dz\phi_m^{pw}(z)\phi_n^{pw}(z) + v\int_{-d/2}^{+d/2} dz\, \phi_m^{well}(z)\phi_n^{well}(z)$$

$$+ \phi_m^{pw}(d_l)\phi_n^{pw}(d_l)\Sigma_l(\omega) + \phi_m^{pw}(d_r)\phi_n^{pw}(d_r)\Sigma_r(\omega), \qquad (3.8)$$

where $\Sigma_l(\omega) = -i\sqrt{\omega/2}$ and $\Sigma_r(\omega) = -i\sqrt{(\omega - w)/2}$, and

$$O_{mn} = \int_{d_l}^{d_r} dz\, \phi_m^{pw}(z)\phi_n^{pw}(z) - \int_{-d/2}^{+d/2} dz\, \phi_m^{pw}(z)\phi_n^{pw}(z) + \int_{-d/2}^{+d/2} dz\, \phi_m^{well}(z)\phi_n^{well}(z).$$

(3.9)

From the Green function (2.43) we can then calculate the local density of states integrated through the well,

$$n_{well}(E) = \int_{-d/2}^{+d/2} dz\, n(z, E), \quad \text{where } n(z, E) = -\frac{1}{\pi}\sum_{mn}\Im G_{mn}(\omega)\phi_m^{well}(z)\phi_n^{well}(z).$$

(3.10)

Figure 3.7 shows our results for $n_{well}(E)$ in the model potential, with parameters given in the figure caption; the pivot energies used to calculate the well solutions (3.7) are $\epsilon_1 = 0.25$ a.u., $\epsilon_2 = 0.35$ a.u. The green curve shows $n_{well}(E)$ calculated with $d_l = -5$ a.u., $d_r = +5$ a.u., in other words, the embedding planes are outside the well as in figure 3.6. The blue curve shows the results calculated with $d_l = 0$, so that the left-hand embedding plane intersects the well—note that we simply change

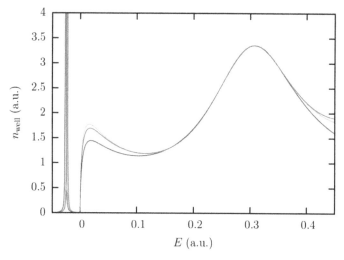

Figure 3.7. $n_{well}(E)$, the local density of states integrated through the square well for the model surface potential (figure 3.6, $v = -0.45$ a.u., $d = 8$ a.u., $w = 0.5$ a.u.). n_{well} is calculated with 16 basis functions, $D = 12$ a.u., augmented by well solutions at $\epsilon_1 = 0.25$ a.u. and $\epsilon_2 = 0.35$ a.u. Green curve, embedding at $d_l = -5$ a.u., $d_r = +5$ a.u.; blue curve, embedding at $d_l = 0$ and $d_r = +5$ a.u.; red curve, 'exact' result from figure 2.4.

the value of d_l in the formulae for H_{mn} and O_{mn}, with the same free-electron embedding potential. This is the interesting case, exactly analogous to the shifted embedding plane intersecting the muffin-tins as in figure 3.5, and there is very little difference between the two curves. For comparison, the red curve shows the exact results taken from figure 2.4—around the pivot energies the agreement is remarkable, but if we stray too far from the pivot energies there are significant differences between the exact results and the results with the augmented basis set. These discrepancies are due to the fact that the functions ϕ_n^{well} are only a good solution to the Schrödinger equation in the well over a limited energy range around the pivot energies. (The energy range can always be split up into bands, with different pivot energies in each band.)

We conclude that shifting the embedding plane into region I really works, and remember—*we are still using the free-electron embedding potential*. There is one proviso, namely that the basis set must be augmented in this region. This is because both ϕ_{pw} and ϕ_{well} contribute to H_{mn} and O_{mn} (equation (3.8)) over an overlap region, when the embedding plane (or planes) lies inside the well; in the case corresponding to the blue curve in figure 3.7, we use both sets of functions over the range -4 a.u. $< z < 0$. We shall present further tests of shifting the embedding plane in realistic LAPW interface calculations in section 3.4.3.

3.1.4 Embedding on to vacuum

In many recent calculations, region I is embedded on to the asymptotic vacuum potential [2]. Assuming a constant vacuum level E_{vac} beyond the vacuum embedding plane S_{vac} (figure 3.2), the vacuum embedding potential over S_{vac} is given by the free-electron expression (2.45), with matrix elements given by

$$\Sigma_{vac}\big(E, \mathbf{k}_\parallel\big)_{\mathbf{gg'}} = -i\delta_{\mathbf{gg'}}\sqrt{\frac{E - E_{vac} - \big|\mathbf{k}_\parallel + \mathbf{g}\big|^2}{2}}, \qquad (3.11)$$

where the term inside the square root is the wave-vector component perpendicular to S_{vac}.

In embedded calculations of image potential-induced surface states (section 4.2) [10], we must use the asymptotic image potential in region II instead of a constant potential. This is given by

$$V(z) \sim E_{vac} - \frac{1}{4|z - z_{im}|}, \qquad (3.12)$$

where E_{vac} is the vacuum zero and z_{im} the image plane [17]. The embedding potential Σ_{im} to replace the asymptotic image region can be found from the outgoing Coulomb function [18],

$$\psi(z) = H_0^-(\eta, \rho), \quad \text{where } \rho = \sqrt{2(E - E_{vac})} \cdot (z - z_{im}), \quad \eta = \frac{1}{4\sqrt{2(E - E_{vac})}}. \qquad (3.13)$$

Conveniently, Thompson and Barnett [19] give a rapidly converging continued fraction expression for $\frac{1}{H_0^-}\frac{\partial H_0^-}{\partial \rho}$, from which we can obtain Σ_{im} using equation (2.16).

3.2 Embedded surface calculations

The full-potential LAPW method provides a way of solving the embedded surface Schrödinger equation which is both accurate and well-adapted to the embedding method [11]. In this section I shall describe calculations in which region I extends indefinitely into the vacuum, with embedding only on to the bulk substrate. This is an adaptation of the layer-LAPW method developed by Krakauer *et al* [14], and was used in the first embedded surface calculations [12]. It is very economical as far as the size of basis set is concerned, and has yielded many useful results [1]. It can be readily modified (and the basis functions simplified) for embedding on to both the crystal substrate and the vacuum (sections 3.1.4 and 3.6).

3.2.1 LAPW basis functions

We see from figure 3.1 that the potential felt by an electron at a surface is typically rather smooth in the interstitial space between the muffin-tins, where the atomic-like potential is predominantly spherically symmetric; beyond the top layer of atoms, the potential tails away fairly smoothly into the vacuum.

In each of these three regions—interstitial, muffin-tins and vacuum, shown schematically in figure 3.8—we use a different expansion of $V(\mathbf{r})$ and a different

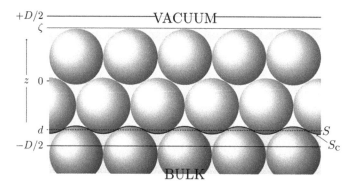

Figure 3.8. Regions of potential in the embedded LAPW surface calculation: the bulk region (brown), is replaced by the embedding potential on the embedding plane S (dashed line). In the surface region I the potential is split up into muffin-tins (green), the interstitial region (grey), and the vacuum (white). S_c (thick solid line) indicates the curvy boundary between muffin-tins in regions I and II. D defines the LAPWs.

form of basis functions [11]. The potential in the interstitial region, shaded grey in figure 3.8, is expanded as a Fourier series,

$$V(\mathbf{r}) = \sum_{\mathbf{g},n} V_{\mathbf{g},n} \exp(i\mathbf{g} \cdot \mathbf{r}_\parallel) \times \begin{cases} \cos(k_n z) & (n \text{ even}) \\ \sin(k_n z) & (n \text{ odd}) \end{cases} + \dots, \tag{3.14}$$

where the sum is over the two-dimensional reciprocal lattice vectors $\mathbf{g}$ and n, which defines the functions in the z-direction with $k_n = n\pi/D$. The dots at the end of equation (3.14) indicate extra terms which arise from fitting the electrostatic boundary conditions (equation (3.30)). Inside each muffin-tin, shaded green in figure 3.8, $V(\mathbf{r})$ is expanded in spherical harmonics $Y_L(\theta, \phi)$,

$$V(\mathbf{r}) = \sum_{l,m} V_{L,\alpha}(r) Y_L(\theta, \phi), \quad \mathbf{r} \in \text{muffin-tin } \alpha, \tag{3.15}$$

where $L = (l, m)$. The spherically symmetric term, with $l = 0$, greatly dominates. In the vacuum region we use a two-dimensional Fourier expansion over the surface reciprocal lattice vectors,

$$V(\mathbf{r}) = \sum_{\mathbf{g}} V_{\mathbf{g}}(z) \exp(i\mathbf{g} \cdot \mathbf{r}_\parallel), \tag{3.16}$$

where $V_{\mathbf{g}}(z)$ is tabulated on a grid.

The LAPW basis functions, which we use to solve the embedded Schrödinger equation in region I, are defined as follows. In the interstitial region between ζ and the embedding plane S at $z = d$, the LAPWs are given by plane-waves with wave-vector $\mathbf{k}_\parallel$ parallel to the surface,

$$\chi_{\mathbf{g},n}(\mathbf{r}; \mathbf{k}_\parallel) = \exp\left[i(\mathbf{k}_\parallel + \mathbf{g}) \cdot \mathbf{r}_\parallel\right] \times \begin{cases} \cos(k_n z), & (n \text{ even}) \\ \sin(k_n z), & (n \text{ odd}) \end{cases} \text{ with } k_n = n\pi/D. \tag{3.17}$$

D is chosen to be greater than the thickness of the surface layer (figure 3.8), to give a range of logarithmic derivatives on the embedding plane, exactly as in section 2.3.

Inside the muffin-tins, the LAPWs are built up of solutions of the Schrödinger equation[1], $u_{l,\alpha}(r)$ and their energy derivatives $\dot{u}_{l,\alpha}(r)$, calculated with the spherically symmetric part of the muffin-tin potential $V_{0,\alpha}(r)$ (equation (3.15)),

$$-\frac{1}{2}\left(\frac{\mathrm{d}^2 u_{l,\alpha}}{\mathrm{d}r^2} - \frac{l(l+1)}{r^2}u_{l,\alpha}\right) + V_{0,\alpha}(r)u_{l,\alpha} = \epsilon_{l,\alpha}u_{l,\alpha}(r). \qquad (3.18)$$

Here, $\epsilon_{l,\alpha}$ is the pivot energy at which $u_{l,\alpha}(r)$ and its energy derivative $\dot{u}_{l,\alpha}(r)$ are evaluated. We should note that equation (3.18) is not an eigenvalue equation—the Schrödinger equation is integrated outwards from the nucleus, at energy $\epsilon_{l,\alpha}$. Inside muffin-tin α we then have

$$\chi_{\mathbf{g},n}(\mathbf{r}; \mathbf{k}_\parallel) = \frac{1}{r}\sum_L \left[A_L^\alpha(\mathbf{g}, n)u_{l,\alpha}(r) + B_L^\alpha(\mathbf{g}, n)\dot{u}_{l,\alpha}(r)\right]Y_L(\theta, \phi), \qquad (3.19)$$

where $A_{l,m}^\alpha$ and $B_{l,m}^\alpha$ are determined by matching χ and $\partial\chi/\partial r$ across the surface of the muffin-tin; typically we go up to $l = 8$ in this expansion.

To extend the LAPW basis functions into the vacuum region, $z > \zeta$ (figure 3.8), we join them on to solutions of the vacuum Schrödinger equation in the same way that we join on to atomic solutions inside the muffin-tins. We solve the Schrödinger equation with the planar-averaged part of the vacuum potential at pivot energy ϵ_v,

$$-\frac{1}{2}\left(\frac{\mathrm{d}^2 w_{\mathbf{g}}}{\mathrm{d}z^2} - \left|\mathbf{k}_\parallel + \mathbf{g}\right|^2 w_{\mathbf{g}}(z)\right) + V_0(z)w_{\mathbf{g}}(z) = \epsilon_v w_{\mathbf{g}}(z), \qquad (3.20)$$

integrating equation (3.20) backwards from a starting point at large z, where $w_{\mathbf{g}}(z)$ decays exponentially. An LAPW in this region is a linear combination of the $w_{\mathbf{g}}(z)$ and its energy derivative $\dot{w}_{\mathbf{g}}(z)$,

$$\chi_{\mathbf{g},n}(\mathbf{r}; \mathbf{k}_\parallel) = \left[A^{\mathrm{vac}}(\mathbf{g}, n)w_{\mathbf{g}}(z) + B^{\mathrm{vac}}(\mathbf{g}, n)\dot{w}_{\mathbf{g}}(z)\right]\exp\left[\mathrm{i}(\mathbf{k}_\parallel + \mathbf{g})\cdot\mathbf{r}_\parallel\right], \qquad (3.21)$$

where the coefficients $A^{\mathrm{vac}}(\mathbf{g}, n)$ and $B^{\mathrm{vac}}(\mathbf{g}, n)$ are chosen so that χ and $\partial\chi/\partial z$ are continuous across the vacuum interface at $z = \zeta$.

These three components of an LAPW we shall call $\phi_{\mathbf{g},n}^{\mathrm{pw}}$, $\phi_{\mathbf{g},n}^{\mathrm{at}}$ and $\phi_{\mathbf{g},n}^{\mathrm{vac}}$, respectively, in our expressions for the Hamiltonian and overlap matrix elements in section 3.2.2.

As they contain u, $\dot{u}$ inside each atom, and w, $\dot{w}$ in the vacuum, the LAPWs represent an accurate solution of the Schrödinger equation in these regions over a range of energies about the pivot energies $\epsilon_{l,\alpha}$, ϵ_v; this is analogous to choosing the two square-well solutions in equation (3.7) so that they describe the solution over the required energy range. Moreover, the plane-wave component of the LAPWs provides a rapidly converging representation of the wave-functions in the interstitial region, where the potential is slowly varying. Previous experience with slab

---

[1] In many applications the scalar relativistic equation is used (section 8.3) [20].

calculations had shown that LAPWs are an excellent basis set for surfaces [21], and they turn out to be ideal for embedded surface calculations—not least because the augmentation of the plane-waves inside the muffin-tins allows us to chop off the caps of the muffin-tins by the embedding plane.

If we decide to terminate region I by the vacuum embedding plane S_{vac}, with Σ_{vac} taking care of the asymptotic behaviour of the wave-function or Green function in region I, we simply extend the interstitial region described by plane-waves in $V(\mathbf{r})$ (equation (3.15)) and $\chi_{\mathbf{g},n}(\mathbf{r}; \mathbf{k}_\parallel)$ (equation (3.17)) up to Σ_{vac}, dispensing with the separate vacuum expansions [2]. As D must now be chosen so that it extends beyond the whole of region I, it is much larger than in the method described above and many more plane-waves will be needed in the expansions; on the other hand, the basis functions and matrix elements are easier to evaluate.

3.2.2 Calculating the Green function

We calculate the Green function $G_{\mathbf{k}_\parallel}(\mathbf{r}, \mathbf{r}'; E)$ at fixed two-dimensional wave-vector $\mathbf{k}_\parallel$ in the surface Brillouin zone, expanding it in terms of the LAPW basis functions $\chi_{\mathbf{g},n}(\mathbf{r}; \mathbf{k}_\parallel)$ (equation (2.37)).

From equation (3.6) the Hamiltonian and overlap matrix elements are given by the following expressions, which remain valid when the embedding plane S intersects the muffin-tins in region I,

$$
H_{\mathbf{g}m,\mathbf{g}'n} = \frac{1}{2} \int_d^\zeta dz \int_{2D \, cell} d\mathbf{r}_\parallel \, \nabla \phi_{\mathbf{g},m}^{pw*} \cdot \nabla \phi_{\mathbf{g}',n}^{pw} + \int_{int} d\mathbf{r} \, \phi_{\mathbf{g},m}^{pw*} V \phi_{\mathbf{g}',n}^{pw}
$$
$$
+ \frac{1}{2} \int_{MT} d\mathbf{r} \, \phi_{\mathbf{g},m}^{pw*} \nabla^2 \phi_{\mathbf{g}',n}^{pw} + \int_{MT} d\mathbf{r} \, \phi_{\mathbf{g},m}^{at*} H \phi_{\mathbf{g}',n}^{at} + \int_\zeta^\infty dz \int_{2D \, cell} d\mathbf{r}_\parallel \, \phi_{\mathbf{g},m}^{vac*} H \phi_{\mathbf{g}',n}^{vac}
$$
$$
+ \int_{2D \, cell} d\mathbf{r}_\parallel \int_{2D \, cell} d\mathbf{r}_\parallel' \, \phi_{\mathbf{g},m}^{pw*}(\mathbf{r}_\parallel, d) \Sigma(\mathbf{r}_\parallel, \mathbf{r}_\parallel') \phi_{\mathbf{g}',n}^{pw}(\mathbf{r}_\parallel', d), \tag{3.22}
$$

and

$$
O_{\mathbf{g}m,\mathbf{g}'n} = \int_d^\zeta dz \int_{2D \, cell} d\mathbf{r}_\parallel \, \phi_{\mathbf{g},m}^{pw*} \phi_{\mathbf{g}',n}^{pw} - \int_{MT} d\mathbf{r} \, \phi_{\mathbf{g},m}^{pw*}(\mathbf{r}; \mathbf{k}_\parallel) \phi_{\mathbf{g}',n}^{pw}(\mathbf{r}; \mathbf{k}_\parallel)
$$
$$
+ \int_{MT} d\mathbf{r} \, \phi_{\mathbf{g},m}^{at*} \phi_{\mathbf{g}',n}^{at} + \int_\zeta^\infty dz \int_{2D \, cell} d\mathbf{r}_\parallel \, \phi_{\mathbf{g},m}^{vac*} \phi_{\mathbf{g}',n}^{vac}. \tag{3.23}
$$

The first term in $H_{\mathbf{g}m,\mathbf{g}'n}$ is an integral over z from the embedding plane at d up to the vacuum boundary at ζ (figure 3.8) and a two-dimensional integral over the surface unit cell. This combines the matrix element of the kinetic energy with the normal derivative term over S (as in equation (2.40)). The second term is over the interstitial region, shaded grey in figure 3.8, and gives the matrix element of $V(\mathbf{r})$, the interstitial potential. The third and fourth terms in $H_{\mathbf{g}m,\mathbf{g}'n}$ are over the muffin-tins, shaded green, removing the plane-wave kinetic energy contribution and adding the contribution from the atomic solutions with the full Hamiltonian H. Next, we have the contribution from the Hamiltonian in the vacuum region, extending from the boundary at $z = \zeta$, in principle out to infinity, but cut off by the rapidly decaying vacuum solutions. The final term in $H_{\mathbf{g}m,\mathbf{g}'n}$ is the matrix element of the embedding potential over the embedding plane. The contributions to $O_{gm,g'n}$ involve similar integrals. These matrix elements are numerically straightforward to evaluate [11].

With the matrix elements we can calculate the Green function (equation (2.42)),

$$G_{\mathbf{k}_\parallel}(\mathbf{r}, \mathbf{r}'; E) = \sum_{gm,g'n} (EO - H)^{-1}_{gm,g'n} \chi_{g,m}(\mathbf{r}; \mathbf{k}_\parallel) \chi^*_{g',n}(\mathbf{r}'; \mathbf{k}_\parallel), \qquad (3.24)$$

hence the surface electronic structure.

3.2.3 Self-consistency—the electron density and potential

The potential $V(\mathbf{r})$ entering the Schrödinger equation (3.1) depends self-consistently on the electron density $\rho(\mathbf{r})$, which can be found from the local density of states $n_{\mathbf{k}_\parallel}(\mathbf{r}, E)$ (equation (2.35)), labelled in surface calculations with the two-dimensional wave-vector $\mathbf{k}_\parallel$. From equations (2.34) and (2.35) we have

$$\rho(\mathbf{r}) = \frac{A}{2\pi^2} \int_{BZ} d\mathbf{k}_\parallel \int_{E_0}^{E_F} dE\, n_{\mathbf{k}_\parallel}(\mathbf{r}, E)$$

$$= -\frac{A}{2\pi^3} \int_{BZ} d\mathbf{k}_\parallel \int_{E_0}^{E_F} dE\, \Im G_{\mathbf{k}_\parallel}(\mathbf{r}, \mathbf{r}; E + i\eta), \quad \eta \to 0+. \qquad (3.25)$$

The first integral is over two-dimensional wave-vectors in the surface Brillouin zone, with A the area of the real-space surface unit cell; the energy integral is over the occupied states up to the Fermi energy E_F, with a lower limit E_0 below the bottom of the energy bands. There is a factor of 2 in these equations to include spin, though in spin-polarized calculations the spin-up and spin-down electrons are treated separately.

The local density of states is typically structured, which makes it difficult to evaluate the energy integral in equation (3.25) as it stands. Fortunately, this integral can be transformed to an integral along a semicircular contour in the upper complex energy plane, where the structure is smeared out. Along the real E-axis, G has poles from discrete states (surface states in the energy gaps) and branch cuts (in the bulk energy bands and above the vacuum zero), but above the real axis G is analytic. This means that we can deform the path of integration in equation (3.25) from the line integral just above the real axis into a semicircle C (figure 3.9) [11],

$$\Im \int_{E_0+i\eta}^{E_F+i\eta} dE\, G_{\mathbf{k}_\parallel}(\mathbf{r}, \mathbf{r}; E) = \Im \int_C dE\, G_{\mathbf{k}_\parallel}(\mathbf{r}, \mathbf{r}; E). \qquad (3.26)$$

G varies smoothly along C, unless there is structure close to E_F, where C drops down to the real axis. But help is at hand even here: if we use the Gauss–Chebyshev method to carry out the numerical integration along C [22], the sampling points are concentrated near the ends of the range, just where structure can appear in the integrand.

The integration over the surface Brillouin zone is invariably replaced by a summation over discrete $\mathbf{k}_\parallel$-points in the irreducible part of the surface Brillouin

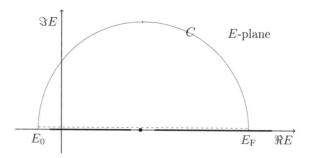

Figure 3.9. Complex E-plane. The singularities of G lie on the real energy axis, with poles at discrete states (filled circle) and branch cuts in energy continua (thick lines). The path of integration can be distorted from the dashed line just above the real axis to contour C.

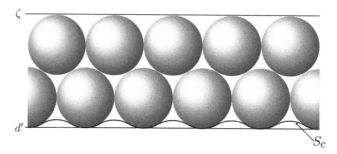

Figure 3.10. Solving Poisson's equation [25]: $\rho_{pw}(\mathbf{r})$, the plane-wave representation of the electron density in the interstitial region (shaded grey in figure 3.8) is extended into the muffin-tins and up to the surface at $z = d'$ (the region shaded blue). To this is added $\rho_{pseudo,\alpha}(\mathbf{r})$, a pseudo-electron density inside the muffin-tins (shaded dark blue), which when added to $\rho_{pw}(\mathbf{r})$ has the same multipole moments as the real muffin-tin charge.

zone ('irreducible' means the part of the Brillouin zone from which the rest can be obtained by symmetry operations),

$$\frac{A}{2\pi^2}\int_{BZ}d\mathbf{k}_{\parallel} \longrightarrow 2\sum_{\mathbf{k}_{\parallel}}w_{\mathbf{k}_{\parallel}}. \tag{3.27}$$

The choice of $\mathbf{k}_{\parallel}$-points is not arbitrary, as they must be adapted to the particular Brillouin zone, but suitable sampling points and their weights $w_{\mathbf{k}_{\parallel}}$ are given in the literature [23].

Knowing the electron density, we next calculate $V_{es}(\mathbf{r})$ (equation (3.1)). This involves solving Poisson's equation with $\rho(\mathbf{r})$ on the right-hand side [24], for which we use a clever method devised by Weinert [11, 25]. We write the electron density in the interstitial region, calculated from the plane-wave part of the wave-functions, as $\rho_{pw}(\mathbf{r})$; this can be continued through the muffin-tins and beyond S_c to $z = d'$ (the blue region in figure 3.10). To this we add a 'pseudo' electron density $\rho_{pseudo,\alpha}(\mathbf{r})$ within muffin-tin α, such that the total density inside the muffin-tin has the same electrostatic multipole moments as the real electron density + ionic or nuclear charge,

$$\tilde{\rho}(\mathbf{r}) = \rho_{\mathrm{pw}}(\mathbf{r}) + \sum_{\alpha} \rho_{\mathrm{pseudo},\alpha}(\mathbf{r}), \quad (\mathbf{r} \in \text{blue region}, \, d' < z < \zeta). \tag{3.28}$$

Weinert [25] showed how to construct a $\rho_{\mathrm{pseudo},\alpha}(\mathbf{r})$ which can be easily Fourier transformed, so that we can write

$$\tilde{\rho}(\mathbf{r}) = \sum_{\mathbf{g},n} \tilde{\rho}_{\mathbf{g},n} \exp(\mathrm{i}\mathbf{g} \cdot \mathbf{r}_{\parallel}) \times \begin{cases} \cos(k_n z) \\ \sin(k_n z). \end{cases} \tag{3.29}$$

The general solution of Poisson's equation in the blue region is then given by

$$V_{\mathrm{es}}(\mathbf{r}) = \sum_{\mathbf{g},n}' \frac{4\pi\tilde{\rho}_{\mathbf{g},n}}{g^2 + k_n^2} \exp(\mathrm{i}\mathbf{g} \cdot \mathbf{r}_{\parallel}) \times \begin{cases} \cos(k_n z) \\ \sin(k_n z) \end{cases} - 2\pi\tilde{\rho}_{0,0}z^2$$
$$+ \sum_{\mathbf{g}}' \left[V_{\mathbf{g}}^+ \exp(gz) + V_{\mathbf{g}}^- \exp(-gz) \right] \exp(\mathrm{i}\mathbf{g} \cdot \mathbf{r}_{\parallel}) + A + Bz. \tag{3.30}$$

In the first line, the term with $\mathbf{g} = 0, n = 0$ is excluded from the summation (indicated by the dash), its contribution being given by the final term on the first line. The terms in the second line are solutions of Laplace's equation in this region, with constants A, B and $V_{\mathbf{g}}^{\pm}$, and we add these explicitly to the Fourier expansion of $V(\mathbf{r})$ in the interstitial region (equation (3.14)). As the ionic or nuclear charge is included in the monopole contribution to $\rho_{\mathrm{pseudo},\alpha}(\mathbf{r})$, equation (3.30) corresponds to the full electrostatic potential V_{es}; that is, $V_{\mathrm{ion}} + V_H$ (equation (3.1)).

In the vacuum region, the electron density is expanded in the same form as equation (3.16),

$$\rho(\mathbf{r}) = \sum_{\mathbf{g}} \rho_{\mathbf{g}}(z)\exp(\mathrm{i}\mathbf{g} \cdot \mathbf{r}_{\parallel}), \tag{3.31}$$

and then each $\mathbf{g}$-component of the electrostatic potential satisfies the one-dimensional Poisson's equation,

$$\frac{\mathrm{d}^2 v_{\mathbf{g}}(z)}{\mathrm{d}z^2} - g^2 v_{\mathbf{g}}(z) = -4\pi\rho_{\mathbf{g}}(z), \tag{3.32}$$

which we integrate back towards ζ, starting from the boundary conditions that $v_{\mathbf{g}}(z) \to 0$, $v_{\mathbf{g}}'(z) \to 0$, at large z. Adding on the solutions to Laplace's equation, the general solution for $z > \zeta$ is given by

$$V_{\mathrm{es}}(\mathbf{r}) = \sum_{\mathbf{g}}' \left[v_{\mathbf{g}}(z) + v_{\mathbf{g}}^- \exp(-gz) \right] \exp(\mathrm{i}\mathbf{g} \cdot \mathbf{r}_{\parallel}) + v_0(z) + C, \tag{3.33}$$

where we have separated out the $\mathbf{g} = 0$ term $v_0(z)$ from the summation. The constants resulting from Laplace are C, and $v_{\mathbf{g}}^-$.

For each surface reciprocal lattice vector there are three unknown constants in equations (3.30) and (3.33), which can be fixed by matching fields across the vacuum and bulk interfaces. At $\mathbf{g} = 0$ these are A and B in the interstitial potential, and C in the vacuum; at non-zero $\mathbf{g}$ the unknowns are $V_{\mathbf{g}}^{+}, V_{\mathbf{g}}^{-}$ and $v_{\mathbf{g}}^{-}$. Two are fixed by the requirement that V_{es} and V_{es}' must match across $z = \zeta$, and the third by matching V_{es} to the bulk electrostatic potential over S_c (figure 3.8). Note that we do not match the electric field across S_c—only the potential—as this extra condition would over-determine $V_{es}(\mathbf{r})$. Having found all the constants, we know the electrostatic potential throughout the interstitial region, including over the surfaces of the muffin-tin spheres. This provides the necessary boundary condition for solving Poisson's equation inside the spheres with the actual electron density $\rho(\mathbf{r})$ rather than $\tilde{\rho}(\mathbf{r})$, thus completing the exercise of finding $V_{es}(\mathbf{r})$ throughout the whole of region I.

We still have to add the exchange-correlation potential $V_{xc}(\mathbf{r})$ to the electrostatic potential to obtain the total effective potential $V(\mathbf{r})$ felt by the electrons (equation (3.1)). There is a huge body of literature on density functional theory (DFT) and on the different approximations for $V_{xc}(\mathbf{r})$ [26], but in many surface calculations the simplest approximation is used, the local density approximation (LDA) [27, 28], in which the exchange-correlation potential just depends on the local electron density,

$$V_{xc}(\mathbf{r}) = \epsilon_{xc}\big(\rho(\mathbf{r})\big) + \rho(\mathbf{r})\frac{\partial \epsilon_{xc}(\rho)}{\partial \rho}\bigg|_{\rho=\rho(\mathbf{r})}. \tag{3.34}$$

Here, $\epsilon_{xc}(\rho)$ is the exchange-correlation energy per unit volume of a uniform free-electron gas with electron density ρ, for which there are various expressions and parametrizations [29, 30]. To evaluate $V_{xc}(\mathbf{r})$ in the interstitial region in a Fourier expansion (equation (3.14)), we perform a least-squares fit of the Fourier coefficients to $V_{xc}(\mathbf{r})$ evaluated at random points in this region; in the muffin-tins and vacuum regions we can linearize $V_{xc}(\mathbf{r})$, giving it directly in the forms given by equations (3.15) and (3.16) [11]. (See [31] for another approach.)

Whether we use the LDA or go beyond it with the generalized gradient approximation for example [28], the exchange-correlation potential in these approximations decays exponentially into the vacuum, whereas it should tend to the image potential (equation (3.12)). We must include the image potential, interpolating between this and the exponential behaviour near the surface, when we study image potential-induced surface states [10]; but in the ground state the wave-functions decay rapidly away from the surface, and we can neglect image potential effects.

Going to self-consistency involves guessing a starting potential, calculating the electron density and from this a new potential, and iterating until the input and output potentials are sufficiently close [5]. In all surface calculations, embedding as well as slab, self-consistency is tricky, as electronic charge tends to slosh out of the surface into the vacuum, setting up a surface dipole with consequent large changes in the potential. The process is unstable, and we have to mix a small fraction of the output potential V_n^{out} with a large fraction of the input potential V_n^{in} to obtain the input for the next iteration V_{n+1}^{in},

$$V_{n+1}^{in} = (1 - \alpha)V_n^{in} + \alpha V_n^{out}. \tag{3.35}$$

Typical values for the mixing parameter α are 0.002. There are other, more sophisticated, mixing schemes [32, 33] but, provided α is small enough, equation (3.35) always works. An important difference between self-consistency in slab and embedding calculations is that in the slab the total charge is fixed and the electronic states are populated by a certain number of electrons, whereas in embedding the Fermi energy is fixed, determined by the bulk. We go to self-consistency only in the surface region, region I, and the embedding potential and the Fermi energy remain the same throughout the calculation.

3.3 First results

3.3.1 Electron density and work-function

One of the first applications of surface embedding was a calculation of the electronic structure of the Al(001) surface [12, 11]. This is a typical s–p bonded metal surface, with a prominent surface state in the band gap of the nearly-free electron band structure. Figure 3.11 shows the surface electron density from this calculation, in which two atomic layers are included in region I. The flat embedding plane is taken halfway between atomic layers, as in figure 3.8, and the embedding potential is calculated from the reflection properties of the semi-infinite bulk, with a muffin-tin potential, as described in section 3.1.1. We should note that the electron density

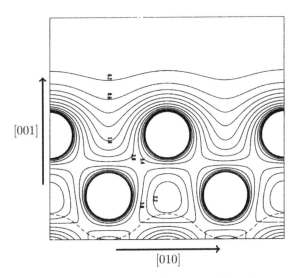

Figure 3.11. Electron density at the Al(001) surface, plotted on a (100) slice, from an embedded surface calculation with two atomic layers included in region I. The dashed line shows the 'natural' boundary S_c between regions I and II, over which $V(\mathbf{r})$ in region I equals that in the substrate. Contours are labelled in units of electrons per bulk unit cell. (From Inglesfield and Benesh [11].)

below the 'natural' boundary S_c is meaningless, as it is just a continuation of the electron density calculated for region I.

A useful quantity to calculate is the work-function: the energy difference between the Fermi energy and the vacuum zero—this gives some measure of the accuracy of the self-consistent electron density, particularly the surface dipole due to charge spilling out into the vacuum. The two-layer calculation gives a work-function of 4.50 eV, in good agreement with the experimental value of 4.41 ± 0.03 eV [34], and with only one layer, the embedded calculation gives 4.63 eV. Many more layers are needed in conventional slab calculations; for example, with nine atomic layers an LAPW slab calculation gave a work-function of 4.53 eV [35]. Embedding works even better on the Al(111) surface—Benesh and Liyanage [3] obtained a work-function of 4.22 eV in an embedded three-layer calculation, and 4.24 eV with only one layer, compared with the experimental value of 4.24 eV [34]. It is worth remarking that usually the more close-packed the metal surface, the greater the work-function, but this does not apply to these surfaces of Al, where the (111) surface is more closely packed than (001) yet has a smaller work-function [6].

Despite these work-function successes, these early embedding calculations do not give a perfect picture of the surface electron density, as we can see from figure 3.11. The contours labelled 3.0 (in units of electrons per bulk unit cell) in the second layer of atoms should be more symmetric; we would expect the electron density to be essentially bulk-like by the second layer. Two factors which contribute to this deficiency are the following. Firstly, the bulk embedding potential is calculated with a muffin-tin potential rather than the full potential of the surface calculations, leading to inaccurate matching with the bulk. Secondly, in going to self-consistency the potential in region I is set equal to the bulk potential over the boundary surface S_c (section 3.2.3), and with a muffin-tin bulk potential this inevitably distorts the potential in region I near the boundary. We shall see in section 3.6 how a full-potential description of the bulk (as well as generally improved computational methods) cures these problems.

3.3.2 Surface states and bulk states at the surface

Figure 3.12 shows the Al(001) surface density of states $n_{s,k_\parallel}(E)$ at $\mathbf{k}_\parallel = 0$, the local density of states $n_{k_\parallel}(\mathbf{r}, E)$ integrated through the surface atoms, from the two-layer embedded calculation [11]. The most striking feature is the discrete state—a surface state—near the bottom of the bulk band gap between 0.204 and 0.243 a.u. (Energies are measured with respect to the bulk muffin-tin zero, and an imaginary part of the energy $\eta = 0.001$ a.u. is added to E to broaden discrete states; relative to the energy zero the bulk Fermi energy is at 0.3085 a.u.) Discrete surface states are a consequence of the symmetry-breaking of the bulk translational symmetry at the surface [6, 36]. They are localized and discrete because, with an energy in a bulk band gap, they decay exponentially with oscillations into the bulk, as well as exponentially into the vacuum. Their importance is that they play a large part in determining the properties of surfaces—for example, the surface reconstruction of W(001) [6]—and they show up as very prominent features in surface electron spectroscopies [37]. Both the absolute

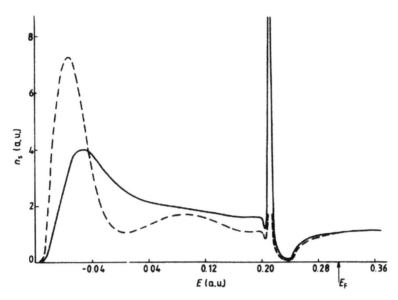

Figure 3.12. Al(001) surface density of states $n_{s,k_\parallel}(E)$ at $k_\parallel = 0$. Full curve, surface layer of atoms; dashed curve, sub-surface layer. Imaginary part of the energy, $\eta = 0.001$ a.u. (From Inglesfield and Benesh [11].)

position of this surface state, 2.65 eV below E_F, and its energy relative to the bottom of the band gap are in good agreement with experiment [38].

The continuum part of the surface density of states comes from states in the bulk energy bands which are reflected from the surface. The embedding method automatically includes these in the surface density of states. In the bulk, the density of states at fixed $k_\parallel$ has an inverse square root singularity near a band edge at E_0,

$$n_{k_\parallel}(E) \sim |E - E_0|^{-\frac{1}{2}}, \tag{3.36}$$

which is rounded off in the surface density of states,

$$n_{s,k_\parallel}(E) \sim |E - E_0|^{\frac{1}{2}}, \tag{3.37}$$

the same as in a one-dimensional system (section 2.5, figure 2.5). Although the square root singularities in $n_{s,k_\parallel}(E)$ are broadened by the imaginary part of the energy, we can see the same general behaviour like equation (3.37) in figure 3.12, especially at the top of the band gap. What is particularly interesting is that the peak above the bottom of the band is much bigger in the second layer than in the top layer, reflecting the fact that as we proceed deeper and deeper into the solid this eventually becomes the inverse square root singularity of equation (3.36). Another interesting feature of the continuum states is the oscillatory behaviour in $n_{s,k_\parallel}(E)$ in the second layer, due to interference between the incident and reflected bulk waves.

For comparison with these embedding results, figure 3.13 shows the results of a seven-layer slab calculation for Al(001) [12]. In this figure we plot the surface weight W (the charge integrated through the surface atoms) of states with $k_\parallel = 0$, as a function of discrete energy E. In a slab calculation all the states are discrete, and

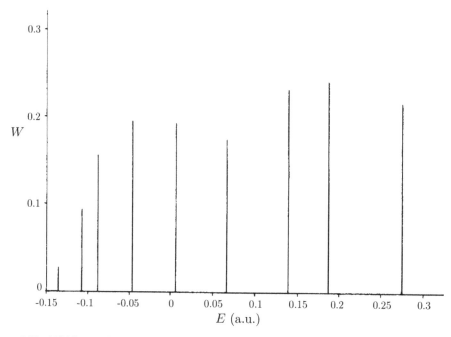

Figure 3.13. Al(001) seven-layer slab calculation at $\mathbf{k}_\parallel = 0$. Weight of states in the surface layer, W, plotted against energy. (From Benesh and Inglesfield [12].)

comparing figure 3.13 with figure 3.12 we see that this gives a poor description of the states at the surface of a real semi-infinite crystal—which is what the embedding calculation provides. The states at $E = 0.14$ a.u. and 0.19 a.u. correspond to the *single* surface state of the isolated surface. These states on the two surfaces of the slab interact through the interior layers, and their energy splits. The Al(001) surface state decays rather slowly into the semi-infinite crystal, with a decay length of about seven atomic layers, so a slab calculation would have to be very thick to give a good description. This is perhaps a rather extreme example, and many surface states—in particular the Tamm surface states [6] in a system with stronger chemical bonds—are well-localized on the surface atoms. But the principle remains the same: it can be difficult to distinguish between bulk and surface states in a slab calculation. Finally, we re-emphasize the important principle of embedding: even though the Al(001) surface state extends well into the bulk, the embedding calculation describes it correctly in the surface region, the region treated explicitly.

3.3.3 Problems

As well as the problems to which we alluded in section 3.3.1, associated with using a muffin-tin potential for calculating the embedding potential, and for matching over S_c in the self-consistency procedure, we encountered a more worrying difficulty in these early embedding calculations (remember the inverse of the old adage: every silver lining has a cloud!). In the case of an Al(001) one-layer calculation, going beyond about 100 basis functions the results suddenly become nonsensical, with the appearance of spurious features ('ghost states') in the density of states[2] and unphysical charge densities. However, increasing the number of basis functions restores sensible behaviour, and with 120 LAPWs the density of states and electron density are very close to the results with less than 100.

This annoying behaviour seems to be connected with the embedding plane intersecting the muffin-tins in region I. Although this is a valid procedure (section 3.1.2), the matrix elements have to be evaluated very accurately, and in more recent calculations attention has been paid, amongst other things, to the number of atomic functions inside the muffin-tins for more accurate matching to the plane-wave part of the LAPWs (equation (3.19)). This greatly improves the stability of the calculations. However, for exceptional stability we have to go to the embedding methods developed by Ishida [2, 31], and described in section 3.5.

3.4 Sub-volume embedding

In this and subsequent sections we shall look at alternative ways of calculating the embedding potential for surface and interface calculations, so that the full potential of the bulk substrate can be represented. We start with a method in which the embedding potential is transferred through a layer of atoms [16].

3.4.1 Adding an atomic layer

Let us suppose that we know the embedding potential Σ_0 on the plane S_0, an embedding interface with the bulk material (figure 3.14). Σ_0 is, of course, independent of whatever we put on top as region I, and we have also seen in section 3.1.2 that the embedding plane can intersect muffin-tins in region I as well as the bulk muffin-tins. We now calculate the Green function for an extra layer of atoms, layer 1, added to the bulk, satisfying the following embedded Schrödinger equation,

$$-\frac{1}{2}\nabla_{\mathbf{r}}^2 G(\mathbf{r}, \mathbf{r}'; E) + \frac{1}{2}\delta(n - n_{S_0})\frac{\partial G(\mathbf{r}_{S_0}, \mathbf{r}'; E)}{\partial n_{S_0}} + V(\mathbf{r})G(\mathbf{r}, \mathbf{r}'; E)$$

$$+ \delta(n - n_{S_0})\int_{S_0} d\mathbf{r}''_{S_0}\Sigma_0(\mathbf{r}_{S_0}, \mathbf{r}''_{S_0}; E)G(\mathbf{r}''_{S_0}, \mathbf{r}'; E) - EG(\mathbf{r}, \mathbf{r}'; E) = -\delta(\mathbf{r} - \mathbf{r}').$$

$$(3.38)$$

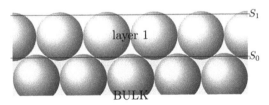

Figure 3.14. A layer of atoms (green), designated layer 1, is embedded on to the bulk (brown) over S_0. The surface inverse of the embedded Green function, with zero-gradient on S_1, gives an embedding potential over S_1, which represents the bulk + layer combined system.

---

[2] Ghost states occur in different electronic structure methods, including pseudopotentials [39].

Σ_0 is added on to the layer Hamiltonian over S_0, and on the top surface of the slab S_1 the embedding potential is set equal to zero. This layer Green function matches on to the bulk Green function over S_0—it extends the bulk Green function into the layer. But over S_1 the zero embedding potential forces G to have zero derivative. However, this is just the Green function, whose surface inverse gives an embedding potential (equation (2.17))

$$\Sigma_1\left(\mathbf{r}_{S_1}, \mathbf{r}'_{S_1}\right) = -G^{-1}\left(\mathbf{r}_{S_1}, \mathbf{r}'_{S_1}\right). \tag{3.39}$$

What is the meaning of Σ_1? It is the embedding potential for the system consisting of semi-infinite bulk + 1 layer, and we have transferred the embedding potential from S_0 to S_1.

This process of transferring the embedding potential through an atomic layer opens up several possibilities for treating surfaces and interfaces. Most importantly, it leads to a very efficient way of calculating the bulk embedding potential, as we shall show in section 3.4.2. It also provides an order-N method of calculating the electronic structure in a surface or interface system in which the perturbation extends a long way into the solid. We can proceed layer-by-layer towards the surface or interface from the unperturbed bulk, calculating the Green function given by equation (3.38) in each layer, hence the embedding potential for adding the next layer. (Of course the 'layer', or better *sub-volume*, can consist of several atomic layers.) Such a procedure is efficient computationally, as the computing time scales linearly with the number of layers, whereas calculating the Green function for N layers treated as one large region scales as N^3 [22]. Whether this is worth the extra programming complication is a different matter.

3.4.2 Bulk embedding potential

To show how we can calculate the bulk embedding potential using these ideas, we re-express equations (3.38) and (3.39) in compact form. To find the surface inverse in equation (3.39), we write $G(\mathbf{r}_S, \mathbf{r}'_S)$[3] in terms of a surface Green function matrix $\mathcal{G}_{\mathbf{gg}'}$,

$$G(\mathbf{r}_S, \mathbf{r}'_S) = \frac{1}{A} \exp\left[i\mathbf{k}_{\parallel} \cdot \left(\mathbf{r}_S - \mathbf{r}'_S\right)\right] \sum_{\mathbf{gg}'} \mathcal{G}_{\mathbf{gg}'} \exp[i(\mathbf{g} \cdot \mathbf{r}_S - \mathbf{g}' \cdot \mathbf{r}'_S)], \tag{3.40}$$

where $\mathbf{k}_{\parallel}$ is the two-dimensional Bloch wave-vector (we suppress this in G and $\mathcal{G}$) and A is the area of the two-dimensional unit cell. Using the expansion of G in terms of basis functions χ_i, the matrix $\mathcal{G}$ is then given by

$$\mathcal{G}_{\mathbf{gg}'} = \sum_{ij} C_{\mathbf{g}i} G_{ij} C^{\dagger}_{j\mathbf{g}'}, \tag{3.41}$$

---

[3] $\mathbf{r}_S$ rather than $\mathbf{r}_{S_0}$, as this refers to any embedding surface rather than the particular surface S_0.

where the rectangular matrix C is defined as

$$C_{gi} = \frac{1}{\sqrt{A}} \int_{\text{2D cell}} \mathrm{d}\mathbf{r}_S \exp\left[-\mathrm{i}(\mathbf{k}_{\parallel} + \mathbf{g}) \cdot \mathbf{r}_S\right] \chi_i(\mathbf{r}_S, z_S), \qquad (3.42)$$

and the dagger indicates the Hermitian conjugate or adjoint of the matrix. The inverse of G gives the matrix form of the embedding potential,

$$\Sigma_{gg'} = -G_{gg'}^{-1}, \qquad (3.43)$$

in terms of which its matrix elements are given by

$$\int_{\text{2D cell}} \mathrm{d}\mathbf{r}_S \int_{\text{2D cell}} \mathrm{d}\mathbf{r}_S' \chi_i^*(\mathbf{r}_S)\Sigma(\mathbf{r}_S, \mathbf{r}_S')\chi_j(\mathbf{r}_S') = \sum_{gg'} C_{ig}^{\dagger}\Sigma_{gg'}C_{g'j}. \qquad (3.44)$$

Using these matrix expressions, we can then combine equations (3.38) and (3.39) to give a matrix equation for the embedding matrix Σ_1 on S_1 in terms of Σ_0 on S_0,

$$\Sigma_1 = -[C_1(EO_1 - H_1 - C_0^{\dagger}\Sigma_0 C_0)^{-1}C_1^{\dagger}]^{-1}, \qquad (3.45)$$

where H_1 and O_1 are the matrix elements of the Hamiltonian (including the normal derivative terms) and overlap in layer 1. C_0 and C_1 are the C-matrices (equation (3.42)) evaluated on S_0 and S_1, respectively. (To simplify notation we neglect any lateral shift in origin of the embedding plane in going from S_0 to S_1.)

This compact equation for Σ_1 in terms of Σ_0, or to generalize, Σ_{N+1} in terms of Σ_N, suggests a direct way of finding the bulk embedding potential. When we add on a layer of bulk material to the semi-infinite bulk, the embedding matrix doesn't change, so taking H_1 as the potential of a bulk layer we have

$$\Sigma = -[C_1(EO_1 - H_1 - C_0^{\dagger}\Sigma C_0)^{-1}C_1^{\dagger}]^{-1}, \qquad (3.46)$$

with the bulk embedding matrix on the left and right. We can solve this equation for Σ by direct iteration, making an initial guess for Σ on the right, getting a new Σ from the left and so on. A good starting point would be Σ calculated from the reflection matrix (equation (3.5)), with the muffin-tin form of potential. Such a process of repeated iteration corresponds physically to adding on bulk layers, and we would hope that after a reasonable number of layers the embedding matrix would stabilize to the accurate bulk value. Although this simple iterative procedure is bound to converge, if the energy E has a small imaginary part, in pratice it is impractical, and hundreds of iterations are necessary. Fortunately, the equation can be solved in a few iterations using an iterative algorithm based on Newton's method [16].

3.4.3 Tests

The iterative algorithm for solving equation (3.46) converges very quickly, and typically only two or three iterations are needed to find the bulk embedding potential, even when E is arbitrarily close to the real energy axis. To show how well it works, we calculate the density of states in a monolayer sandwiched between metal substrates on each side, with the geometry shown in figure 3.15.

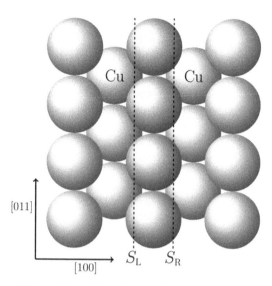

Figure 3.15. Embedding a (100) metal monolayer (brown spheres) in Cu (green spheres): the figure shows a slice through the $(0\bar{1}1)$ plane, and the [100] and [011] directions are indicated. The Cu substrates are replaced by embedding potentials on planes S_L and S_R.

First we consider a Cu(100) monolayer embedded into Cu, using embedding potentials calculated from equation (3.46) on embedding surfaces S_L and S_R (figure 3.15). An LAPW basis set was used in the layer, with a muffin-tin potential neglecting non-spherical terms and assuming a constant interstitial potential. The results for the density of states in the Cu muffin-tin at $\mathbf{k}_{\parallel} = (0.1,\ 0.2)$ a.u. are shown in figure 3.16 [16], and agreement between the embedding calculation (indicated by open diamonds) and results calculated using the layer-Korringa–Kohn–Rostoker (KKR) scattering method [40] (the solid line) is practically perfect. The number of iterations needed to calculate the bulk Cu embedding matrix is shown by crosses (right-hand axis). The initial estimate of Σ at the starting energy of the calculation, the bottom of the band, was the zero matrix, and only three iterations were needed to obtain the converged embedding potential matrix. At subsequent energies, Σ at the previous energy point was taken as the starting guess, and over nearly the whole of the energy range, with an energy interval of 0.005 a.u., only two iterations were needed. More iterations were only necessary in the d-bands, where the electronic structure changes rapidly. Halving the energy interval to 0.0025 a.u., and again using Σ from the previous energy as the starting point, reduces the number of iterations in most cases to just one—we are getting more information for no extra cost.

As a second example, we take a Ni monolayer embedded in Cu, this time calculating the Ni muffin-tin density of states with shifted embedding planes [16]. To find the Cu embedding potential on a shifted plane we first calculate the embedding potential for bulk Cu on a plane half-way between atomic layers (S_0 in figure 3.14), and add on a Cu layer with a shifted plane S_1. We then use the embedding potential at S_1 on the left- and right-hand embedding surfaces S_L and S_R, on either side of the

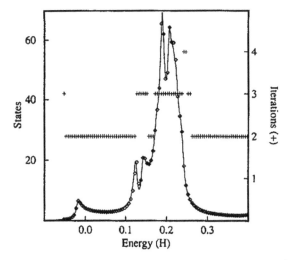

Figure 3.16. Cu muffin-tin density of states at $k_{\parallel} = (0.1, 0.2)$ a.u. in a Cu(100) layer embedded into bulk Cu on both sides, with muffin-tin potentials. The full curve is the layer-KKR result, and the diamonds show the embedding results, using LAPW basis functions. The embedding potential is obtained by solving equation (3.46) using Newton's method, and the number of iterations needed for convergence is indicated by crosses. The imaginary part of the energy is $\eta = 0.005$ a.u. (From Crampin *et al* [16].)

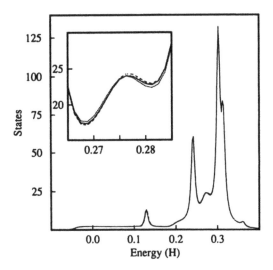

Figure 3.17. Ni muffin-tin density of states at $k_{\parallel} = 0$ in a Cu/Ni/Cu (100) sandwich. The full curve is the layer-KKR result and the broken curves are the embedding results, calculated with embedding planes at (short dash to long dash) 0.50, 0.55 and 0.60 × the interlayer spacing. (From Crampin *et al* [16].)

Ni monolayer (figure 3.15). Figure 3.17 shows the density of states at $k_{\parallel} = (0, 0)$ in the Ni muffin-tin (dashed lines), calculated for different positions of the embedding planes, at 0.50, 0.55 and 0.60 × the interlayer spacing [16]. These are compared with the density of states calculated using layer-KKR (solid line), and we see that at the scale of the larger figure the results are indistinguishable: only on the scale shown in the inset can we see any difference between the results. This is a convincing demonstration not only that we can use an embedding plane which intersects muffin-tins, but also that the embedding plane can be shifted.

In these examples, the LAPW basis used, both in calculating the embedding potentials and in calculating the density of states of the embedded monolayer, consisted of 160 plane-waves, and Σ was expanded in 29 surface reciprocal lattice vectors (equation (3.43)).

3.4.4 Applications

An early application of sub-volume embedding, building up the surface region sub-volume by sub-volume (section 3.4.1), was a self-consistent calculation of the work-function of a stepped Al(111) surface in a jellium model, in which the ion cores are smeared out into a positive background [41]. The whole embedded region extended 12 a.u. into the solid and 12 a.u. into the vacuum, divided into four sub-volumes, with free-electron embedding potentials appropriate to bulk jellium and vacuum at each end. The results for the change in work-function $\Delta\phi$ compared with the flat surface, as a function of step density, are shown in figure 3.18; the integer n by each data point gives the number of atomic rows in the terrace. We see that the stepped surfaces have a lower work-function than the flat surface ($n = \infty$), as a result of the surface electron density being smoother than the stepped background potential (Smoluchowski smoothing [6]). A surprising result is that the minimum work-function occurs not for the narrowest terraces, but for $n = 3$. This calculation is an example where sub-volume embedding is particularly economical, with the large surface unit cells of the widest terraces.

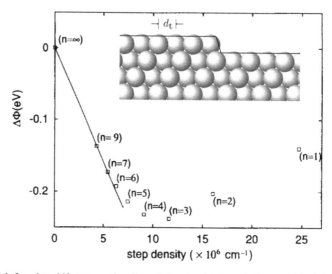

Figure 3.18. Work-function shift $\Delta\phi$ as a function of step density for a jellium model of the stepped Al(111) surface. The inset shows the atomic structure of the steps, with the jellium edge indicated by the solid line. The calculated values of $\Delta\phi$ are indicated by squares, with n giving the terrace width in units of d_t, the width between atomic rows. (Adapted from Crampin *et al* [41].)

Sub-volume embedding has been developed by Van Hoof [42] into a full-potential method for calculating surface and interface electronic structure, going to self-consistency with LAPW basis functions (Van Hoof uses transfer matrix and reflection matrix methods rather than equation (3.46) to find the bulk embedding potential). In his first applications he used three sub-volumes to describe the surface region—the top two layers of atoms and a vacuum sub-volume, embedded on one side on to the self-consistent bulk and on the other side on to vacuum. In this way, he obtained a work-function for Al(001) of 4.4 eV, in precise agreement with experiment, and for Cu(001) a work-function of 5.0 eV, in satisfactory agreement with the experimental value of 4.6 eV [42]. The computer programs have been successfully used to calculate the transport properties of interfaces such as magnetic multilayer systems [43] and domain walls in ferromagnets [44, 45], though with a simplified description of the potential near the interface.

Despite these successes, the main legacy of transferring the embedding potential through an atomic layer is its use in calculating the bulk embedding potential, as we shall see in the following sections. Modern computing power makes it very easy to find the Green function in a single region I consisting of many layers, rather than splitting it up into sub-volumes. But the idea of sub-volume embedding may still find application in semiconductor interfaces, for example, where the screening is much less effective than in metals.

3.5 Embedding with buffer regions

The embedding potential for replacing the semi-infinite bulk can be found very accurately and easily by introducing a buffer region between the 'natural' embedding surface and a plane embedding surface which does not intersect the muffin-tins in region I. This corresponds to shifting the embedding potential from S_c to the surface S shown in figure 3.4, and Ishida [2, 31] has shown how this can be done using the full potential of the bulk crystal without explicitly defining S_c at all. The method draws on some of the results of section 3.4.1, in particular the way that the embedding potential can be transferred through an atomic layer, but it avoids using an embedding surface which intersects muffin-tins (compare figure 3.14). This gives rise to a very stable method for embedding surface region I, without ghost states (section 3.3.3), and one which can be used with *any* electronic structure method— not just LAPWs (section 3.1.3).

We start with a layer of the bulk material, region Ω, between left and right 'natural' embedding surfaces S_L and S_R (figure 3.19, left). If Σ_{S_L} is the embedding potential for embedding on to semi-infinite bulk to the left of S_L, and Σ_{S_R} is the embedding potential for embedding to the left of S_R, we know from the arguments of section 3.4.2 that $\Sigma_{S_R} = \Sigma_{S_L}$ (there is a lateral shift implicit in this). Starting with an estimate of Σ_{S_L} we can integrate this through Ω (*transfer* it, in the language of section 3.4.1) to obtain Σ_{S_R}, and if these satisfy $\Sigma_{S_R} = \Sigma_{S_L}$ they are the actual embedding potentials for embedding on to semi-infinite bulk to the left. Unfortunately, this procedure is too complicated with the curvy 'natural' embedding surfaces.

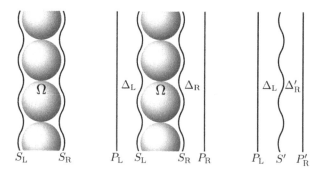

Figure 3.19. Left: bulk layer, region Ω, with 'natural' embedding surfaces S_L and S_R to the left and right. Middle: buffer regions Δ_L and Δ_R are added to the layer Ω to give a region Ω' between planes P_L and P_R. Right: Δ_L and Δ'_R (laterally translated Δ_R) are joined together to give an auxiliary region Ω''.

The next stage is to add buffer regions Δ_L and Δ_R on either side of region Ω, bounded by plane surfaces P_L and P_R, respectively (figure 3.19, middle); any convenient potential can be used in the buffer region, for example, a smooth continuation of $V(\mathbf{r})$ in Ω. This whole region is designated Ω'. We also construct a third region Ω'' (figure 3.19, right), consisting of Δ_R laterally translated to give Δ'_R which joins on to Δ_L over S'. The embedding potential over the flat surface P_L equivalent to Σ_{S_L} over the curvy surface S_L can be found by transferring it through Ω' and Ω''.

Suppose that we have a trial embedding potential Σ_{P_L} over P_L. We transfer this, using the formalism of section 3.4.1, through region Ω' to the right-hand plane surface P_R, giving Σ_{P_R}. We also transfer Σ_{P_L} through region Ω'' to P'_R (figure 1.19) to give $\Sigma_{P'_R}$. Then the condition for Σ_{P_L} to be the actual embedding potential is [31]

$$\Sigma_{P'_R} = \Sigma_{P_R}. \tag{3.47}$$

To show this, we integrate $\Sigma_{P'_R}$ back through Δ'_R to give $\Sigma_{S'}$ on S' (this is shorthand for saying that $\Sigma_{S'}$ transfers through Δ'_R to give $\Sigma_{P'_R}$, equivalent to an actual integration of the Schrödinger equation). Similarly, we integrate Σ_{P_R} back through Δ_R to give Σ_{S_R}. As the potentials in Δ'_R and Δ_R are the same (apart from a shift), we see that

$$\Sigma_{P'_R} = \Sigma_{P_R} \Longrightarrow \Sigma_{S'} = \Sigma_{S_R}. \tag{3.48}$$

We also see from figure 3.19 that $\Sigma_{S'} = \Sigma_{S_L}$, as both can be obtained by transferring Σ_{P_L} through Δ_L, and equation (3.48) becomes

$$\Sigma_{P'_R} = \Sigma_{P_R} \Longrightarrow \Sigma_{S_L} = \Sigma_{S_R}. \tag{3.49}$$

This means that Σ_{S_L} obtained by transferring Σ_{P_L} through Δ_L is the bulk embedding potential over S_L, and consequently Σ_{P_L} is the embedding potential over the convenient planar surface P_L. The remarkable point about this procedure is that the condition (3.47) satisfied by Σ_{P_L} only involves transfers through regions Ω' and Ω'', without any reference to the 'natural' embedding surfaces. These are only relevant by defining boundaries to the potentials in Δ_L and Δ_R. To find the Σ_{P_L} satisfying (3.47), an accelerated iterative procedure is used [31].

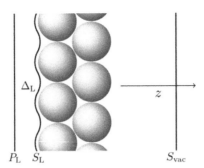

Figure 3.20. Embedded surface calculations with a buffer region. Region I lies between P_L and S_{vac}, with the bulk embedding potential Σ_{P_L} on P_L, and the vacuum embedding potential on S_{vac}. Between S_L and S_{vac} the electrons feel a self-consistent potential, and in buffer region Δ_L the same potential as that used to find Σ_{P_L}.

The geometry for using Σ_{P_L} in surface calculations is shown in figure 3.20. The surface region is padded out with the buffer region Δ_L between the 'natural' embedding surface S_L and the planar surface P_L; the same potential must be used in Δ_L as in the evaluation of Σ_{P_L} (figure 3.19). On the right-hand side, the asymptotic vacuum region is replaced by Σ_{vac} or Σ_{im} on surface S_{vac}, which lies beyond the surface charge density (section 3.1.4). Region I, where the embedded Schrödinger equation is solved, then comprises the whole region between P_L and S_{vac}, with the self-consistent potential between S_L and S_{vac} and the fixed buffer potential in Δ_L.

3.5.1 Applications

This method of constructing the embedding potential, with an embedding plane which does not overlap the muffin-tins in region I (figure 3.20), leads to great stability in computation (no ghost states!). The accuracy of the method is illustrated by Ishida's calculations of the electronic structure of the Ag(001), Pd(001) and Rh(001) surfaces [2], for which the electron density is shown in figure 3.21. In this work, the embedding potential was constructed using the method described above, with the potential in the atomic layer taken from a separate self-consistent bulk calculation. Region I consists of the top two layers of atoms, extending to the vacuum embedding plane at about 10 a.u. above the surface; an LAPW basis set was used in region I (section 3.2.1) and the self-consistent potential was matched on to the bulk potential over the dashed lines shown in figure 3.21 (roughly, the 'natural' embedding surface) (section 3.2.3). In figure 3.21 the electron density below the dashed line is taken from the bulk calculation, and what is remarkable is the quality of the matching across this line, with the electron density contours in the surface region having exactly the right shape. The inaccuracies encountered in figure 3.11, at least in part associated with using the muffin-tin approximation to construct the embedding potential, have gone completely.

Because the embedding plane P_L (figure 3.20) does not intersect the atomic spheres of the neighbouring plane of atoms in region I (compare with figure 3.5), we do not need an augmented basis set inside the atoms for embedding to work. (Remember the tests described in section 3.1.3—these would not have worked

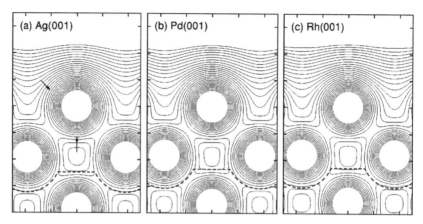

Figure 3.21. Charge density plotted on a (100) slice for (a) Ag(001), (b) Pd(001), and (c) Rh(001) surfaces, from embedded surface calculations with two atomic layers included in region I and full-potential embedding potentials. The lowest contour level is 5×10^{-4} a.u., increasing successively by a factor of $10^{1/8}$, and the arrows indicate the directions of increasing electron density. Dashed lines indicate the surface over which the potential is matched to the bulk; the bulk electron density is plotted below this surface. (From Ishida [2].)

without the augmentation functions inside the well.) In fact, the first results published with this approach were for the Al(111) surface calculated using a plane-wave basis with pseudopotentials to describe the Al atoms [31]. This gives results in excellent agreement with the embedded LAPW calculations of Benesh and Liyanage [3], who used the methods described in sections 3.1 and 3.2 with an embedding potential derived from the reflection matrix of the semi-infinite crystal. Figure 3.22 shows the $\mathbf{k}_\parallel$-resolved surface density of states at the $\bar{\mathrm{K}}$-point (the corner of the hexagonal surface Brillouin zone). n_s is shown for the three atomic layers included in the surface region, and we see that there is excellent agreement between the two sets of results. We saw in section 3.3.1 that the LAPW calculation of Benesh and Liyanage [3] gave an excellent value for the Al(111) work-function; 4.22 eV compared with the experimental value of 4.24 eV. Ishida'a embedded pseudopotential calculation gives a similarly excellent work-function of 4.25 eV [31].

3.6 The transfer matrix and embedding

Another way of evaluating the embedding potential shifted through the buffer region Δ_L to P_L is to use the *transfer matrix* [46]. This is closely related to the method described in section 3.5, and has the advantage that we do not have to solve equation (3.47) by an iterative procedure.

There are several definitions of the transfer matrix, but the most useful for our purposes is one that relates the amplitude and derivative of the wave-function on one side of an atomic layer with that on the other (compare this with the definition in [44]) (figure 3.23, left). For the moment we shall take the surface normals in the same direction on both surfaces, into the layer Ω over S_L and out of the layer over S_R.

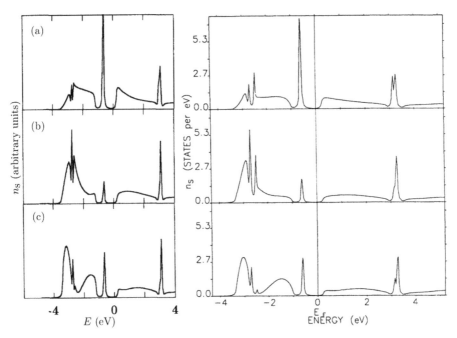

Figure 3.22. Al(111) $\mathbf{k}_\parallel$-resolved surface density of states at $\bar{K}$. The left-hand panel shows pseudopotential results with full-potential embedding, and the right-hand panel LAPW results with an embedding potential derived from the reflection matrix. In both calculations the surface region consists of three atomic layers, and the figure shows n_s in the atomic spheres of (a) the top layer, (b) the sub-surface layer and (c) the third layer. Energies are measured relative to E_F. (From Ishida [31] and Benesh and Liyanage [3].)

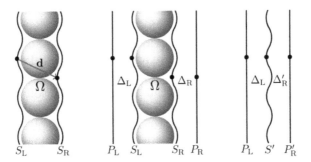

Figure 3.23. Left: bulk layer, region Ω, the red arrow indicating lattice vector $\mathbf{d}$. Middle: buffer regions Δ_L and Δ_R are added to give region Ω'. Right: Δ_L and Δ'_R (laterally translated Δ_R) are joined together to give Ω''. The dots indicate the origins on each surface.

Writing the wave-function over the left-hand surface as the vector Ψ_L and the normal derivative as Ψ'_L, and similarly over S_R, we have

$$\begin{pmatrix} \Psi_R \\ \Psi'_R \end{pmatrix} = T_\Omega \begin{pmatrix} \Psi_L \\ \Psi'_L \end{pmatrix}, \tag{3.50}$$

where T_{Ω} is the transfer matrix. The origins on S_L and S_R, indicated by the black dots in figure 3.23 (left), are displaced by the lattice vector $\mathbf{d}$, which is the displacement of the layers when stacked together to give the bulk crystal.

When we stack the layers in this way, S_R of one layer becomes S_L of the next, and T_{Ω} describes the way that the wave-function changes as we go from layer to layer. So the eigenvalue equation

$$T_{\Omega}\begin{pmatrix} \Psi_i \\ \Psi_i' \end{pmatrix} = \lambda_i \begin{pmatrix} \Psi_i \\ \Psi_i' \end{pmatrix} \tag{3.51}$$

has eigenvectors which are the amplitude and derivative over S_L (or S_R) of the Bloch solutions of the bulk Schrödinger equation, and eigenvalues which are the corresponding Bloch phase factors in going from one layer to the next,

$$\lambda_i = \exp(i\mathbf{k} \cdot \mathbf{d}), \tag{3.52}$$

where $\mathbf{k}$ is the Bloch wave-vector. The transfer matrix has a size of $2\times$ the number of basis vectors N_g in the expansions over the surfaces, so fixing $\mathbf{k}_{\parallel}$, the component of $\mathbf{k}$ parallel to the surface, these equations give $2N_g$ values of the z-component k_z^i. As the wave-vectors come in pairs, $\pm k_z^i$, there are N_g values of k_z^i [46]: some of the k_z^is are real, corresponding to the real energy bands in the band structure, but others are complex, corresponding to the complex bands which weave their way between the real bands—only the real bands correspond to allowed solutions in the infinite bulk crystal, but the complex bands correspond to solutions of the Schrödinger equation allowed in a truncated crystal [47].

Figure 3.24 shows the complex bands at real energy for Cu built up out of (001) layers, evaluated at $\mathbf{k}_{\parallel} = 0$. This was in fact calculated from the eigenvalues of the transfer matrix of the Dirac equation, so it includes spin–orbit splitting which doubles the size of the transfer matrix (see section 8.2.1) [48]. We see that the complex bands join the real bands at extrema and provide a continuous path through the band structure as E varies. At real energy, these are sometimes called *real lines* [47].

The general solution of the Schrödinger equation of a semi-infinite crystal, in the bulk region and away from the surface, can be built up from solutions with complex wave-vector: if the semi-infinite crystal lies at $z < 0$, we take those solutions with $\Im k_z \leqslant 0$ [6]. So the wave-function of a surface state, for example, consists of a linear combination of bulk solutions which decay as $z \rightarrow -\infty$, matched on to the surface solution which decays into vacuum. In an embedding calculation of surface states or surface electronic structure, like the one shown in figure 3.12, this is all taken care of by the embedding potential; this suggests that there is an intimate connection between the general solutions of the bulk Schrödinger equation, complex band theory, and the embedding potential.

The link is that Σ can be constructed from the eigenvectors of equation (3.51) [31]. Taking Ω (figure 3.23, left) as a layer of semi-infinite bulk, extending to the left, the general solution of the Schrödinger equation over S_L is given by a linear combination of the amplitude part of the eigenfunctions of equation (3.51), with the derivative given by the linear combination of the Ψ_i'. But the relationship

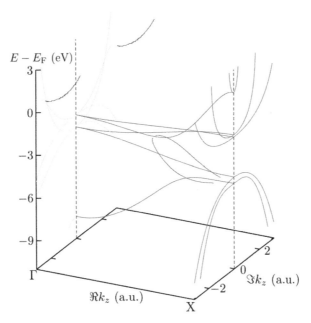

Figure 3.24. Cu(001) complex band structure at $\mathbf{k}_\parallel = 0$, including spin–orbit splitting, in the Γ-X direction in the bulk Brillouin zone. Red lines correspond to real bands, green lines to complex bands with $\Re k_z^i$ at Γ (the centre of the bulk Brillouin zone), blue lines to complex bands with $\Re k_z^i$ at X (at the Brillouin zone boundary) and brown lines to general complex bands. (Figure courtesy of S Crampin.)

between amplitude and derivative over S_L gives the embedding potential (equation (2.16)). To satisfy outgoing or decaying boundary conditions, we must take the eigenvectors for which $\Re k_z^i < 0$ and $\Im k_z^i < 0$; if we work at complex energy with $\Im E > 0$, both of these conditions are satisfied at the same time.

To formalise this we write the N_g components of the N_g allowed amplitudes and derivatives as $N_\mathrm{g} \times N_\mathrm{g}$ matrices, $\Psi_{\nu i}$ and $\Psi'_{\nu i}$, where ν labels the basis vector $\mathbf{g}$, and i the eigenvector of the transfer matrix. In embedding theory (section 2.1) we know the amplitude ϕ of the trial function over S_L, which we can expand in terms of the basis vectors with expansion coefficients ϕ_ν. We then write the identity

$$\phi_\nu = \sum_{i,\nu'} \Psi_{\nu i} \Psi_{i\nu'}^{-1} \phi_{\nu'}. \tag{3.53}$$

It follows (in the notation of section 2.1) that

$$\psi'_\nu = \sum_{i,\nu'} \Psi'_{\nu i} \Psi_{i\nu'}^{-1} \phi_{\nu'}, \tag{3.54}$$

and comparing with equation (2.16), we see that the embedding potential, which describes the semi-infinite bulk to the left of S_L, is given by [31, 46, 48]

$$\left(\Sigma_{S_\mathrm{L}}\right)_{\nu\nu'} = \frac{1}{2} \sum_i \Psi'_{\nu i} \Psi_{i\nu'}^{-1}. \tag{3.55}$$

(We have taken account of the direction of the normal derivative, which in equation (3.54) is directed over S_L into Ω, whereas in the definition of Σ_{S_L} (equation (2.16)) it is in the opposite direction.)

3.6.1 Transferring the transfer matrix

As in section 3.5, we can work with the much more convenient flat surfaces P_L and P_R by introducing buffer regions Δ_L and Δ_R (figure 3.23, middle) [46], with the origins on each surface shown by the black dots.

We first consider region Ω' between the flat surfaces P_L and P_R, consisting of the atomic layer Ω extended by buffer regions Δ_L and Δ_R. The transfer matrix $T_{\Omega'}$, taking the solution of the Schrödinger equation from P_L to P_R, is the product of the transfer matrices for the separate regions [46], given by

$$T_{\Omega'} = T_{\Delta_R} T_\Omega T_{\Delta_L}, \tag{3.56}$$

where T_{Δ_L} and T_{Δ_R} are the transfer matrices of Δ_L and Δ_R. Similarly, the transfer matrix through Ω'', consisting of Δ_L and Δ_R joined together with a lateral shift in Δ_R (figure 3.23, right), is given by

$$T_{\Omega''} = T_{\Delta_R} T_{\Delta_L} \tag{3.57}$$

(from the origins shown in figure 3.23 we see that $T_{\Delta_R'} = T_{\Delta_R}$). We can combine this with equation (3.56) to give a new transfer matrix T defined as

$$T = T_{\Omega''}^{-1} T_{\Omega'} = T_{\Delta_L}^{-1} T_\Omega T_{\Delta_L}. \tag{3.58}$$

As T and T_Ω are related by a similarity transformation they have the same eigenvalues, and we can find the complex band structure without using the curvy surfaces S_L and S_R: T is constructed from $T_{\Omega'}$ and $T_{\Omega''}$, both of which are defined between planar surfaces, with correspondingly straightforward plane-wave expansions.

The eigenvectors of T satisfy the equation

$$T \begin{pmatrix} \Upsilon_i \\ \Upsilon_i' \end{pmatrix} = \lambda_i \begin{pmatrix} \Upsilon_i \\ \Upsilon_i' \end{pmatrix}, \tag{3.59}$$

so from equation (3.58) the eigenvectors of T and T_Ω with the same eigenvalue are related by

$$\begin{pmatrix} \Psi_i \\ \Psi_i' \end{pmatrix} = T_{\Delta_L} \begin{pmatrix} \Upsilon_i \\ \Upsilon_i' \end{pmatrix}. \tag{3.60}$$

Hence, (Ψ_i, Ψ_i') can be found by propagating $(\Upsilon_i, \Upsilon_i')$ from P_L through the buffer region Δ_L to S_L. The embedding potential shifted through Δ_L to P_L is then given by [46]

$$\left(\Sigma_{P_L} \right)_{\nu\nu'} = \frac{1}{2} \sum_i \Upsilon_{\nu i}'' \Upsilon_{i\nu'}^{-1}. \tag{3.61}$$

which is entirely equivalent to equation (3.55). As in section 3.5, we have constructed the embedding potential on the planar surface P_L, which fully describes the semi-infinite crystal to the left of S_L; to the right of P_L we can add any region I—but we must preserve the buffer layer Δ_L.

This embedding potential is the same as Σ_{P_L} as derived in section 3.5, and indeed there are similar transfers through Ω' and Ω'' in both derivations. In fact, this transfer matrix approach is the method of choice nowadays for calculating the embedding potential for surface and interface calculations: not only does the construction of Σ_{P_L} using the eigenvectors eliminate the iterative solution of equation (3.47), but it also gives the outgoing embedding potential even for real E. It has proved to be extremely stable and accurate, and has been generalized to the relativistic case [48].

3.6.2 The transfer matrix and Green functions

The two transfer matrices which we need, $T_{\Omega'}$ (3.56) and $T_{\Omega''}$ (3.57), can be found using Green function methods [46, 48]. Let us take the case of Ω', and assume we know the Green function $G_0(\mathbf{r}, \mathbf{r}')$ in this volume, with zero derivative on the planes P_L and P_R on either side (figure 3.23, middle). Then from equation (2.12) the solution of the Schrödinger equation inside Ω' is given in terms of the derivatives over P_L and P_R by

$$\psi(\mathbf{r}) = \frac{1}{2}\int_{P_L} d\mathbf{r}_L\, G_0(\mathbf{r}, \mathbf{r}_L)\frac{\partial\psi(\mathbf{r}_L)}{\partial n_L} - \frac{1}{2}\int_{P_R} d\mathbf{r}_R\, G_0(\mathbf{r}, \mathbf{r}_R)\frac{\partial\psi(\mathbf{r}_R)}{\partial n_R}, \qquad (3.62)$$

where $\mathbf{r}_{L/R}$ indicates the surface coordinate on $P_{L/R}$ and the surface derivative is taken into Ω' on P_L and out of Ω' on P_R (the same convention as in the definition of the transfer matrix). The surface integrals in equation (3.62) are over one surface unit cell if we take G_0 as the Bloch Green function with wave-vector $\mathbf{k}_\parallel$ parallel to the surface, and ψ a wave-function with the same $\mathbf{k}_\parallel$ (ψ is *any* solution of the Schrödinger equation in the z-direction). We now put $\mathbf{r}$ on P_L, and expand $\psi(\mathbf{r}_L)$, $G_0(\mathbf{r}_L, \mathbf{r}'_L)$ and the other terms in equation (3.62) in terms of two-dimensional plane-waves, summed over surface reciprocal lattice vectors (as in equation (3.40)),

$$\psi(\mathbf{r}_L) = \frac{1}{\sqrt{A}}\exp\left(i\mathbf{k}_\parallel \cdot \mathbf{r}_L\right)\sum_g \psi_{L;g}\exp(i\mathbf{g}\cdot\mathbf{r}_L),$$

$$G_0(\mathbf{r}_L, \mathbf{r}'_L) = \frac{1}{A}\exp\left[i\mathbf{k}_\parallel\cdot(\mathbf{r}_L - \mathbf{r}'_L)\right]\sum_{gg'}\mathcal{G}_{LL;gg'}\exp[i(\mathbf{g}\cdot\mathbf{r}_L - \mathbf{g}'\cdot\mathbf{r}'_L)], \quad (3.63)$$

with similar expansions on P_R.[4] In abstract notation we then have

$$\psi_L = \frac{1}{2}\mathcal{G}_{LL}\psi'_L - \frac{1}{2}\mathcal{G}_{LR}\psi'_R,$$

$$\psi_R = \frac{1}{2}\mathcal{G}_{RL}\psi'_L - \frac{1}{2}\mathcal{G}_{RR}\psi'_R. \qquad (3.64)$$

---

[4] We must remember the origin shift shown in figure 3.23 (middle) in expanding $G_0(\mathbf{r}_L, \mathbf{r}'_R)$.

This can be rearranged to give ψ_R and ψ'_R in terms of ψ_L and ψ'_L,

$$\psi_R = \mathcal{G}_{RR}\mathcal{G}_{LR}^{-1}\psi_L + \frac{1}{2}(\mathcal{G}_{RL} - \mathcal{G}_{RR}\mathcal{G}_{LR}^{-1}\mathcal{G}_{LL})\psi'_L,$$

$$\psi'_R = -2\mathcal{G}_{LR}^{-1}\psi_L + \mathcal{G}_{LR}^{-1}\mathcal{G}_{LL}\psi'_L, \qquad (3.65)$$

and comparing this with equation (3.50) we see that the transfer matrix is given by

$$T_{\Omega'} = \begin{pmatrix} \mathcal{G}_{RR}\mathcal{G}_{LR}^{-1} & \frac{1}{2}(\mathcal{G}_{RL} - \mathcal{G}_{RR}\mathcal{G}_{LR}^{-1}\mathcal{G}_{LL}) \\ -2\mathcal{G}_{LR}^{-1} & \mathcal{G}_{LR}^{-1}\mathcal{G}_{LL} \end{pmatrix}. \qquad (3.66)$$

The zero-derivative Green function G_0 which enters this expression can be found using embedding ideas. G_0 satisfies an embedding-type Schrödinger equation in Ω', with the embedding potentials on P_L and P_R set equal to zero,

$$-\frac{1}{2}\nabla_{\mathbf{r}}^2 G_0(\mathbf{r}, \mathbf{r}'; E) - \frac{1}{2}\delta(n - n_L)\frac{\partial G_0(\mathbf{r}_L, \mathbf{r}'; E)}{\partial n_L} + \frac{1}{2}\delta(n - n_R)\frac{\partial G_0(\mathbf{r}_R, \mathbf{r}'; E)}{\partial n_R}$$
$$+ V(\mathbf{r})G_0(\mathbf{r}, \mathbf{r}'; E) - EG_0(\mathbf{r}, \mathbf{r}'; E) = -\delta(\mathbf{r} - \mathbf{r}'), \quad \mathbf{r}, \mathbf{r}' \in \Omega'. \qquad (3.67)$$

As in (equation (3.38)), omitting the embedding potentials but keeping the surface derivative terms forces G_0 to have zero derivative on P_L and P_R. (Again the derivatives are taken in the directions shown in figure 3.23 (middle) rather than the usual embedding convention.) This can be solved using a basis-set expansion, typically LAPWs [46]. The potential in the atomic layer Ω (the filling, if we think of Ω' as a sandwich) is taken from a separate self-consistent bulk calculation, and $V(\mathbf{r})$ in the buffer layers Δ_L and Δ_R (the bread of the sandwich) is taken as constant or a smooth extension of the interstitial potential in Ω.

So starting from a full-potential bulk calculation, we find G_0 in Ω', hence $T_{\Omega'}$ and, using exactly the same methods $T_{\Omega''}$. We then find T from equation (3.58), and construct the embedding potential on P_L from its eigenvectors using equation (3.61).

3.6.3 Model calculation

To demonstrate using transfer matrix methods for calculating the embedding potential on a shifted surface, we shall consider a one-dimensional model potential, devised by Chulkov and co-workers [49] to describe the Al(001) surface (figure 3.25). This consists of a sinusoidal potential in the bulk, and a parametrized surface potential joined on smoothly to the asymptotic image potential which an electron feels deep in the vacuum (equation (3.12)); the parameters of the model potential are fitted to the observed band gap and surface state energies of Al(001).

First let us embed the system shown in figure 3.25, adding an embedding potential Σ_b to replace the bulk at $z = -10$ a.u. and Σ_{im} to replace the Coulomb tail of the image potential at $z = +10$ a.u. We can find Σ_b from the eigenvector of the (2×2) transfer matrix T across one atomic layer (here, one oscillation of the bulk potential) which corresponds to the wave-function decaying/travelling away into the bulk

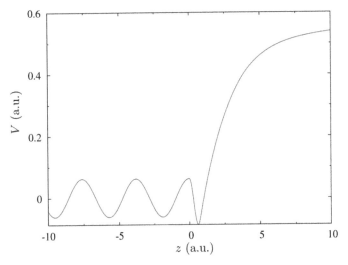

Figure 3.25. One-dimensional model potential for the Al(001) surface [49], showing the periodic bulk potential, the potential step at the surface and in the vacuum the asymptotic image potential. The top layer of atoms is at $z = 0$, and $V = 0$ is the average bulk potential.

(equation (3.55)).[5] The right-hand embedding potential Σ_{im} can be found from the outgoing Coulomb wave-functions (equation (3.13), section 3.1.4). The density of states for this embedded Al(001) surface is shown by the red dotted curve in figure 3.26. The most prominent feature is the Shockley surface state close to the bottom of the band gap, qualitatively similar to the Al(001) results shown in figure 3.12 from the full three-dimensional embedded surface calculation.

We now shift the bulk embedding potential in the slightly artificial system shown in figure 3.27: Σ_b evaluated at $z_1 = -3.8$ a.u. (a maximum of the bulk potential) is shifted through a buffer region of constant potential Δ_L to $z_2 = -10$ a.u. (In this model there is no right-hand buffer layer.) The embedding potential at z_1 is given in terms of the solution of the bulk Schrödinger equation by the usual expression (equation (3.55)),

$$\Sigma_b(z_1; E) = \frac{1}{2}\frac{\psi'(z_1)}{\psi(z_1)}, \qquad (3.68)$$

where $(\psi(z_1)\psi'(z_1))$ is the eigenvector of T travelling or decaying into the bulk. This solution of the Schrödinger equation at z_1 can then be integrated back through Δ_L to give $(\psi(z_2)\psi'(z_2))$, where the two vectors are related by the transfer matrix T_{Δ_L},

$$\begin{pmatrix} \psi(z_2) \\ \psi'(z_2) \end{pmatrix} = T_{\Delta_L}^{-1}\begin{pmatrix} \psi(z_1) \\ \psi'(z_1) \end{pmatrix}. \qquad (3.69)$$

---

[5] Another simple way of finding the one-dimensional bulk embedding potential is given in section 10.4.1.

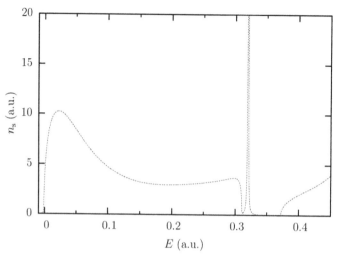

Figure 3.26. Al(001) surface density of states $n_s(E)$, integrated between $z = -3.8$ a.u. and 10 a.u. Red dotted curve, calculated with the potential shown in figure 3.25, and embedded at $z = -10$ a.u. on to the bulk embedding potential and at $z = +10$ a.u. on to the Coulomb embedding potential. Green curve, calculated with the potential shown in figure 3.27 and embedded at $z = -10$ a.u. on to the embedding potential shifted from -3.8 a.u. to -10 a.u. Fifty basis functions are used, and the imaginary part of the energy is $\eta = 10^{-4}$ a.u.

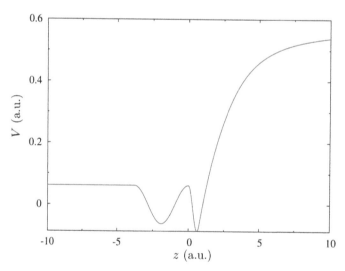

Figure 3.27. Al(001) surface with bulk embedding potential shifted from $z = -3.8$ a.u. through a buffer region with constant potential to $z = -10$ a.u.

T_{Δ_L} is given by the free-electron expression [46],

$$T_{\Delta_L} = \begin{pmatrix} \cos kd & \dfrac{1}{k}\sin kd \\ -k\sin kd & \cos kd \end{pmatrix},$$ (3.70)

where $d = z_1 - z_2$, and k is the wave-vector of the free-electron solutions of the Schrödinger equation in the buffer region. The bulk embedding potential, shifted

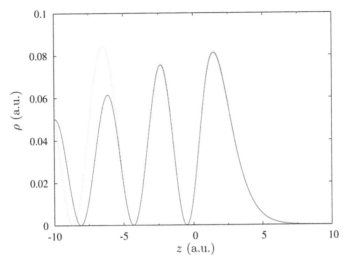

Figure 3.28. Electron density $\rho(z)$ of Al(001) surface state at $E = 0.312$ a.u. Red curve, embedded at $z = -10$ a.u. on to the bulk embedding potential, and at $z = +10$ a.u. on to the Coulomb embedding potential. Green curve, embedded at $z = -10$ a.u. on to the embedding potential shifted from -3.8 a.u. to -10 a.u. Fifty basis functions are used.

from z_1 to z_2 through Δ_L, is then given by substituting $(\psi(z_2)\psi'(z_2))$ into equation (3.68).

We now solve the Schrödinger equation using the potential of figure 3.27, embedded on to the shifted bulk embedding potential at -10 a.u. and Σ_{im} at $+10$ a.u. As we emphasised in section 3.6.1, we must include the potential of the buffer region Δ_L in the Schrödinger equation, though it is only the solution to the right of $z_1 = -3.8$ a.u. which has physical meaning. The resulting surface density of states $n_s(E)$ integrated between -3.8 a.u. and $+10$ a.u. is shown by the green curve in figure 3.26; we see almost perfect agreement with $n_s(E)$ (red dotted line) calculated with the potential shown in figure 3.25 and the unshifted bulk embedding potential.

The electron density of the Shockley surface state, evaluated with the original and shifted embedding potentials, is shown in figure 3.28. The red curve evaluated with the original embedding potential at -10 a.u. and $V(z)$ shown in figure 3.25 is accurate across the entire range; the green curve evaluated with the embedding potential shifted from -3.8 a.u. to -10 a.u. and $V(z)$ of figure 3.27 is meaningful for $z > -3.8$ a.u. In this physical range we see that the two curves are in excellent agreement, and the slight difference which we can just detect in the peaks is due to numerical inaccuracies. (We did not try to optimize the computation or the parameters used in the calculation.)

3.7 Embedding an isolated adsorbate

Although most surface applications of the embedding method assume two-dimensional periodicity parallel to the surface, it has also been applied to the isolated adsorbate problem, though (so far) only on a model surface [7, 50]. Chemisorption is

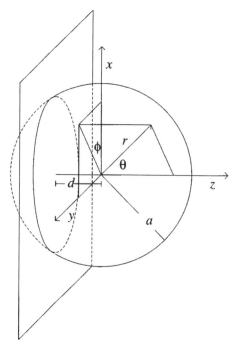

Figure 3.29. The embedding sphere and coordinate system used for the isolated adsorbate. The semi-infinite background of the jellium model fills the left-hand region of space, with the jellium edge (the geometrical surface) indicated by the plane at $z = -d$. (From Trioni *et al* [7].)

an important problem [51, 52] for which numerous methods have been developed, based for the most part on Dyson's equation (see, for example, [52, 53, 54, 55, 56]) or the Lippmann-Schwinger equation [57]. We shall discuss the classic tight-binding/ Dyson's equation approach in section 6.2.

To apply embedding to a single adsorbate, a spherical embedding surface is used, large enough so that region I includes the perturbed substrate (where the potential differs significantly from the clean surface) as well as the adsorbate (figure 3.29). In region II, outside the sphere, it is assumed that the electrons feel the potential of the unperturbed surface, so the embedding potential over the surface of the sphere can be found from the Green function for the clean surface. In principle, we could find this for any surface, including the full self-consistent potential, but then we have the problem of how to treat substrate atoms which are *partially* included in region I. If we describe the substrate atoms with local pseudopotentials this might not be a problem, but dealing with a chopped-off non-local pseudopotential or muffin-tin becomes non-trivial (though it could probably be handled as in section 3.5). To avoid these complications, we can use a one-dimensional model potential to describe the substrate, the jellium model[6] in the original paper [7] or the one-dimensional Chulkov potential [50].

---

[6] In the jellium model of a surface, the electrons feel the potential of a semi-infinite positive background, plus their self-consistent field calculated in DFT [58].

3.7.1 The adsorbate embedding potential

We start with $G_{M,k_\parallel}(\mathbf{r}, \mathbf{r}'; E)$, the Green function for the clean metal surface with Bloch wave-vector $\mathbf{k}_\parallel$ (section 3.2.2). With a one-dimensional potential this factorizes,

$$G_{M,k_\parallel}(\mathbf{r}, \mathbf{r}'; E) = \exp\left[i\mathbf{k}_\parallel \cdot \left(\mathbf{r}_\parallel - \mathbf{r}'_\parallel\right)\right] g_{M,k_\parallel}(z, z'; E), \qquad (3.71)$$

where $g_{M,k_\parallel}$ satisfies the one-dimensional Schrödinger equation,

$$\left[-\frac{1}{2}\frac{\partial^2}{\partial z^2} + V(z) - \left(E - \frac{k_\parallel^2}{2}\right)\right] g_{M,k_\parallel}(z, z'; E) = -\delta(z - z'). \qquad (3.72)$$

The full surface Green function, without the restriction of Bloch form, is then given by integrating over all $\mathbf{k}_\parallel$ (in other words Fourier transforming $g_{M,k_\parallel}$),

$$G_M(\mathbf{r}, \mathbf{r}'; E) = \frac{1}{4\pi^2}\int d\mathbf{k}_\parallel \exp\left[i\mathbf{k}_\parallel \cdot \left(\mathbf{r}_\parallel - \mathbf{r}'_\parallel\right)\right] g_{M,k_\parallel}(z, z'; E) \qquad (3.73)$$

(with a three-dimensional substrate potential, $G_{M,k_\parallel}$ is integrated over the surface Brillouin zone [7]). $g_{M,k_\parallel}$ in the range of z which spans the embedding sphere (figure 3.29) is readily calculated using embedding, but to improve convergence of the $\mathbf{k}_\parallel$ integral it is convenient to rewrite equation (3.73) as

$$G_M(\mathbf{r}, \mathbf{r}'; E) = \frac{1}{4\pi^2}\int d\mathbf{k}_\parallel \exp\left[i\mathbf{k}_\parallel \cdot \left(\mathbf{r}_\parallel - \mathbf{r}'_\parallel\right)\right]\left(g_{M,k_\parallel}(z, z'; E) + \frac{1}{\gamma}\exp(-\gamma|z - z'|)\right)$$

$$-2ik\sum_{l,m} j_l\left(kr_<\right)h_l\left(kr_>\right)Y_L(\theta, \phi)Y_L^*(\theta', \phi'), \qquad (3.74)$$

where $\gamma = \sqrt{k_\parallel^2 - 2E}$ and $k = \sqrt{2E}$. The second term in the brackets in the integral in equation (3.74) subtracts the free-electron solution of equation (3.72) and the final term is the spherical expansion of the free-electron Green function [59]; $r_</r_>$ are the lesser/greater of $|\mathbf{r}|$, $|\mathbf{r}'|$. Writing G_M in this way builds in the discontinuity in the derivative of the Green function as $\mathbf{r}$ passes through $\mathbf{r}'$.

The embedding potential is given by the surface inverse of the Green function G_0 satisfying the zero-derivative boundary condition over the embedding sphere (equation (2.17)). To find this from G_M, we use Green's theorem, which gives [60]

$$G_0(\mathbf{r}, \mathbf{r}') = G_M(\mathbf{r}, \mathbf{r}') - \frac{1}{2}\int_S d\mathbf{r}_S \frac{\partial G_M(\mathbf{r}, \mathbf{r}_S)}{\partial n_S} G_0(\mathbf{r}_S, \mathbf{r}'), \quad \mathbf{r}, \mathbf{r}' \in \text{region II}, \qquad (3.75)$$

where the integral is over the surface of the embedding sphere, and the surface normal is directed outwards. To solve this we expand the Green functions over the surface in terms of spherical harmonics, for example,

$$G_M(\mathbf{r}_S, \mathbf{r}'_S) = \frac{1}{a^2}\sum_{L,L'} G_{M;L,L'} Y_L(\theta, \phi)Y_L^*(\theta', \phi'), \qquad (3.76)$$

where a is the radius of the embedding sphere[7]. The integral equation (3.75) then becomes the matrix equation

$$\mathcal{G}_0 = \mathcal{G}_{\mathrm{M}} - \frac{1}{2}\mathcal{G}'_{\mathrm{M}}\mathcal{G}_0, \tag{3.77}$$

so $\mathcal{G}_0$ is given by

$$\mathcal{G}_0 = \left(I + \frac{1}{2}\mathcal{G}'_{\mathrm{M}}\right)^{-1}\mathcal{G}_{\mathrm{M}}, \tag{3.78}$$

and the embedding potential expansion coefficients by

$$\Sigma = -\mathcal{G}_{\mathrm{M}}^{-1}\left(I + \frac{1}{2}\mathcal{G}'_{\mathrm{M}}\right). \tag{3.79}$$

3.7.2 Adsorbate calculations: density of states and energy

We can now find the Green function for the adsorbate system by solving the Schrödinger equation (2.36) within the embedding sphere, using the embedding potential given by equation (3.79). Inside the sphere, the potential $V(\mathbf{r})$ in equation (2.36) comprises the adsorbate atom potential, the electrostatic potential due to the jellium background and the self-consistent potential, calculated within DFT, due to the electron charge density; as the changes in potential are confined entirely to the sphere (this is our assumption), the electrostatic contribution from region II can be found by once-and-for-all matching $V(\mathbf{r})$ over the surface of the sphere.

In the original paper [7] the adsorbate atom was described by a pseudopotential, of the norm-conserving (hence energy-independent) Bachelet–Hamann–Schlüter form [61]. This meant that spherical plane-wave basis functions could be used within region I,

$$\chi_{n,l,m}(r, \theta, \phi) = j_l(k_n r)\, Y_{l,m}(\theta, \phi). \tag{3.80}$$

The wave-vector k_n in the argument of the spherical Bessel function is given by $k_n = n\pi/\tilde{a}$, where (as usual) $\tilde{a} > a$, so that the basis functions have no particular boundary conditions on the embedding surface. The only complication in evaluating the matrix elements (equation (2.40)) comes from the non-locality (that is, l-dependence) of the pseudopotential. In [50], a spherical version of LAPWs was used with a muffin-tin form of adsorbate potential.

As well as the adsorbate density of states and charge density, Trioni *et al* [7] calculated the adsorbate binding energy $\mathcal{E}$ as a function of adsorbate–substrate distance. An important contribution to this calculation is the change in the *total* density of states due to the adsorbate, which we can find directly from the matrix

---

[7] We introduce the factor of $1/a^2$ because $\frac{1}{a^2}\sum_L Y_L(\theta, \phi) Y_L^*(\theta', \phi') = \delta(\mathbf{r}_S - \mathbf{r}'_S)$, where $\mathbf{r}_S$, $\mathbf{r}'_S$ lie on the surface of the sphere. This simplifies the form of the matrix equations.

elements in region I. The total density of states (sometimes just called the density of states) is defined as

$$n(E) = \int_{\text{I+II}} d\mathbf{r}\, n(\mathbf{r}, E) \qquad (3.81)$$

(compare with equation (2.43)), and the integrated density of states as

$$N(E) = \int^{E} dE\, n(E). \qquad (3.82)$$

Then the change in $N(E)$ on adsorption is given by [7]

$$\Delta N(E) = -\frac{1}{\pi} \Im \ln \det\left(\frac{H + \Sigma - EO}{H_{\text{M}} + \Sigma - EO}\right), \qquad (3.83)$$

where H is the Hamiltonian of region I in the adsorbate system, from which we have separated out Σ, and H_{M} the Hamiltonian of the clean surface; note that Σ and the overlap matrix O are the same in both the adsorbate system and the clean surface. Equation (3.83) is based on a 'generalized phase shift' result based on Lloyd's formula [62] and derived in [63]. A similar generalized phase shift is used in Dyson's equation treatments of the impurity and adsorbate problems [55, 64]. By numerically differentiating $\Delta N(E)$ we can find $\Delta n(E)$, hence the total change in one-electron energy on adsorption,

$$\Delta U_{1-\text{e}} = \int^{E} dE\, E\, \Delta n(E). \qquad (3.84)$$

To find the binding energy itself, we must use the grand canonical formulation of DFT [64], as the number of electrons is not constrained to be constant. This gives

$$\mathcal{E} = \Delta T + \Delta U - E_{\text{F}} \Delta Q, \qquad (3.85)$$

where ΔT is the change in electron kinetic energy, ΔU is the change in electrostatic and exchange-correlation energy, and ΔQ is the deviation in electron number from charge neutrality. The change in electron kinetic energy is given by

$$\Delta T = \Delta U_{1-\text{e}} - \Delta\left(\int_{\text{I}} d\mathbf{r}\, \rho(\mathbf{r}) V(\mathbf{r})\right), \qquad (3.86)$$

where ρ is the electron density. ΔU can be readily found, as our assumption is that the changes in charge density are all restricted to region I. The most interesting part of equation (3.85) is the final term, which corresponds to subtracting (or adding) excess (or missing) electron charge at the substrate Fermi energy—this acts as an electron reservoir [7]. From equation (3.82) we have

$$\Delta Q = N(E_{\text{F}}) - Z, \qquad (3.87)$$

where Z is the valency of the adsorbate atom.

3.7.3 Adsorption results

As an example of this embedding approach to chemisorption, we consider a N atom adsorbed on jellium, with the electron density corresponding to Al [7]. (We shall present more recent studies in section 4.4.2.) Here the sphere radius a was taken to be 7 a.u, and in the basis functions (equation (3.80)) the maximum values $l = 8$ and $n = 16$ were used.

The change in electron density due to the adsorbed atom is shown in figure 3.30, calculated at an atom–surface distance of 1 a.u., and we see an enhancement of electronic charge in the atomic region. This is indicative of an ionic contribution to the bonding. The adsorption energy as a function of N–jellium distance is shown in figure 3.31, giving an equilibrium separation of 1 a.u. The inset to this figure gives ΔQ (equation (3.87)) as a function of distance, and at equilibrium $\Delta Q = 0.169$ electrons, compared with $Z = 5$; this is reduced by about 10% when a is increased to 7.5 a.u. This value of ΔQ still seems rather large, despite the energy correction in equation (3.85), and it would be useful to repeat the energy calculation with a considerably larger value of sphere radius.

Here, we have a rare example of a total energy calculation within the framework of embedding, and we must ask the question: why don't embedded surface calculations, like those we have discussed earlier in this chapter, include evaluations of the energy and structural optimization? The main reason is that slab calculations for surfaces, based on solutions of the Schrödinger equation for three-dimensional crystals, are very well-developed for calculating energy, forces on atoms, structural optimization and so on. The surface energy is a local quantity, much more so than the local density of states, which depends on distant boundary conditions. This is because the total energy involves integrals of the density of states over occupied states, and this leads to greater locality. In this situation it is easier to take structural parameters from separate slab calculations, rather than optimize them within embedding.

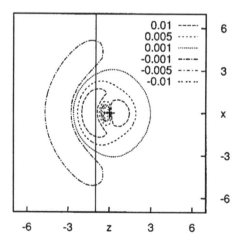

Figure 3.30. N adsorbed on jellium (Al): electron density difference plot. The N atom is 1 a.u. from the jellium edge, indicated by the vertical line. The embedding sphere radius is $a = 7$ a.u. (From Trioni *et al.* [7].)

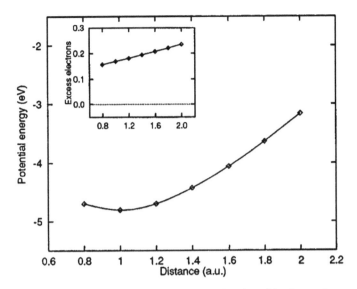

Figure 3.31. N adsorbed on jellium (Al): adsorption energy as a function of the distance between the atom and the jellium edge. The inset shows the excess charge ΔQ as a function of distance. The embedding sphere radius is $a = 7$ a.u. (From Trioni *et al* [7].)

Another, more fundamental, reason is that the embedded energy calculation is based on the assumption that changes in the charge density and the potential are confined to region I. Although the grand canonical correction term in equation (3.85) takes care (to first order) of deviations from overall charge neutrality, it does not take care of these changes in potential outside region I. Moreover, is the first-order correction good enough? To answer these questions requires more research.

References

[1] Inglesfield J E 2001 Embedding at surfaces *Comput. Phys. Commun.* **137** 89–107
[2] Ishida H 2001 Surface-embedded Green-function method: A formulation using a linearized-augmented-plane-wave basis set *Phys. Rev.* B **63** 165409
[3] Benesh G A and Liyanage L S G 1994 Surface-embedded Green-function method for general surfaces: Application to Al(111) *Phys. Rev.* B **49** 17264–72
[4] Callaway J 1991 *Quantum Theory of the Solid State* 2nd edn (San Diego: Academic Press)
[5] Thijssen J M 2007 *Computational Physics* 2nd edn (Cambridge: Cambridge University Press)
[6] Inglesfield J E 1982 Surface electronic structure *Rep. Prog. Phys.* **45** 223–84
[7] Trioni M I, Brivio G P, Crampin S and Inglesfield J E 1996 Embedding approach to the isolated adsorbate *Phys. Rev.* B **53** 8052–64
[8] Feldbauer G, Wolloch M, Bedolla P O, Mohn P, Redinger J and Vernes A 2015 Adhesion and material transfer between contacting Al and TiN surfaces from first principles *Phys. Rev.* B **91** 165413
[9] Ishida H 2014 Rashba spin splitting of Shockley surface states on semi-infinite crystals *Phys. Rev.* B **90** 235422
[10] Nekovee M, Crampin S and Inglesfield J E 1993 Magnetic splitting of image states at Fe(110) *Phys. Rev. Lett.* **70** 3099–102

[11] Inglesfield J E and Benesh G A 1988 Surface electronic structure: Embedded self-consistent calculations *Phys. Rev.* B **37** 6682–700

[12] Benesh G A and Inglesfield J E 1984 An embedding approach for surface calculations *J. Phys. C: Solid St. Phys.* **17** 1595–606

[13] Andersen O K 1975 Linear methods in band theory *Phys. Rev.* B **12** 3060–83

[14] Krakauer H, Posternak M and Freeman A J 1979 Linearized augmented plane-wave method for the electronic band structure of thin films *Phys. Rev.* B **19** 1706–19

[15] Pendry J B 1974 *Low Energy Electron Diffraction: The Theory and Its Application to Determination of Surface Structure* (London: Academic Press)

[16] Crampin S, van Hoof J B A N, Nekovee M and Inglesfield J E 1992 Full-potential embedding for surfaces and interfaces *J. Phys.: Condens. Matter* **4** 1475–88

[17] Lang N D and Kohn W 1973 Theory of metal surfaces: Induced surface charge and image potential *Phys. Rev.* B **7** 3541–50

[18] Thompson I J 2010 Coulomb functions In ed F W J Olver, D W Lozier, R F Boisvert and C W Clark *NIST Handbook of Mathematical Functions* chapter 33 (Cambridge: Cambridge University Press) pp 741–56

[19] Thompson I J and Barnett A R 1986 Coulomb and Bessel functions of complex arguments and order *J. Comput. Phys.* **64** 490–509

[20] Koelling D D and Harmon B N 1977 A technique for relativistic spin-polarised calculations *J. Phys. C: Solid St. Phys.* **10** 3107–14

[21] Posternak M, Krakauer H, Freeman A J and Koelling D D 1980 Self-consistent electronic structure of surfaces: Surface states and surface resonances on W(001) *Phys. Rev.* B **21** 5601–12

[22] Press W H, Flannery B P, Teukolsky S A and Vetterling W T 1989 *Numerical Recipes* 1st edn (Cambridge: Cambridge University Press)

[23] Cunningham S L 1974 Special points in the two-dimensional Brillouin zone *Phys. Rev.* B **10** 4988–94

[24] Zangwill A 2013 *Modern Electrodynamics* (Cambridge: Cambridge University Press)

[25] Weinert M 1981 Solution of Poisson's equation: beyond Ewald-type methods *J. Math. Phys.* **22** 2433–9

[26] Jones R O and Gunnarsson O 1989 The density functional formalism, its applications and prospects *Rev. Mod. Phys.* **61** 689–746

[27] Kohn W and Sham L J 1965 Self-consistent equations including exchange and correlation effects *Phys. Rev.* **140** A1133–A8

[28] Jones R O 2006 Introduction to density functional theory and exchange-correlation functionals In ed J Grotendorst, S Blügel and D Marx *Computational Nanoscience: Do It Yourself!* (Jülich: John von Neumann Institute) pp 45–70

[29] Perdew J P and Zunger A 1981 Self-interaction correction to density-functional approximations for many-electron systems *Phys. Rev.* B **23** 5048–79

[30] Ceperley D M and Alder B J 1980 Ground state of the electron gas by a stochastic method *Phys. Rev. Lett.* **45** 566–9

[31] Ishida H 1997 Surface-embedded Green function calculation using non-local pseudopotentials *Surf. Sci.* **388** 71–83

[32] Dederichs P H and Zeller R 1983 Self-consistency iterations in electronic-structure calculations *Phys. Rev.* B **28** 5462–72

[33] Srivastava G P 1984 Broyden's method for self-consistent field convergence acceleration *J. Phys. A: Math. Gen.* **17** L317–L21

[34] Grepstad J K, Gartland P O and Slagsvold B J 1976 Anisotropic work-function of clean and smooth low-index faces of aluminium *Surf. Sci.* **57** 348–62

[35] Wimmer E, Weinert M, Freeman A J and Krakauer H 1981 Theoretical $2p$-core-level shift and crystal-field splitting at the Al(001) surface *Phys. Rev.* B **24** 2292–4

[36] Borstel G and Inglesfield J E 2000 Electronic states on metal surfaces In ed K Horn and M Scheffler *Handbook of Surface Science volume 2, Electronic Structure* (Amsterdam: Elsevier) pp 209–45

[37] Kevan S D and Eberhardt W 1992 Surface states on metals In ed S D Kevan *Angle-Resolved Photoemission* (Amsterdam: Elsevier) pp 99–143

[38] Plummer E W 1985 Deficiencies in the single particle picture of valence band photoemission *Surf. Sci.* **152-153** 162–79

[39] Holzwarth N A W, Matthews G E, Dunning R B, Tackett A R and Zeng Y 1997 Comparison of the projector augmented-wave, pseudopotential, and linearized augmented-plane-wave formalisms for density-functional calculations of solids *Phys. Rev.* B **55** 2005–17

[40] MacLaren J M, Crampin S, Vvedensky D D and Pendry J B 1989 Layer Korringa-Kohn-Rostoker technique for surface and interface electronic properties *Phys. Rev.* B **40** 12164–75

[41] Crampin S, Nekovee M, van Hoof J B A N and Inglesfield J E 1993 Subvolume embedding for interfacial electronic structure *Surf. Sci.* **287/288** 732–5

[42] van Hoof J 1997 *The Embedding Method* PhD thesis (Nijmegen: Katholieke Universiteit)

[43] Schep K M, van Hoof J B A N, Kelly P J, Bauer G E W and Inglesfield J E 1997 Interface resistances of magnetic multilayers *Phys. Rev.* B **56** 10805–8

[44] van Hoof J B A N, Schep K M, Kelly P J and Bauer G E W 1998 Ab initio magneto-resistance in magnetic domain walls *J. Magn. Magn. Mater.* **177-181** 188–92

[45] van Hoof J B A N, Schep K M, Brataas A, Bauer G E W and Kelly P J 1999 Ballistic electron transport through magnetic domain walls *Phys. Rev.* B **59** 138–41

[46] Wortmann D, Ishida H and Blügel S 2002 *Ab initio* Green-function formulation of the transfer matrix: Application to complex band structures *Phys. Rev.* B **65** 165103

[47] Heine V 1963 On the general theory of surface states and scattering of electrons in solids *Proc. Phys. Soc.* **81** 300–10

[48] James M and Crampin S 2010 Relativistic embedding method: The transfer matrix, complex band structures, transport, and surface calculations *Phys. Rev.* B **81** 155439

[49] Chulkov E V, Silkin V M and Echenique P M 1999 Image potential states on metal surfaces: binding energies and wave-functions *Surf. Sci.* **437** 330–52

[50] Achilli S, Trioni M I, Chulkov E V, Echenique P M, Sametoglu V, Pontius N, Winkelmann A, Kubo A, Zhao J and Petek H 2009 Spectral properties of Cs and Ba on Cu(111) at very low coverage: Two-photon photoemission spectroscopy and electronic structure theory *Phys. Rev.* B **80** 245419

[51] Grimley T B 1967 The electron density in a metal near a chemisorbed atom or molecule *Proc. Phys. Soc.* **92** 776–82

[52] Grimley T B and Pisani C 1974 Chemisorption theory in the Hartree-Fock approximation *J. Phys. C: Solid St. Phys.* **7** 2831–48

[53] Gunnarsson O, Hjelmberg H and Lundqvist B I 1976 Binding energies for different adsorption sites of hydrogen on simple metals *Phys. Rev. Lett.* **37** 292–5

[54] Scheffler M, Droste Ch, Fleszar A, Máca F, Wachutka G and Barzel G 1991 A self-consistent surface-Green-function (SSGF) method *Physica* B **172** 143–53

[55] Bormet J, Neugebauer J and Scheffler M 1994 Chemical trends and bonding mechanisms for isolated adsorbates on Al(111) *Phys. Rev.* B **49** 17242–52

[56] Dederichs P H, Lounis S and Zeller R 2006 The Korringa-Kohn-Rostocker (KKR) Green function method II. Impurities and clusters in the bulk and on surfaces In ed J Grotendorst, S Blügel and D Marx *Computational Nanoscience: Do It Yourself!* (Jülich: John von Neumann Institute) pp 279–98

[57] Lang N D and Williams A R 1978 Theory of atomic chemisorption on simple metals *Phys. Rev.* B **18** 616–36

[58] Lang N D and Kohn W 1971 Theory of metal surfaces: Work function *Phys. Rev.* B **3** 1215–23

[59] Morse P M and Feshbach H 1953 *Methods of Theoretical Physics* (New York: McGraw-Hill)

[60] Inglesfield J E 1970 Green functions, surfaces, and impurities *J. Phys. C: Solid St. Phys.* **4** L14–8

[61] Bachelet G B, Hamann D R and Schlüter M 1982 Pseudopotentials that work: From H to Pu *Phys. Rev.* B **26** 4199–228

[62] Lloyd P 1967 Wave propagation through an assembly of spheres: II. The density of single-particle eigenstates *Proc. Phys. Soc.* **90** 207–16

[63] Inglesfield J E and Benesh G A 1988 Bulk and surface states on Al(001) *Surf. Sci.* **200** 135–43

[64] Drittler B, Weinert M, Zeller R and Dederichs P H 1989 First-principles calculation of impurity-solution energies in Cu and Ni *Phys. Rev.* B **39** 930–9

IOP Publishing

The Embedding Method for Electronic Structure

John E Inglesfield

Chapter 4

Electrons at surfaces

In this chapter we will discuss several applications of surface embedding calculations—surface states, the screening charge induced by an external electric field, and electronic states in adsorbate systems. The examples I choose in this chapter are where embedding has enabled new physics to emerge, mainly because the embedding potential contains the effects of the semi-infinite substrate on the electronic wave-functions in the surface region.

4.1 Surface states and surface resonances

The embedding method is ideal for studying surface states, electronic states which are localized at the surface: with an energy lying within a bulk band gap they decay into the solid, and as their energy lies below the vacuum zero they also decay into the vacuum [1]. The advantage of the embedding method for calculating surface states is that the embedding potential ensures that the decay of the surface state wave-function into the bulk is correctly treated. In an embedding calculation of the surface density of states, the surface states show up as discrete states at fixed wave-vector $\mathbf{k}_{\parallel}$ parallel to the surface, whereas the bulk states at the surface form a continuum (section 3.3.2). This enables us to distinguish between localized surface states and surface resonances; although resonances show up as a peak in the surface density of states, their energy overlaps with the continuum and they couple to bulk states.

The surface electronic structure of Ta(011) provides a good example, as angle-resolved photoemission experiments show an apparent transition between a surface state and a surface resonance with increasing $\mathbf{k}_{\parallel}$ [2–4]. The experimental results are summarised in figure 4.1(a), showing the surface state close to $\mathbf{k}_{\parallel} = 0$ just below the bulk continuum, disappearing close to the wave-vector at which a surface resonance appears in the middle of the continuum. This was associated by Kneedler *et al* [2] with an 'avoided crossing' between these surface features and bulk features in the spectrum (the surface features were identified by their sensitivity to H-adsorption). To understand this behaviour, the surface electronic structure was calculated using a

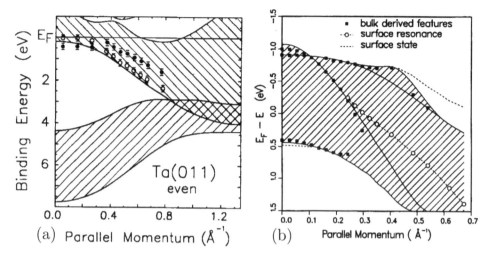

Figure 4.1. (a) Ta(011) photoemission results as a function of wave-vector along the $\bar{\Delta}$ symmetry direction. Filled circles are surface features and open circles are bulk features, both with even symmetry. Shaded regions are the calculated bulk continua. (b) Calculated surface bands of Ta(011) with even symmetry along $\bar{\Delta}$. Dashed lines show surface states, open circles surface resonances and filled circles features associated with the bulk band structure. Solid lines indicate the extrema of projected bulk bands. (From Kneedler *et al* [2] and Van Hoof, Crampin, and Inglesfield [4].)

single surface layer embedded on to the bulk [4]. LAPW basis functions were used with the full potential (section 3.2.1), and the bulk embedding potential was calculated using reflection matrix methods (section 3.1.1). The resulting surface band structure shown in figure 4.1(b) is in qualitative agreement with experiment: the almost dispersionless surface state enters the continuum close to the wave-vector at which a surface resonance appears in the continuum.

To study this in more detail, figure 4.2 shows the surface density of states calculated at different wave-vectors along the $\bar{\Delta}$ symmetry direction of the surface Brillouin zone. In figure 4.2(a), showing $n_s(E)$ at $\bar{\Gamma}$ (the centre of the surface Brillouin zone), we see a surface state pulled from the continuum of bulk states at the surface; the weak peaks at the extrema of the bulk bands correspond to the 'bulk-derived' features shown in figure 4.1. As we move through these figures, in figure 4.2(c) the surface state has just entered the bulk continuum, giving rise to an enhanced peak at the continuum edge. This is exactly analogous to the behaviour we saw in figure 2.5 in section 2.5, where the continuum of states is enhanced just before a bound state is pulled off. Figure 4.2(d) shows a large peak in the middle of the bulk states, associated with the bulk extremum which drops sharply down in figure 4.1(b), and in (e) and (f) a surface resonance is pulled up from this extremum, weakening the bulk feature in the process. This is a surface resonance rather than a surface state, as it overlaps and interacts with the bulk continuum, leading to a broadening of the peak. It seems to be a coincidence that the surface resonance appears close to the wave-vector at which the surface state disappears, and as they have different orbital symmetries the notion of an avoided crossing does not seem to be justified [4].

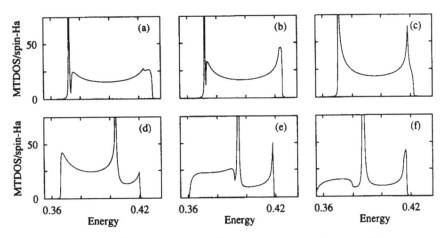

Figure 4.2. Density of states with even symmetry within the surface atom muffin-tin at different wave-vectors along the $\bar{\Delta}$ symmetry direction. (a) at $\bar{\Gamma}$ ($k_\parallel = 0$); (b) at 0.06 $\bar{\Gamma}\bar{N}$ (0.08 Å^{-1}); (c) at 0.10 $\bar{\Gamma}\bar{N}$ (0.13 Å^{-1}); (d) at 0.16 $\bar{\Gamma}\bar{N}$ (0.21 Å^{-1}); (e) at 0.20 $\bar{\Gamma}\bar{N}$ (0.27 Å^{-1}); (f) at 0.24 $\bar{\Gamma}\bar{N}$ (0.32 Å^{-1}). The imaginary part of the energy $\eta = 5 \times 10^{-5}$ a.u. (From Van Hoof, Crampin and Inglesfield [4].)

This is an early study of the distinction between surface states and resonances, and the way that they interact with bulk continuum edges. We shall meet a much more recent study of these effects with spin–orbit induced surface states in relativistic embedded calculations (section 8.3.2).

4.2 Image states

An interesting class of surface states occurs when a band gap overlaps the vacuum zero of energy: the Coulombic image potential felt by electrons outside the surface (equation (3.12)) gives rise to a Rydberg series of *image states*, (sometimes called Rydberg surface states), with energies just below the vacuum zero [5, 6]. As these states lie above E_F, it was the development of techniques for studying unoccupied states like inverse photoemission [7] and two-photon photoemission [8] which led to a wealth of experimental data.

Embedding provides a very accurate method of calculating image states [9, 10]. In the original embedding calculations for image states [9, 11], region I consisted of the *near-surface region*, extending typically 10 a.u. from just outside the surface layer of atoms into the vacuum (figure 4.3), with the metal surface and substrate replaced by an embedding potential Σ_{surf} over S_{surf}, and the image potential beyond S_{vac} replaced by Σ_{im}. Σ_{surf} can be found from a preliminary surface calculation: having gone to self-consistency using the programs and methods described in section 3.2, we then calculate the Green function for the embedded surface satisfying the zero-derivative boundary condition on S_{surf}, and from this we find Σ_{surf} (equation (3.39)).

The reason why the calculation was broken up in this way was that the LAPW basis functions used in the early programs (equation (3.21)) were clearly unsuitable for describing the image states, which consequently had to be calculated separately from the surface electronic structure. Moreover, splitting the calculation into an initial calculation of Σ_{surf}, then an embedded calculation for the near-surface region,

near-surface region

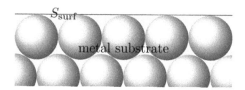

Figure 4.3. Embedding calculation for image states: the near-surface region, region I, is embedded on to the metal surface at S_{surf} and the vacuum at S_{vac}.

meant that relatively small basis sets could be used at each stage—significant at a time when computer resources were comparatively limited! Modern calculations of image states (as in [10]) do not make this separation, with region I comprising the surface layers of atoms as well as the near-surface region, using the geometry shown in figure 3.20 and the embedding methods described in sections 3.5 and 3.6.

On the vacuum side of the near-surface region where the electrons feel the asymptotic Coulomb image potential, Σ_{im} may be found from the Coulomb functions (section 3.1.4). In the near-surface region, where the Schrödinger equation is solved explicitly, a three-dimensional surface barrier potential is used, taken from the self-consistent surface calculation (which uses the LDA for exchange-correlation); an interpolation scheme is used to join the planar average of this potential with the asymptotic image potential. This potential is smooth, so plane-waves can be used to expand the Green function.

4.2.1 Threshold behaviour

As a first example of embedding calculations of image states we consider their threshold behaviour, in particular the local density of states around the vacuum zero. The example we take is a non-magnetic calculation of image states on Ni(001), for which the vacuum zero E_{vac} lies in the middle of the $X_4' - X_1$ bulk band gap [11].

The density of states at $\mathbf{k}_\| = 0$, integrated through the near-surface region, is shown in figure 4.4 as a function of energy, measured relative to E_{vac}. With an imaginary part to the energy $\eta = 0.001$ a.u., the first two image states in the Rydberg series are well-resolved. What is remarkable is the constant density of states across the threshold at $E = E_{vac}$—we cannot tell where the vacuum zero is! The constant density of states just below E_{vac} consists of an infinite number of Rydberg states, smeared out by η; the constant density of states above consists of a true continuum of states, incident on the surface from the vacuum and unable to penetrate into the bulk. Reducing η enables more states in the Rydberg series to be resolved

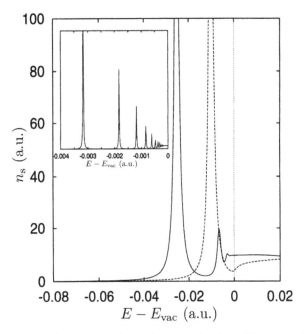

Figure 4.4. The density of states in the near-surface region n_s as a function of the energy relative to the vacuum zero for Ni(001) at $\mathbf{k}_\parallel = 0$. Solid line, image potential interpolation in the near-surface region. Dashed line, LDA surface potential. The imaginary part of the energy $\eta = 0.001$ a.u. The inset shows results with reduced η. (From Nekovee [12].)

(inset, figure 4.4), but we see exactly the same behaviour (though much closer to threshold)—smeared-out image states giving a step-like density of states below the vacuum zero, which joins *continuously* through E_{vac} with the adjacent continuum.

This contrasts remarkably with the behaviour when the potential in the near-surface region is taken as the exponentially decaying surface potential from the initial self-consistent surface calculation, without the image potential (in this case Σ_{vac} is taken as the free-electron embedding potential). The results of this calculation are shown by the dashed line in figure 4.4, showing a single surface state split off from the vacuum continuum. Immediately above E_{vac}, the vacuum continuum has a $(E - E_{vac})^{1/2}$ singularity, identical with the singularity in the surface density of states at a band edge (equation (3.37)).

The continuous behaviour of the density of states in the near-surface region (or indeed *any* finite region of integration) across E_{vac} is a consequence of the Coulombic form of the image potential. It is, in fact, well-known in scattering theory, and in the context of optical absorption by excitons [13]. But this calculation seems to be the first time that it was explored in the context of image states. An intuitive way of understanding this behaviour is that the imaginary part of the energy η, as an inverse lifetime, restricts the distance over which an electron can propagate near the surface. The infinite range of the Coulomb potential means that the electron still feels a negative potential at the maximum distance, and this represents an effective threshold at which the step in n_s commences.

What happens near threshold in a metal without a band gap at the vacuum zero, and which consequently does not show image states? In this case, image resonances occur with similar threshold behaviour to what we have just found. Figure 4.5 shows the density of states of states in the near-surface region at the Al(001) surface [11], a system without a band gap at E_{vac}. The first feature to notice is the surface state near the bottom of the bulk band gap, at $E - E_{vac} \approx -0.28$ a.u.—this is the Shockley surface state we found in section 3.3.2 (figure 3.12), rather smaller in amplitude in this calculation as most of the weight of this state is *inside* the surface. The large feature just below E_{vac} is an image resonance, a state which can leak into the bulk, hence forming part of the continuum of bulk states at the surface. With $\eta = 0.001$ a.u. we can just make out a second resonance, but above this we have a density of states continuous across E_{vac}. Reducing η we see many more resonances (inset, figure 4.5)—in fact, there is an infinite series of resonances—but once again there is smooth behaviour, without any singularity, across E_{vac}. A difference from the previous case of surface states, is that with surface resonances n_s, though smoothly varying around E_{vac}, is not constant. This is because the resonances are superimposed on a background of bulk states. Again, we contrast the image potential results with the density of states calculated with just the LDA exchange-correlation potential (dashed line, figure 4.5)—with no image potential n_s has the singularity of an infinite slope at E_{vac}.

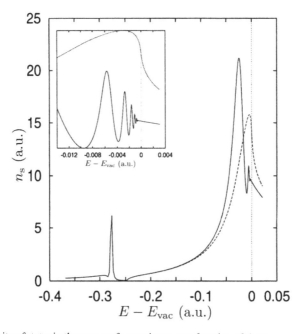

Figure 4.5. The density of states in the near-surface region n_s as a function of the energy relative to the vacuum zero for Al(001) at $\mathbf{k}_{\parallel} = 0$. Solid line, image potential interpolation in the near-surface region; dashed line, LDA surface potential. The imaginary part of the energy $\eta = 0.001$ a.u. The inset shows results with reduced η. (From Nekovee [12].)

4.2.2 Magnetic image states

The image states at the surface of a ferromagnetic metal are split in energy, an effect which can be probed directly using spin-polarized inverse photoemission. Nekovee *et al* [9] have used the near-surface embedding method to study these states on the surface of ferromagnetic Fe(110). The modifications to the method described above are firstly, the embedding potential Σ_{surf} to replace the metal substrate is spin-dependent, and must be calculated from a self-consistent *spin-polarized* surface calculation; and secondly, the LDA exchange-correlation potential from this initial calculation, which enters the surface barrier potential in the near-surface region, is also spin-polarized.

The initial spin-polarized self-consistent calculation of the Fe(110) surface was made using a single Fe layer embedded on to the ferromagnetic bulk, with the von Barth–Hedin local spin-density approximation for the spin-dependent exchange-correlation potential [14]. This gave a magnetic moment on the surface Fe atoms of $2.75 \, \mu_B$, in good agreement with other calculations. From this preliminary calculation the spin-dependent embedding potential Σ_{surf} and spin-dependent surface barrier potential in the near-surface region can be found.

Figure 4.6(a) shows the spin-up (bulk majority spin) and spin-down (minority) density of states in the near-surface region at $\mathbf{k}_{\parallel} = 0$, and we see that the image states are split, with the spin-up states having slightly lower energy than the spin-down. The splitting is greatest for the first state ($n = 1$) in the Rydberg series, and further analysis shows that this varies as

$$\Delta E_n \sim n^{-3}, \quad n \text{ large}, \tag{4.1}$$

the same as the variation in many-body lifetime broadening effects. In this figure, we can just see the threshold effect discussed in section 4.2.1, with the smeared-out image states giving rise to a constant density of states across the threshold.

The two contributions to the spin-splitting—the spin-dependence of the surface barrier potential and the spin-dependence of Σ_{surf}—are separated in the lower panels of the figure. Figure 4.6(b) shows the density of states when Σ_{surf} is replaced by an infinite barrier potential, so that an electron in an image state cannot penetrate into the Fe surface and simply feels the spin-polarized surface barrier potential: we see that the splitting is reversed, so that the spin-up image states have *higher* energy than the spin-down! On the other hand, if we turn off the spin-polarization of the surface barrier potential, and include only the spin-polarized Σ_{surf}—which represents the effect of the substrate on the image states—we obtain the results shown in figure 4.6(c), with spin-up states having lower energy and increased spin-splitting. We thus have two competing effects: a negative spin-splitting from the surface barrier, and a positive splitting from Σ_{surf}, for the first Rydberg state -14 meV and $+63$ meV, respectively. The combined effect is the positive spin-splitting of $+55$ meV seen in figure 4.6(a).

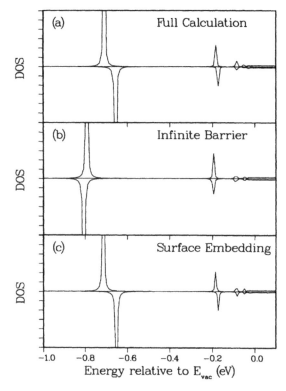

Figure 4.6. The spin-polarized density of states in the near-surface region as a function of energy relative to the vacuum zero for Fe(110) at $\mathbf{k}_\parallel = 0$. Upper curves show spin-up (bulk majority spin) and lower curves spin-down (bulk minority spin). (a) Full calculation; (b) calculated with an infinite barrier at the surface, showing the effect of spin-polarization of the surface barrier potential in the near-surface region; (c) as in the full calculation, but with no spin-polarization of the surface barrier potential. The imaginary part of the energy $\eta = 0.0001$ a.u. (From Nekovee, Crampin and Inglesfield [9].)

The negative splitting with the surface barrier potential is a consequence of a change in sign of the magnetization in the tail of the ground state electron density just outside the Fe(110) surface. Figure 4.7(a) shows the spin-up and spin-down electron densities in the near-surface region together with the net magnetization, and we see that the spin-polarization changes sign beyond about 1 a.u. outside the surface. This results in majority spin electrons feeling a less attractive potential than minority as we move out into the near-surface region, as shown in figure 4.7(b). The inset in figure 4.7(b) shows the probability densities of the $n = 1$ and $n = 2$ image states and we see that their weight is mainly in the region of potential with reversed spin-polarization. The reversal in spin-polarization seems to be due to the minority spin states predominating over majority spin at E_F in Fe: it is the electrons near E_F which penetrate further into the vacuum.

The happy conclusion to this embedding calculation of image states was a later measurement of the spin-splitting on Fe(110), using spin-polarized inverse

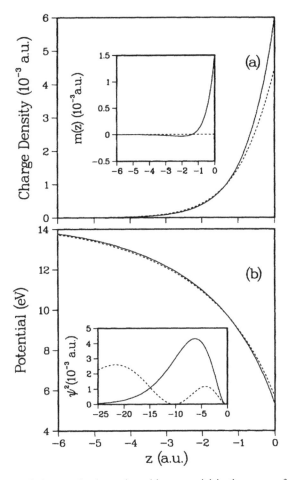

Figure 4.7. Planar averaged electron density and resulting potential in the near-surface region outside the Fe(110) surface. (a) Ground state electron density: solid line, spin-up; dashed line, spin-down. The inset shows the net magnetization. (b) Interpolated surface potential felt by spin-up electrons (solid line) and spin-down electrons (dashed line). The inset shows the probability density of the $n = 1$ and $n = 2$ image states. Note that $z = 0$ corresponds to the boundary of the near-surface region with the Fe(110) surface, and negative z corresponds to moving into vacuum. (From Nekovee, Crampin and Inglesfield [9].)

photoemission [15]: the measured splitting of the $n = 1$ state was 57 ± 5 meV. Convincing agreement, I think!

4.3 Screening of an external field

The screening of an external electric field at a metal surface is interesting in its own right—in classical electrostatics, the field $\mathcal{E}$ is perfectly screened by a surface charge density of $\mathcal{E}/4\pi$, but what is the spatial distribution of this charge? The distribution is important for interpreting experiments such as field emission, field evaporation, scanning tunnelling microscopy, electrochemistry and so on. On a basic level, the centre of gravity of the screening charge corresponds to the position of the image plane z_{im} from which the image potential is measured, and the plane from which the separation between capacitor plates is measured [16]. The energy of the image states, which we have discussed in section 4.2, depends on z_{im}. Moreover, the variation in shape of the screening charge with electric field—a non-linear effect—contributes to second-harmonic generation, a sensitive optical probe of surfaces [17]. Surface embedding is particularly suited to calculating the screening of an external field, as the bulk is properly treated and acts as an electron reservoir as we go to self-consistency, with constant E_F allowing for changes in the number of electrons in the surface region with the applied field [18, 19]. The embedding calculations, performed with the methods described in chapter 3, were the first calculations of surface screening to include atoms: Lang and Kohn [16, 20] treated the screening at the surface of semi-infinite jellium, in which the atoms are smeared out into a uniform positive background cut off at the geometrical surface, and the same model was used by Weber and Liebsch [17] in their work on non-linear effects and second-harmonic generation. This early work captured the physics of screening, but we shall see that atoms *are* important!

The embedding calculations for screening simply involve a change in the boundary conditions on the electrostatic potential in going to self-consistency (section 3.2.3). To do this, we go back to Poisson's equation for the $\mathbf{g} = 0$ component of the electrostatic potential in the vacuum region (equation (3.32)),

$$-\frac{d^2 v_0(z)}{dz^2} = 4\pi\rho_0(z),$$

(4.2)

which we now solve with the boundary condition that

$$v_0'(z) \to \mathcal{E}, \quad z \to \infty.$$

(4.3)

$\mathcal{E}$ is the applied electric field, and our convention is that a positive field leads to a reduction in the number of electrons at the surface—the standard definition of electric field. The second boundary condition on v_0 is the same as in section 3.2.3, $v_0(z) \to 0$ at large z, any shift in potential being absorbed by the constant C in equation (3.33). We impose the boundary conditions at some cut-off distance in the vacuum, beyond the range of the vacuum solutions in the basis functions (equation (3.21)): there is a limit to the strength of negative field—one which attracts electrons to the surface—the potential at the cut-off must not drop below E_F, otherwise electrons will be drawn out of the surface *ad infinitum*.

4.3.1 Screening at the Ag(001) surface

The screening charge $\delta\rho(\mathbf{r})$ is the difference in electron density with and without the field, and this has been calculated for Ag(001) using the LAPW methods and programs described in section 3.2, with two atomic layers embedded on to bulk Ag [19]. The results are shown in the left-hand contour plot of figure 4.8 with an electric field $\mathcal{E} = +0.01$ a.u. (5.14×10^9 V m^{-1}), a field screened by a depletion in the

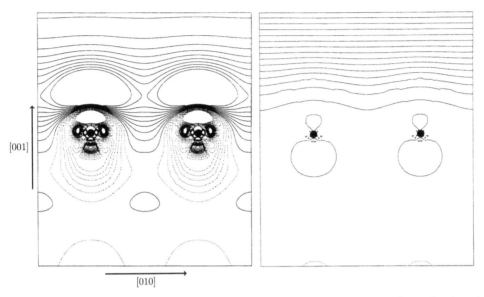

Figure 4.8. Screening of an electric field $\mathcal{E} = +0.01$ a.u. at Ag(001). Plots are on a (100) slice passing through atoms in the surface layer, indicated by heavy dots. The left-hand plot shows the screening charge, solid lines showing contours of decreased electron density and dashed lines increased electron density; the right-hand plot shows the change in potential. (From Aers and Inglesfield [19].)

number of electrons. The corresponding change in potential $\delta V(\mathbf{r})$ is shown in the right-hand plot. The most striking feature of these results is that most of the screening charge is located on top of the surface atoms, so that the field barely penetrates the solid. There is atomic structure apparent in the screening charge, with polarization effects inside the ion cores; however, the main effect of the ion cores is to exclude the screening charge, which bends over the tops of the atoms into the region between.

The plot of $\delta V(\mathbf{r})$ shows just how effective the screening charge is in excluding the external field. This is relevant to processes like field evaporation, as the force on an atom in an applied field is the screened field at the nucleus, from the Hellmann–Feynman theorem.

The planar averaged screening charge $\delta\bar{\rho}(z)$ is shown in figure 4.9 as a function of the distance from the geometrical surface (half an inter-layer spacing beyond the top layer of atoms). $\delta\bar{\rho}(z)$ is actually calculated as the planar average of the change in $\tilde{\rho}(\mathbf{r})$ (equation (3.28)), the electron density including the pseudo-charge density inside the muffin-tins, which has the same multipole moments as the actual charge density; this gives a smoother representation. This figure shows the large peak in the screening charge near the geometrical surface, with Friedel oscillations extending into the solid. Although the total screening charge is given by $\mathcal{E}/4\pi$ per unit area—this two-layer embedded calculation is accurate to within 1–2%—we see from figure 4.9 that the shape of $\delta\bar{\rho}(z)$ varies with field. In particular, a negative field tends

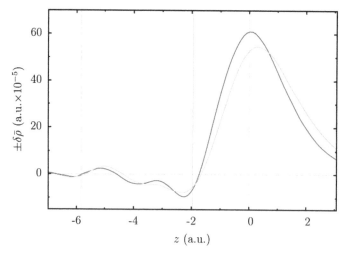

Figure 4.9. The planar-averaged screening charge $\pm\delta\bar{\rho}(z)$ at Ag(001) for fields $\mathcal{E} = +0.02$ a.u. (red line) and $\mathcal{E} = -0.02$ a.u. (green line). The positive field gives a decrease in electron number at the surface and the negative field an increase, so taking ρ as the electron density, the red curve is $-\delta\bar{\rho}(z)$. z is measured from the geometrical surface, and the blue line indicates the position of the first layer of atoms and the gold line the second layer.

to pull the screening charge into the vacuum. From our results, we find that the centre of gravity of the screening charge can be fitted quite well by [19]

$$z_0 = 0.97 - 8.83\mathcal{E} \quad \text{(a.u.).} \tag{4.4}$$

The zero-field value of z_0 is the electrostatic origin of the surface z_{im}, from which the asymptotic image potential is measured [16]. So, at Ag(001) the image plane lies at 0.97 a.u. on the vacuum side of the geometrical surface. This compares with $z_{im} = 1.35$ a.u. for jellium of density $r_s = 3$ a.u. [17], which has often been used to model the response of Ag surfaces to external fields.

By fitting a model potential to Rydberg surface state energies, Chulkov *et al* [21] obtained $z_{im} = 0.13$ a.u.—much closer to the geometrical surface than our value. However, we should be cautious about such a comparison as they fit parameters of a model potential with the same form as the potential shown in figure 3.25. In fact, the comparison may not be meaningful anyway! Our value of z_{im}, while appropriate to the image potential felt by a static or quasi-static charge, may not be valid for a dynamic object like an electron in a Rydberg surface state.

The field-dependence of the shape of the screening charge, represented by the coefficient -8.83 a.u. in equation (4.4), gives a second-order current normal to the surface. Assuming that the frequency of perturbation ω is much less than the plasma frequency ω_p (which determines how quickly the metal responds), the screening charge per unit area is given by $\mathcal{E}\cos\omega t/4\pi$ and its centre of gravity varies like $-8.83\mathcal{E}\cos\omega t$, giving a second-order current normal to the surface proportional to $8.83\,\mathcal{E}^2\cos^2\omega t$. This gives rise to second-harmonic radiation, with an intensity proportional to $(8.83)^2$. The Ag(001) surface screening is in fact much stiffer than the equivalent jellium with $r_s = 3$ a.u., for which the coefficient of $\mathcal{E}$ in equation (4.4) is 30 a.u. So, we predict a second-harmonic intensity that is a factor of 12 smaller. This is, apparently, in the right direction to improve agreement with experiments on Ag electrodes.

4.3.2 Screening at stepped surfaces

Screening at stepped (or vicinal) surfaces is particularly interesting, because we know that on the macroscopic scale the electric field is concentrated at steps and points (think of lightning conductors). It also has pratical relevance, because in practice all surfaces have steps, and these must affect the screening of the electric field in contexts like electrochemical cells[1] and field emission. The difficulty with doing calculations for stepped surfaces is that the surface unit cell is large, and the open atomic layers parallel to the surface are close together; however, the efficient screening at a metal surface means that we need only include a few layers in the surface region in an embedding calculation and the methods for calculating the shifted embedding potential described in section 3.6 can cope with open atomic layers.

Palladium is a metal widely used for electrodes in electrochemistry, and Merrick *et al* [23] have calculated the screening of an electric field at the Pd(311) surface. This surface is made up of (111) and (100) terraces and steps, with rows of atoms in the [0$\bar{1}$1] direction. The surface region of three atomic layers plus vacuum was embedded on to bulk Pd using the programs of Ishida (section 3.6) [24]. The surface electron density in zero field is shown in figure 4.10, and we see that the charge density tends to fill up the gaps between the outermost atoms, smoothing

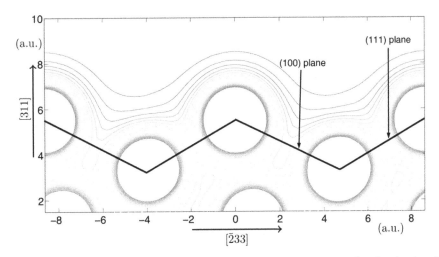

Figure 4.10. Valence electron density at the Pd(311) surface linearly averaged in the [0$\bar{1}$1] direction (parallel to the rows of atoms and perpendicular to the figure). Electron density contour spacings are 5×10^{-3} a.u., with a cut-off of 0.1 a.u. in the atomic cores. (From Merrick [22].)

---

[1] In an electrochemical cell the electric field is greatest at the electrode surface, across the Helmholtz double layer formed by the electrolyte.

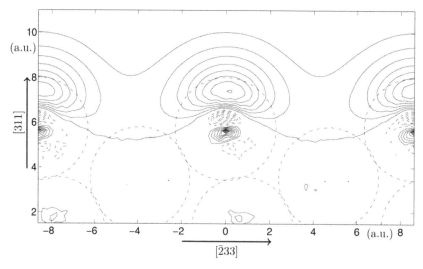

Figure 4.11. Screening charge at the Pd(311) surface linearly averaged in the [0$\bar{1}$1] direction, applied electric field $\mathcal{E} = 0.0014$ a.u. Solid contour lines represent positive charge, with a contour interval of 10^{-5} a.u. The circles represent atomic muffin-tins projected on to the (0$\bar{1}$1) plane. (From Merrick [22].)

the surface profile. This Smoluchowski smoothing effect reduces the work-function compared with close-packed surfaces [1], and the calculation (which uses LDA for exchange-correlation) gives a work-function of 5.64 eV compared with work-functions within the same computational scheme of 5.86 eV for Pd(111) and 5.80 eV for Pd(001) [23].

The screening charge density at Pd(311) in an applied electric field $\mathcal{E} = 0.0014$ a.u. (7×10^8 V m^{-1}) is shown in figure 4.11. Again, we see that screening is extremely effective, with the screening charge in this case sitting on top of the outermost atoms at the step edges of this vicinal surface. Although this is what we would expect from macroscopic electrostatics, at the microscopic level it is remarkable that the smoothing charge between the atoms barely contributes to the screening.

4.4 Adsorbates

In this section we shall discuss some applications of the embedding method to adsorbates, starting with overlayer systems in which the two-dimensional periodicity of the overlayer means that state-of-the-art programs like those of Ishida [24] can be used without modification. As we discussed in section 4.2, in modern programs region I extends well into the vacuum, so that these calculations deliver image states as well as the wide variety of surface states which adsorbate systems display. We shall also consider an isolated adsorbate atom, for which the embedding method described in section 3.7 can be applied; breaking the two-dimensional periodicity means that the treatment of the underlying surface and substrate is not as sophisticated as in the overlayer case, but nevertheless some useful physics emerges.

4.4.1 Alkali metal overlayers on Cu(111) quantum-well systems

Alkali metal overlayers on Cu(111) are classic examples of quantum-well systems, in which the band gap at $\bar{\Gamma}$ in the bulk Cu band structure behaves like a potential barrier to electrons in the overlayer [25]. As an example of this, Butti *et al* [10] have used surface embedding to study the electron states on Na/Cu(111). In this calculation, region I consisted of the top two Cu layers plus the overlayer of Na, together with the near-surface region extending 12 a.u. into the vacuum, embedded on to the bulk embedding potential Σ_{P_L} over P_L and Σ_{im} over S_{vac} (figure 3.20). In the near-surface region, the potential was interpolated between the self-consistent density-functional potential, calculated in the generalized gradient approximation for exchange-correlation [26], and the asymptotic image potential. The $p(3 \times 3)$ surface structure of 1 ML Na on Cu(111) (referred to in [25] as $(3/2 \times 3/2)$) was taken from experiment [27, 28].

Results for the density of states integrated through a Na muffin-tin, plotted for different wave-vectors in the surface Brillouin zone, are shown in figure 4.12, with the single peak lying between the solid lines corresponding to the quantum-well state, and the first three image states appearing as peaks between the dashed lines. The quantum-well state, which is localized close to the Na overlayer with p_z symmetry, has an energy at $\bar{\Gamma}$ of -0.173 eV relative to E_F, compared with a value of -0.127 eV from scanning tunnelling spectroscopy measurements [29]. The energies at $\bar{\Gamma}$ of the three image states are shown in table 4.1: the agreement with experiment is excellent, the differences mostly lying within experimental error. The broad peak between $E \approx 1$ and 3 eV above E_F in the density of states at $\bar{\Gamma}$ (figure 4.12) is a resonance coming from the unoccupied Na 3s valence state overlapping with the 3p level. The image states overlap with this resonance [10], but despite this they remain sharp, presumably because of minimum spatial overlap. The advantages of embedding compared with conventional slab calculations are apparent from this figure; whereas the localized states on each surface in a slab calculation interact and split [25], here they are uniquely identified with a well-defined width.

An embedding study of a $p(2 \times 2)$ monolayer of Cs on Cu(111) reveals an even more complicated behaviour of surface states and resonances [30]. As well as the quantum-well state and the image states, this system shows a resonance with considerable weight on the Cs overlayer dispersing across E_F, and more remarkably, a very sharp state (the 'gap state') localized on the overlayer with an energy at $\bar{\Gamma}$ of 2.7 eV above E_F (0.9 eV above the vacuum zero). Unlike the case of $(3/2 \times 3/2)$ Na/Cu(111), here the quantum-well state is unoccupied at $\bar{\Gamma}$ with an energy of about 40 meV above E_F, in agreement with scanning tunnelling spectroscopy measurements [31]. The overlayer resonance, which shows up in photoemission experiments, can be interpreted either as the Cu(111) surface state pushed down in energy below the bulk Cu band edge by Cs adsorption, or as arising from a free-electron state in a free-standing Cs monolayer [32]. With the latter interpretation, there are *two* quantum-well states associated with the Cs layer: the unoccupied quantum-well state and the partially occupied overlayer resonance.

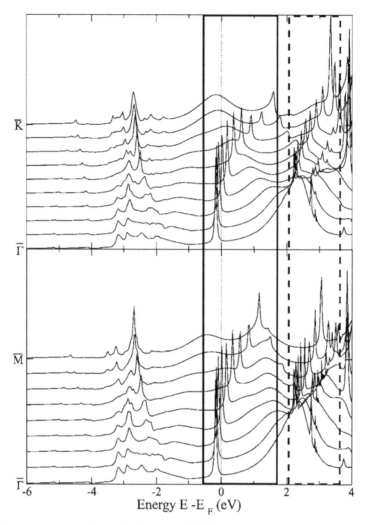

Figure 4.12. Density of states in Na muffin-tins in Na/Cu(111) as a function of energy relative to E_F for wave-vectors in the $\bar{\Gamma}$-$\bar{M}$ and $\bar{\Gamma}$-$\bar{K}$ directions in the surface Brillouin zone. The dispersion of the quantum-well peak lies between the solid lines and of the image states between the dashed lines. (From Butti *et al* [10].)

Table 4.1. Calculated and experimental energies of image states at $\bar{\Gamma}$ for the overlayer system Na/Cu(111) (1 ML of Na in a (3/2 × 3/2) structure), measured from the vacuum zero [10].

Image state index	Calc. energy (eV)	Exptl. energy (eV)
$n = 1$	−0.742	−0.72 ± 0.03
$n = 2$	−0.223	−0.27 ± 0.03
$n = 3$	−0.099	−0.13 ± 0.04

4.4.2 Alkali and alkaline-earth atoms adsorbed on Cu(111)

The electronic structure of isolated Cs and Ba atoms adsorbed on Cu(111) has been calculated for comparison with photoemission and two-photon photoemission studies of the occupied and unoccupied states in this system [33]. These calculations used the adsorbate embedding method given in section 3.7. The Cu(111) substrate was described by the one-dimensional Chulkov potential [21], from which the embedding potential over the spherical embedding surface can be calculated; a full-potential method was used for the adsorbate, with a spherical version of LAPWs inside the embedding sphere. The adatom–surface distances were taken from separate calculations, with an embedding sphere radius of $a = 6.35$ Å about twice as large as a in section 3.7.3; a maximum value of $l = 18$ was used in the spherical expansions of the basis functions and the embedding potential.

The symmetry-resolved densities of states on the adsorbate atoms are shown in figure 4.13, integrated through a sphere of radius 2.65 Å centred on the atoms [33]. The main feature in the density of states for the Cs adatom is a very sharp resonance peak with σ-symmetry, at 3.0 eV above E_F. This resonance, which is mostly s-like but with an appreciable p_z component, corresponds to the fully ionized 6s atomic level. This resonance is very narrow, mainly because the Cu(111) substrate has a band gap at this energy over a large part of the surface Brillouin zone. In the occupied states with σ-symmetry, the weak peak just below E_F corresponds to the Cu(111) surface state interacting with the Cs$^+$ ion. The unoccupied σ-resonance and a broader π-resonance with an energy at 4.1 eV above E_F, both correspond to features seen in

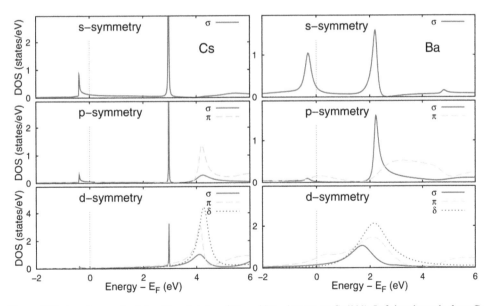

Figure 4.13. l- and m-resolved densities of states of Cs and Ba adatoms on Cu(111). Left-hand panels show Cs and right-hand panels Ba, with densities of states integrated over a sphere of radius 2.65 Å. Top panels: $l = 0$, σ-symmetry. Middle panels: $l = 1$, σ-symmetry ($m = 0$), π-symmetry ($m = 1$). Lower panels: $l = 2$, σ-symmetry ($m = 0$), π-symmetry ($m = 1$), δ-symmetry ($m = 2$). (From Achilli *et al* [33].)

the two-photon photoemission spectra [33]. Although not discussed in [33], we can plausibly identify the unoccupied σ-resonance with the gap state on $p(2 \times 2)$-Cs/ Cu (111), and the occupied σ-peak with the overlayer resonance (section 4.4.1, [30]).

The unoccupied σ-resonance on the Ba adatom (figure 4.13, [33]) at an energy of 2.2 eV above E_F is much broader than in the case of Cs, showing greater interaction with the Cu substrate. The energy of this state agrees with experiment, though the calculated width of 0.22 eV (without any inelastic contribution) is significantly less than the experimental value of about 0.4 eV [33]. This resonance comes from the atomic 6s level, as does the occupied σ-resonance just below E_F; unlike the occupied σ-resonance on Cs, it has considerable weight on the Ba atom. The p-states contribute to the unoccupied σ-resonance, and also to a broad π-resonance at higher energy—two-photon photoemission shows a corresponding π-resonance at 2.7 eV above E_F. There is no experimental evidence for the δ-resonance which is seen in the calculated density of states, presumably because of matrix element effects.

These results for the density of states suggest that the 6s atomic level in Ba interacts with Cu(111) to give two 'molecular' levels, the occupied and unoccupied σ-resonances. As a result, the 6s level is partially ionized, giving Ba$^+$. In the case of the Cs adsorbate, on the other hand, the density of states results imply complete ionization, giving Cs$^+$. This interpretation of the bonding received additional support from the electron density plots shown in figure 4.14 [33]. (These plots

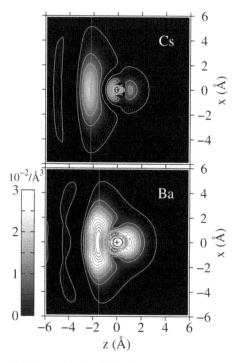

Figure 4.14. Electron density of Cs (upper plot) and Ba (lower plot) adsorbates on Cu(111). The crosses mark the position of the adsorbate and the thin vertical lines the image plane of Cu(111), from which the image potential is measured. (From Achilli *et al* [33].)

show the difference in electron density—the density of the chemisorbed system minus the density of the clean surface, as in section 3.7.3.) The Cs/Cu(111) results show that charge is transferred from the Cs to the Cu surface, accumulating on the image plane: the image charge screens the Cs^+ ion. The Ba/Cu(111) results also show charge on the image plane, but there remains valence charge density on the Ba. This corresponds to partial ionization.

An advantage of the embedding method, both for overlayers and isolated adsorbates, is that the proper description of the substrate means that the (elastic) linewidths of resonances are accurately described. This can lead to greater appreciation of the physics: in the case of adsorbates on Cu(111), an important factor in the greater linewidth of the unoccupied σ-resonance on Ba compared with Cs is the energy of the resonance relative to the surface-projected Cu band gap [33]. The effect of the substrate band gap is particularly striking when the adsorption of an isolated Na atom on Cu(111) (with the wide band gap crossing E_F) is compared with adsorption on the jellium-like Al(111): the empty Na s-level on Cu(111) has a linewidth of 0.11 eV compared with 1.4 eV on Al(111) [34].

4.4.3 Silicene on Ag(111)

There is much current interest in the electronic properties of silicene as it is the Si analogue of graphene. DFT calculations of a free-standing layer of silicene show some buckling of the layer, but two-dimensional Dirac bands crossing E_F persist, corresponding to the existence of massless fermions [35]. However, experiments and theoretical calculations indicate that the Dirac bands do not occur in a silicene layer supported on Ag(111) [36, 37].

To gain more insight into this system, in particular, to see how the electronic states on the silicene layer mix in with the continuum of bulk states in the Ag(111) substrate, Ishida *et al* [38] have calculated its electronic structure using the surface embedding method. This is a good example of the complicated surface structures which can now be handled by modern versions of surface embedding [24]. In their calculation, in which region I consisted of the top two layers of Ag(111) and the silicene overlayer, the atomic structure was taken from a separate slab calculation using the Vienna *ab initio* simulation package (VASP) [39, 40]. (As we discussed in section 3.7, energetics and structure optimization have rarely featured in embedding, and do not form part of Ishida's surface embedding programs [38].) The optimized structure forms the (4 × 4) unit cell shown in figure 4.15, with the green-shaded Si atoms displaced outwards.

The embedding results for the 'band structure' in the silicene overlayer on Ag (111) in the (4 × 4) structure are shown in figure 4.16 [38]. This actually maps $\bar{n}_{Si}(E, \mathbf{k}_\parallel)$, the local density of states integrated through the muffin-tin of a Si atom and averaged over all the Si atoms, as a function of energy and wave-vector in the $\bar{\Gamma}$-$\bar{M}_2$ and $\bar{\Gamma}$-$\bar{K}_2$ symmetry directions. The different maps indicate (b) the density of states with σ-symmetry (s + $p_{x,y}$), (c) states with π-symmetry (p_z), and (a) the total. The brightness corresponds to the size of $\bar{n}_{Si}(E, \mathbf{k}_\parallel)$, so the maps not only trace peaks in the density of states as a function of wave-vector, but they also indicate the size

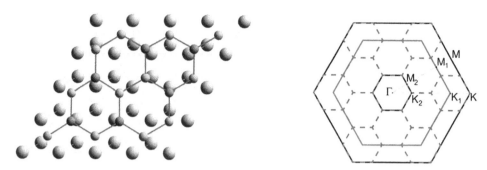

Figure 4.15. Left: top view of the optimized structure of silicene/Ag(111) showing the (4×4) unit cell, with the top two layers of Ag atoms (grey spheres) and the Si atoms (green and blue spheres). Right: surface Brillouin zones. The outer hexagon is the Brillouin zone of (1×1) Ag(111) and the inner hexagon shaded pink is the Brillouin zone of (4×4) silicene/Ag(111). (From Ishida *et al* [38].)

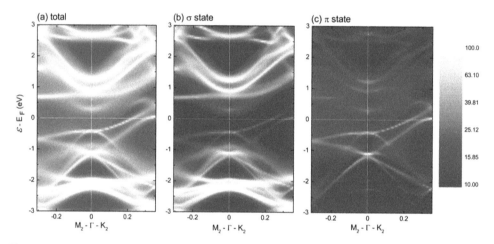

Figure 4.16. The density of states in the silicene overlayer on Ag(111) mapped as a function of $(E - E_F)$, and wave-vector in the $\bar{\Gamma}$-$\bar{M}_2$ and $\bar{\Gamma}$-$\bar{K}_2$ symmetry directions. The brightness gives the size of the density of states, with the scale shown on the right. The density of states is shown with (b) σ-symmetry, (c) π-symmetry and (a) the total. (From Ishida *et al* [38].)

and width of the peaks. These results immediately confirm that there are no Dirac bands at E_F (figure 4.16(a)).

Turning to the symmetry decomposition, the main features in $\bar{n}_{Si}$ with σ-symmetry are relatively dispersionless bands lying between -2 and -3 eV below E_F, and bands dispersing upwards from about 1 eV, with little σ-weight at E_F. As Mahatha *et al* [37] have pointed out, these bands correspond to energy-shifted states in a free-standing silicene layer, but their width, apparent in figure 4.16(b), shows that there must be considerable interaction with the substrate [38]. This is clear from figure 4.17, which maps the density of states in the top-layer Ag and second-layer Ag as well as in the silicene: we see that these σ-bands have substantial weight on the top layer of Ag atoms adjacent to the silicene, but much less on the second layer.

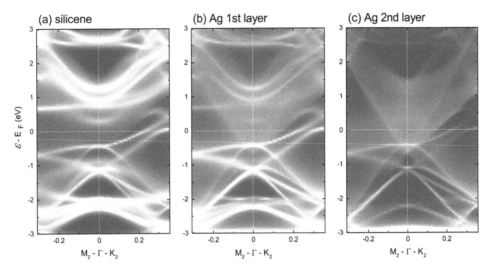

Figure 4.17. The density of states in (a) the silicene overlayer, (b) top-layer Ag and (c) second-layer Ag mapped as a function of $(E - E_F)$ and wave-vector in the $\bar{\Gamma}\text{-}\bar{M}_2$ and $\bar{\Gamma}\text{-}\bar{K}_2$ symmetry directions. The brightness scale is the same as in figure 4.16. (From Ishida *et al* [38].)

Though not particularly evident in figure 4.16(c), the π-contribution to $\bar{n}_{Si}(E, \mathbf{k}_{\|})$ consists of surface resonance structure on top of a nearly energy-independent background [38]. (The background doesn't show up because of the plotting: $\bar{n}_{Si}(E, \mathbf{k}_{\|}) < 10$ a.u. appears as a uniform dark blue.) The featureless background comes from the π-states on the silicene hybridizing with the sp bands of the Ag substrate. There are two rather sharp resonance bands which start off at $\bar{\Gamma}$ at $E = -0.4$ eV and -1.1 eV and cross E_F along $\bar{\Gamma}\text{-}\bar{K}_2$, the upper of which has been identifies by Mahatha *et al* [37] as an interface resonance mainly localized on the top Ag layer. Ishida *et al* [38] have suggested that this state is pulled off the band edge of the surface-projected Ag 5s-5p bands by the perturbation of the silicene layer. It is the Si π-states which give rise to the Dirac bands in free-standing silicene, but we conclude from figures 4.16 and 4.17 that all the π-structure in silicene/Ag(111) has considerable weight on the top-layer Ag as well in the silicene; this is a result of the strong hybridization which ultimately destroys the Dirac properties.

References

[1] Inglesfield J E 1982 Surface electronic structure *Rep. Prog. Phys.* **45** 223–84
[2] Kneedler E, Skelton D, Smith K E and Kevan S D 1990 Surface-state–surface-resonance transition on Ta(011) *Phys. Rev. Lett.* **64** 3151–4
[3] Kneedler E, Smith Kevin E, Skelton D and Kevan S D 1991 Surface electronic structure and dynamical interactions on Ta(011) and H/Ta(011) *Phys. Rev.* B **44** 8233–42
[4] van Hoof J B A N, Crampin S and Inglesfield J E 1992 The surface state-surface resonance transition on Ta(011) *J. Phys.: Condens. Matter* **4** 8477–88
[5] Echenique P M and Pendry J B 1978 The existence and detection of Rydberg states at surfaces *J. Phys. C: Solid St. Phys.* **11** 2065–75

[6] Echenique P M and Pendry J B 1989 Theory of image states at metal surfaces *Prog. Surf. Sci.* **32** 111–72

[7] Dose V, Altmann W, Goldmann A, Kolac U and Rogozik J 1984 Image-potential states observed by inverse photoemission *Phys. Rev. Lett.* **52** 1919–21

[8] Giesen K, Hage F, Himpsel F J, Riess H J and Steinmann W 1985 Two-photon photo-emission via image-potential states *Phys. Rev. Lett.* **55** 300–3

[9] Nekovee M, Crampin S and Inglesfield J E 1993 Magnetic splitting of image states at Fe(110) *Phys. Rev. Lett.* **70** 3099–102

[10] Butti G, Caravati S, Brivio G P, Trioni M I and Ishida H 2005 Image potential states and electronic structure of Na/Cu(111) *Phys. Rev.* B **72** 125402

[11] Nekovee M and Inglesfield J E 1992 Threshold behaviour of surface density of states at the vacuum level *Europhys. Lett.* **19** 535–40

[12] Nekovee M 1995 *Image Potential States and Dielectric Response at Metal Surfaces* PhD thesis (Katholieke Universiteit Nijmegen)

[13] Elliott R J 1957 Intensity of optical absorption by excitons *Phys. Rev.* **108** 1384–9

[14] von Barth U and Hedin L 1972 A local exchange-correlation potential for the spin polarized case: I *J. Phys. C: Solid St. Phys.* **5** 1629–42

[15] Passek F, Donath M, Ertl K and Dose V 1995 Longer living majority than minority image state at Fe(110) *Phys. Rev. Lett.* **75** 2746–9

[16] Lang N D and Kohn W 1973 Theory of metal surfaces: Induced surface charge and image potential *Phys. Rev.* B **7** 3541–50

[17] Weber M and Liebsch A 1987 Density-functional approach to second-harmonic generation at metal surfaces *Phys. Rev.* B **35** 7411–6

[18] Inglesfield J E 1987 The screening of an electric field at an Al(001) surface *Surf. Sci.* **188** L701–7

[19] Aers G C and Inglesfield J E 1989 Electric fields and Ag(001) surface electronic structure *Surf. Sci.* **217** 367–83

[20] Lang N D and Kohn W 1971 Theory of metal surfaces: Work function *Phys. Rev.* B **3** 1215–23

[21] Chulkov E V, Silkin V M and Echenique P M 1999 Image potential states on metal surfaces: binding energies and wave functions *Surf. Sci.* **437** 330–52

[22] Merrick Ian 1995 *Embedding at Electrode Surfaces* PhD thesis (Cardiff University)

[23] Merrick I, Inglesfield J E and Attard G A 2005 Local work-function and induced screening effects at stepped Pd surfaces *Phys. Rev.* B **71** 085407

[24] Ishida H 2001 Surface-embedded Green-function method: A formulation using a linearized-augmented-plane-wave basis set *Phys. Rev.* B **63** 165409

[25] Carlsson J M and Hellsing B 2000 First-principles investigation of the quantum-well system Na on Cu(111) *Phys. Rev.* B **61** 13973–82

[26] Perdew J P, Burke K and Ernzerhof M 1996 Generalized gradient approximation made simple *Phys. Rev. Lett.* **77** 3865–8

[27] Diehl R D and McGrath R 1996 Structural studies of alkali metal adsorption and coadsorption on metal surfaces *Surf. Sci. Rep.* **23** 43–171

[28] Kliewer J and Berndt R 2001 Low temperature scanning tunneling microscopy of Na on Cu(111) *Surf. Sci.* **477** 250–8

[29] Kliewer J and Berndt R 2001 Scanning tunneling spectroscopy of Na on Cu(111) *Phys. Rev.* B **65** 035412

[30] Chis V, Caravati S, Butti G, Trioni M I, Cabrera-Sanfelix P, Arnau A and Hellsing B 2007 Two-dimensional localization of fast electrons in $p(2 \times 2)$-Cs/Cu(111) *Phys. Rev. B* **76** 153404

[31] Corriol C, Silkin V M, Sánchez-Portal D, Arnau A, Chulkov E V, Echenique P M, von Hofe T, Kliewer J, Kröger J and Berndt R 2005 Role of elastic scattering in electron dynamics at ordered alkali overlayers on Cu(111) *Phys. Rev. Lett.* **95** 176802

[32] Breitholtz M, Chis V, Hellsing B, Lindgren S-Å and Walldén L 2007 Overlayer resonance and quantum well state of Cs/Cu(111) studied with angle-resolved photoemission, LEED, and first-principles calculations *Phys. Rev. B* **75** 155403

[33] Achilli S, Trioni M I, Chulkov E V, Echenique P M, Sametoglu V, Pontius N, Winkelmann A, Kubo A, Zhao J and Petek H 2009 Spectral properties of Cs and Ba on Cu(111) at very low coverage: Two-photon photoemission spectroscopy and electronic structure theory *Phys. Rev. B* **80** 245419

[34] Trioni M I, Achilli S and Chulkov E V 2013 Key ingredients of the alkali atom—metal surface interaction: Chemical bonding versus spectral properties *Prog. Surf. Sci.* **88** 160–70

[35] Cahangirov S, Topsakal M, Aktürk E, Şahin H and Ciraci S 2009 Two- and one-dimensional honeycomb structures of silicon and germanium *Phys. Rev. Lett.* **102** 236804

[36] Lin C-L, Arafune R, Kawahara K, Kanno M, Tsukahara N, Minamitani E, Kim Y, Kawai M and Takagi N 2013 Substrate-induced symmetry breaking in silicene *Phys. Rev. Lett.* **110** 076801

[37] Mahatha S K, Moras P, Bellini V, Sheverdyaeva P M, Struzzi C, Petaccia L and Carbone C 2014 Silicene on Ag(111): A honeycomb lattice without Dirac bands *Phys. Rev. B* **89** 201416

[38] Ishida H, Hamamoto Y, Morikawa Y, Minamitani E, Arafune R and Takagi N 2015 Electronic structure of the 4×4 silicene monolayer on semi-infinite Ag(111) *New J. Phys.* **17** 015013

[39] Kresse G and Hafner J 1993 *Ab initio* molecular dynamics for liquid metals *Phys. Rev. B* **47** 558–61

[40] Kresse G and Furthmüller J 1996 Efficiency of ab-initio total energy calculations for metals and semiconductors using a plane-wave basis set *Comput. Mater. Sci.* **6** 15–50

Chapter 5

Confined electrons and embedding

Especially with advances in nanotechnology, there are many systems in which electrons are confined by a boundary, in one, two or three dimensions [1]. We need only think of quantum dots, nanoclusters, carbon nanotubes, atoms encapsulated by the C_{60} molecule, quantum corrals constructed on surfaces and so on [2]. Especially when the boundary has a complicated shape, and the Schrödinger equation is non-separable, this can be a difficult problem. Perhaps surprisingly, embedding enables us to tackle the problem of confinement by a wall of arbitrary shape, in a straightforward and efficient way [3]. The method also sheds some light on the embedding method in general.

5.1 Variational principle for confined systems

A typical confined system is represented by region I in figure 5.1—we want to solve the Schrödinger equation subject to the Dirichlet boundary condition that the wave-function vanishes on the boundary S,

$$H\phi = E\phi, \quad \text{with} \quad \phi(\mathbf{r}_S) = 0, \ \mathbf{r}_S \text{ on } S. \tag{5.1}$$

Alternatively, we can think of region I as a cavity in region II, where the potential $\mathcal{V}$ is effectively infinite—so far, so trivial you might think, but this is the form in which we can solve the problem using the embedding method, replacing region II by an embedding potential over S. Although we cannot treat infinite $\mathcal{V}$, we can use such large values that to all intents and purposes we have solved the true confinement problem [3].

To find the embedding potential to replace region II we use equation (2.15), relating the derivative over S of the solution in region II to its amplitude. Let us consider a small portion of S, locally flat, and use z for the normal coordinate (directed according to our convention *into* S) and $\mathbf{r}_\parallel$ as the parallel coordinate. Then

doi:10.1088/978-0-7503-1042-0ch5

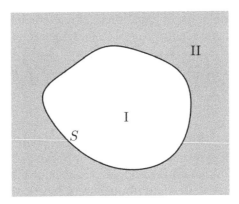

Figure 5.1. Electrons in region I are confined by boundary S, on which we require $\phi(\mathbf{r}_S) = 0$. This is the same problem as region I joined on to region II in which the potential $\mathcal{V} \to \infty$.

because $\mathcal{V}$ is very large, the general solution in II, in the vicinity of this portion of S can take the form

$$\psi(\mathbf{r}) = f(\mathbf{r}_{\parallel})\exp(-\gamma z), \quad \gamma \approx \sqrt{2\mathcal{V}}, \tag{5.2}$$

and f is any smoothly varying function. So, on this portion of S we have

$$\psi(\mathbf{r}_S) = f(\mathbf{r}_{\parallel}), \quad \frac{\partial \psi(\mathbf{r}_S)}{\partial n_S} = -\gamma f(\mathbf{r}_{\parallel}), \tag{5.3}$$

and from equation (2.15) the embedding potential is given by the local, energy-independent form

$$\Sigma(\mathbf{r}_S, \mathbf{r}'_S) = \sqrt{\frac{\mathcal{V}}{2}} \, \delta(\mathbf{r}_S - \mathbf{r}'_S). \tag{5.4}$$

The embedded variational expression (2.22) then takes the simple form

$$E = \frac{\int_I d\mathbf{r} \, \phi H\phi + \int_S d\mathbf{r}_S \left\{ \frac{1}{2}\phi\frac{\partial\phi}{\partial n_s} + \sqrt{\frac{\mathcal{V}}{2}}\phi^2 \right\}}{\int_I d\mathbf{r} \, \phi^2}. \tag{5.5}$$

Minimizing E gives a wave-function ϕ which not only satisfies the Schrödinger equation in I, but also

$$\frac{\partial\phi(\mathbf{r}_S)}{\partial n_s} = -\sqrt{2\mathcal{V}}\,\phi(\mathbf{r}_S), \tag{5.6}$$

which for large $\mathcal{V}$ and well-behaved functions means

$$\phi(\mathbf{r}_S) \approx 0, \tag{5.7}$$

as we require. We can make use of equation (2.28), to combine the kinetic energy in H with the normal derivative term over S so that equation (5.5) becomes

$$E = \frac{\int_I \mathrm{dr} \left\{ \frac{1}{2} \nabla\phi \cdot \nabla\phi + \phi V\phi \right\} + \sqrt{\frac{\mathcal{V}}{2}} \int_S \mathrm{dr}_S \, \phi^2}{\int_I \mathrm{dr} \, \phi^2}. \tag{5.8}$$

This is usually simpler to use than equation (5.5) as it removes the need to evaluate the normal derivative of ϕ over the (possibly complicated) boundary.

We have obtained a minimum variational principle for the confinement problem. However, because $\mathcal{V}$ is always finite (though very large), there is inevitably some leakage of the wave-function outside S. This means that minimizing equation (5.5) leads to an energy which is slightly smaller than the true energy of the confined electron. It is straightforward to show that the error varies as $1/\sqrt{\mathcal{V}}$, and knowing this error behaviour eliminates the problem of leakage. This work was actually inspired by a paper by Brownstein [4], who obtained a stationary (not minimum) variational method for the confined Hamiltonian. His variational expression has one advantage over ours—it only involves volume integrals through region I, whereas in equation (5.8) we also have the integral $\int_S \mathrm{dr}_S \, \phi^2$ over S. However, as our examples will show, having the minimum principle is worth quite a lot.

As a first example we look at confinement in a one-dimensional square well of width d (figure 5.2): we have an infinite barrier at $z = 0$, and at $z = d$ a very large barrier of height $\mathcal{V}$, which is replaced by an embedding potential $\Sigma = \sqrt{\mathcal{V}/2}$. As basis functions for expanding the wave-functions in the well—region I—we use

$$\chi_n(z) = \sin(n\pi z/D), \quad n = 1, 2, 3..., \tag{5.9}$$

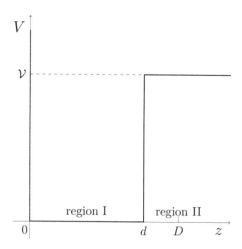

Figure 5.2. Square well with an infinite barrier at $z = 0$ and a very large barrier, height $\mathcal{V}$, at $z = d$. The well constitutes region I, and the right-hand barrier region II, replaced by an embedding potential at $z = d$. The basis functions in region I are defined by D.

where (as usual) we take $D > d$, to give the flexibility in boundary conditions at the embedding plane. Taking $d = 5$ a.u., $D = 7$ a.u., the results we obtain for the lowest energy eigenvalues for different values of $\mathcal{V}$ and basis set size N are shown in table 5.1, together with the exact results in the hard-wall limit,

$$E_m = \frac{1}{2}\left(\frac{m\pi}{d}\right)^2. \tag{5.10}$$

We see that the convergence properties of the eigenvalues are excellent, and the results with 10 basis functions are converged to better than 10^{-6} a.u. (results for $N = 15$ are identical to $N = 10$ and are not shown in the table). For $\mathcal{V} = 10^{10}$ a.u., the eigenvalues of the embedded problem agree with the hard-wall limit to $\sim 10^{-6}$ a.u.

It is interesting to look at the properties of the wave-functions as the number of basis functions increases. Figure 5.3 shows $\phi(z)$ for the third eigenvalue for

Table 5.1. First three energy eigenvalues (in a.u.) for the square well, for different basis-set size N and different values of $\mathcal{V}$, the barrier height, compared with the exact (hard wall) results. Width of well $d = 5$ a.u., and the parameter defining the basis function $D = 7$ a.u.

$\mathcal{V} = 10^6$ a.u.		$\mathcal{V} = 10^8$ a.u.		$\mathcal{V} = 10^{10}$ a.u.		
$N = 5$	10	5	10	5	10	Hard wall
0.197426	0.197336	0.197476	0.197387	0.197481	0.197392	0.197392
0.789434	0.789345	0.789635	0.789546	0.789655	0.789566	0.789568
1.776593	1.776026	1.777049	1.776479	1.777094	1.776524	1.776529

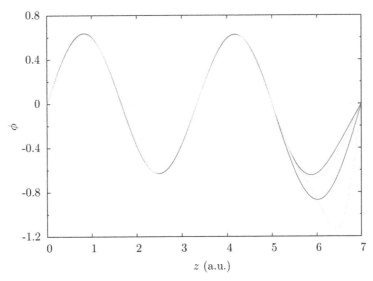

Figure 5.3. Third eigenstate $\phi(z)$ for an electron confined in a square well, width $d = 5$ a.u., height of potential barrier $\mathcal{V} = 10^{10}$ a.u. (figure 5.2), replaced by an embedding potential. Red curve, five basis functions; green, 10; blue, 15; yellow, 20. The basis functions are defined by equation (5.9) with $D = 7$ a.u.

$\mathcal{V} = 10^{10}$ a.u., and we see that within the well $0 < z < d$ (where we are trying to solve the problem), there is perfect convergence. But if we use the basis set expansion to evaluate ϕ beyond d, in the region $d < z < D$ where the basis functions are still defined, we see that there is no hint of convergence. This suggests that although $\phi(z)$ converges in the embedded region, the individual amplitudes a_n in the expansion (2.24) do not converge. In some sense, it is the freedom in the a_n, or in ϕ outside region I, which allows the method to work so well within the embedded region.

One-dimensional examples are in some sense trivial, as they do not emphasise the real power of this method for treating confinement—the fact that we can use basis functions which need not satisfy the zero-amplitude boundary condition on S. We shall consider some examples which exploit this flexibility in the next section.

5.2 Confined H atom

The ground-state energy of a H atom placed off-centre in a hard-wall spherical cavity was calculated by Crampin et al [3] in their original paper on embedding a confined system. This may seem somewhat artificial, but it is relevant to the energy levels of impurities in quantum dots [5]—something which cannot have been imagined when this type of problem was first studied in the 1980s [6]. There is an extensive literature on this type of problem using a wide range of methods, and work continues to this day.

In the embedding calculation the ground state H wave-function was expanded in terms of the basis functions

$$u_{\alpha,\beta}(r, \theta) = e^{-r} r^\alpha \cos^\beta(\theta), \quad \alpha, \beta = 0, 1,...(N-1), \quad (5.11)$$

where r and θ are radial and angular coordinates relative to the atom centre. Table 5.2 shows the results for an atom placed 0.5 a.u. off-centre in a cavity of radius 3 a.u., using $\mathcal{V} = 1.8 \times 10^9$ a.u. in the embedding potential, with different sizes of basis set. These are compared with the energy calculated using Brownstein's variational method [4], with the same basis set. We see that the embedding method converges uniformly to the same value as Brownstein's method. However, embedding has the advantage of giving a true minimum principle, without the ghost states below the true ground state which are found in Brownstein's variational principle (see $N = 4, 6$).

Table 5.2. Ground state energy (in a.u.) of a H atom displaced 0.5 a.u. off-centre in a hard-wall spherical cavity of radius 3 a.u., for different basis set size (N^2). The embedding results were calculated taking $\mathcal{V} = 1.8 \times 10^9$ a.u. and are compared with Brownstein's results [4]. The number in brackets after each energy gives its position in the ordered eigenvalues in each method.

Method	$N = 2$	$N = 4$	$N = 6$
Embedding	$-0.31730(1)$	$-0.41323(1)$	$-0.41389(1)$
Brownstein	$-0.44906(1)$	$-0.41013(4)$	$-0.41389(7)$

The ground state energy determined by the embedding method is in perfect agreement with the energy of -0.41389 a.u. found by Diamond *et al* [7], working with a trial wave-function constrained to vanish at a finite number of points over the boundary. More recent papers have used trial functions for confined atoms and molecules made up of B-splines—piecewise continuous functions joined at *knots*—which are constructed to vanish on the boundary. For the off-centre H atom in the geometry used by Crampin *et al*, this method gives a ground state energy of -0.41388 a.u. [8], again almost perfect agreement.

5.3 Surface state confinement by islands on Ag(111)

Electrons occupying the Shockley surface state on the (111) surface of noble metals behave like two-dimensional electrons, localized in the direction perpendicular to the surface, but moving freely over the surface. They can be confined in their motion over the surface, either by natural features such as surface steps or atomic islands, or by artificial structures like quantum corrals, atoms deposited on the surface using scanning microscopy techniques, in a ring structure for example [2].

A remarkable example of surface state confinement is on the Ag(111) surface, where hexagonal islands of Ag atoms, a monolayer thick and of nanometre dimensions, could be grown by Ag deposition [9]. Scanning tunnelling microscopy (STM) measurements were made of these islands, recording the current I with respect to bias voltage V (dI/dV is approximately proportional to the local density of states, see equation (2.34)), and results are shown in the upper row of figure 5.4. We can clearly see the hexagonal shape of the island, which has an estimated area of about 94 nm^2. As the bias voltage increases, the local density of states is probed at higher energies and increasing structure appears with more nodal lines.

This structure can be modelled by two-dimensional free electrons confined by an hexagonal box. The electrons in the surface state band have a free-electron dispersion relation (energy E as a function of wave-vector k) given by

$$E = E_0 + \frac{k^2}{2m^*}, \tag{5.12}$$

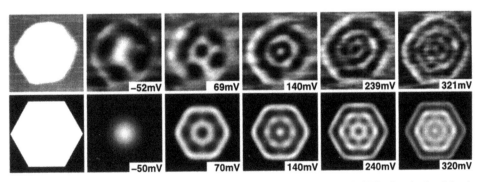

Figure 5.4. Upper row: topographic STM images of an hexagonal Ag island on Ag(111), showing local density of states at different bias voltages. Lower row: calculated local density of states at corresponding energies for free electrons confined by a two-dimensional hexagon. (From Li *et al* [9].)

where E_0 is the energy of the bottom of the band, $E_0 = -67$ meV relative to the Fermi energy, and m^* is the effective mass of the surface states, $m^* = 0.42$. Putting these parameters into the Schrödinger equation, the Green function—hence, the local density of states—of electrons confined inside the hexagon can be calculated using the embedding method. It is the Green function which is calculated, rather than the wave-functions, because there are several broadening processes at work. Firstly, the surface state electrons can leak through the barrier, as confinement is not perfect; secondly, the walls of the barrier, or the edge of the island, can scatter electrons from the surface state into bulk states with the same energy—a surface state is only well-defined over a surface of infinite extent [10]; and thirdly, there are many-body effects. These effects were modelled by evaluating the Green function at an energy with a small imaginary part. The results for the local density of states at different bias energies are shown in the lower row in figure 5.4—remarkable agreement with the experimental results [9].

An example like this—electrons moving under a simple Hamiltonian but confined by a complicated (non-separable) boundary—could also be solved by finite element techniques [11], for example. Here we are demonstrating the flexibility of the embedding method.

5.4 Electron transport through nanostructures

Electron transport through nanostructures is important both from a practical point of view, advances in fabrication techniques making it possible to construct complicated structures [12], but also theoretically, with studies of quantum conductance [13]. The embedding method for confined systems provides a straightforward way of calculating the Green function, hence the transport properties, of electrons going through nanostructures [14].

An interesting problem is the way that electrons propagate round corners in nanostructures, and figure 5.5 shows a model structure for studying this, a kink in a

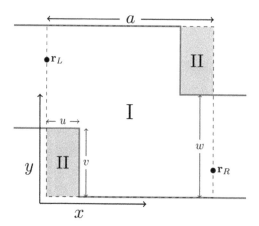

Figure 5.5. Kink in a two-dimensional electron waveguide, shown by the red lines. The square box bounded by the dashed lines is treated by embedding, with electrons confined to region I by potential $\mathcal{V}$ in the shaded regions II.

two-dimensional waveguide [14, 15]. The waveguide, with straight sections extending to the left and right of the kink, is shown by the red lines, and we study the square box of side a indicated by the dashed lines. The electrons are confined to region I by a very large potential $\mathcal{V}$ in the shaded regions II, which we handle by the embedding method.

5.4.1 Eigenstates

Before we calculate the electron transport through the kink, we can test the confinement method by calculating the eigenstates of the Hamiltonian for region I, with an embedding potential along the sides of regions II and a zero-derivative boundary condition at the open ends of the kink. Following equation (2.24), we expand the wave-function ϕ in equation (5.8) as

$$\phi(\mathbf{r}) = \sum_{mn} a_{mn}\chi_{mn}(\mathbf{r}), \qquad (5.13)$$

with the basis functions given by

$$\chi_{mn}(\mathbf{r}) = \cos(m\pi x/a)\, \sin(n\pi y/a), \qquad (5.14)$$

where a is the side of the box (figure 5.5). This choice of basis function gives zero derivative at the ends of the kink, and zero amplitude at the top and bottom. The wave-function expansion coefficients then satisfy the matrix eigenvalue equation

$$\sum_{mn}(H_{kl;mn} - EO_{kl;mn})a_{mn} = 0, \qquad (5.15)$$

with eigenvalue E. The overlap matrix O is the usual integral through region I (equation (2.27)), that is, the square with regions II removed, and the Hamiltonian matrix is given by

$$H_{kl;mn} = \frac{1}{2}\int_I d\mathbf{r}\, \nabla\chi_{kl}(\mathbf{r}) \cdot \nabla\chi_{mn}(\mathbf{r}) + \sqrt{\frac{\mathcal{V}}{2}}\int_S d\mathbf{r}_S\, \chi_{kl}(\mathbf{r}_S)\chi_{mn}(\mathbf{r}_S), \qquad (5.16)$$

where the second embedding potential integral is over the sides of the two shaded regions II (figure 5.5). (We can see from this expression how easy it is to adapt the method to any obstruction in the waveguide, such as a corner or stub.)

Results for several eigenvalues, showing their convergence with basis-set size[1], are shown in table 5.3 for a kink with $a = 5$ a.u., $u = 1$ a.u., $v = 2$ a.u., waveguide width $w = 3$ a.u. (figure 5.5), for two values of confining potential, $\mathcal{V} = 10^5$ a.u. and 10^7 a.u. Convergence is much slower than for the one-dimensional square well, and even with the 41×40 basis the eigenvalues have only converged to $\sim 10^{-3}$ a.u. for $\mathcal{V} = 10^7$ a.u.; the sixth eigenvalue seems to converge particularly slowly. This is presumably because

---

[1] An $M \times N$ basis corresponds to m in equation (5.14) varying between 0 and $M - 1$, and n between 1 and N.

Table 5.3. Eigenvalues (in a.u.) for the kink with zero-derivative boundary conditions at the open ends, for different values of $\mathcal{V}$ and basis set size. The number of eigenvalue ordered in size is shown in the left-hand column. The kink is defined by $a = 5$ a.u., $u = 1$ a.u., $v = 2$ a.u., waveguide width $w = 3$ a.u. (figure 5.5).

	$\mathcal{V} = 10^5$ a.u.			$\mathcal{V} = 10^7$ a.u.		
	21×20	31×30	41×40	21×20	31×30	41×40
1	0.44293	0.43860	0.43689	0.45524	0.44560	0.44167
4	1.37818	1.37456	1.37311	1.38853	1.38099	1.37780
6	2.07448	2.06033	2.05475	2.11455	2.08267	2.06981
8	2.62351	2.62214	2.62160	2.62953	2.62666	2.62547
10	3.64146	3.63668	3.63464	3.65211	3.64436	3.64047

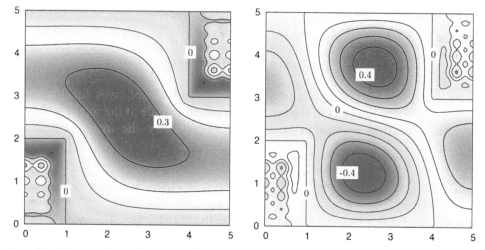

Figure 5.6. Wave-functions $\phi(\mathbf{r})$ for the kink, with zero-derivative boundary conditions at the open ends. Left-hand figure, first eigenstate; right-hand figure, fourth eigenstate. The contours $\phi(\mathbf{r}) = 0$ and the extremal contours are labelled, with a contour interval of 0.1. The calculation is for $\mathcal{V} = 10^7$ a.u. with a basis-set size of 41×40, and the kink is defined by $a = 5$ a.u., $u = 1$ a.u., $v = 2$ a.u., waveguide width $w = 3$ a.u. (figure 5.5).

the basis functions (equation (5.14)) satisfy the zero derivative boundary condition at $x = 0$ and a, reducing the variational freedom needed for embedding. To see how well the wave-functions satisfy the boundary conditions we have imposed, figure 5.6 shows $\phi(\mathbf{r})$ for the first and fourth eigenstate, calculated with the 41×40 basis for a confining potential $\mathcal{V} = 10^7$ a.u. We see that the $\phi(\mathbf{r}) = 0$ contours follow the boundaries of the kink—the edges of the two regions II—remarkably well. Figure 5.6 also illustrates very clearly the point we made about figure 5.3, the fact that there is complete freedom in the behaviour of the wave-functions when they are extended into region II. As the basis-set size increases, there is convergence within region I, but the wave-functions in region II do not converge to anything.

5.4.2 Transmission

The basic problem in transport through the kink is to find the reflection and transmission amplitudes in the expressions for the wave-function in the input (left-hand) and the output (right-hand) straight sections [14],

$$\psi(\mathbf{r}) = \sum_q \left[\delta_{pq} \exp(ik_q x) + r_{pq} \exp(-ik_q x) \right] \sin[q\pi(y - v)/w], \quad \mathbf{r} \text{ in left-hand straight,}$$

$$\psi(\mathbf{r}) = \sum_q t_{pq} \exp(ik_q x) \sin(q\pi y/w), \quad \mathbf{r} \text{ in right-hand straight.} \tag{5.17}$$

The sums are over channels q, with the incident wave in the pth channel, with r_{pq} and t_{pq} the reflection and transmission amplitudes. The wave-vector in the qth channel is given by

$$k_q = \begin{cases} \sqrt{2E - q^2\pi^2/w^2}, & 2E > q^2\pi^2/w^2, \\ i\sqrt{q^2\pi^2/w^2 - 2E}, & 2E < q^2\pi^2/w^2, \end{cases} \tag{5.18}$$

and we only consider incident waves in open channels—those for which the wave-vector is real.

Now, any wave-function $\psi(\mathbf{r})$ and Green function $G(\mathbf{r}, \mathbf{r}')$ satisfying the Schrödinger equation in region I are related by equation (2.11). Integrating through the kink region and using Green's theorem gives

$$\psi(\mathbf{r}) = \frac{1}{2} \int_R d\mathbf{r}_R \left[\psi(\mathbf{r}_R) \frac{\partial G(\mathbf{r}, \mathbf{r}_R)}{\partial n_R} - G(\mathbf{r}, \mathbf{r}_R) \frac{\partial \psi(\mathbf{r}_R)}{\partial n_R} \right]$$
$$- \frac{1}{2} \int_L d\mathbf{r}_L \left[\psi(\mathbf{r}_L) \frac{\partial G(\mathbf{r}, \mathbf{r}_L)}{\partial n_L} - G(\mathbf{r}, \mathbf{r}_L) \frac{\partial \psi(\mathbf{r}_L)}{\partial n_L} \right], \tag{5.19}$$

where the surface (here line) integrals are over the the right-hand and left-hand ends of the kink, with coordinates $\mathbf{r}_R$ and $\mathbf{r}_L$ shown in figure 5.5. The normal derivatives in this case are taken from left to right in the positive x-direction—outside to inside at the left-hand end, and inside to outside at the right-hand end. This equation simplifies if we use G_0, the Green function satisfying zero-derivative boundary conditions at both ends of the kink, and putting $\mathbf{r}$ at each end we obtain the integral equations

$$\psi(\mathbf{r}_L) = \frac{1}{2} \int_L d\mathbf{r}'_L \, G_0(\mathbf{r}_L, \mathbf{r}'_L) \frac{\partial \psi(\mathbf{r}'_L)}{\partial n_L} - \frac{1}{2} \int_R d\mathbf{r}'_R \, G_0(\mathbf{r}_L, \mathbf{r}'_R) \frac{\partial \psi(\mathbf{r}'_R)}{\partial n_R}$$
$$\psi(\mathbf{r}_R) = \frac{1}{2} \int_L d\mathbf{r}'_L \, G_0(\mathbf{r}_R, \mathbf{r}'_L) \frac{\partial \psi(\mathbf{r}'_L)}{\partial n_L} - \frac{1}{2} \int_R d\mathbf{r}'_R \, G_0(\mathbf{r}_R, \mathbf{r}'_R) \frac{\partial \psi(\mathbf{r}'_R)}{\partial n_R}. \tag{5.20}$$

We can find this Green function using our confinement method. Expanding $G_0(\mathbf{r}, \mathbf{r}')$ in terms of the basis set (equation (5.14)),

$$G_0(\mathbf{r}, \mathbf{r}') = \sum_{mn} G_{kl;mn} \chi_{kl}(\mathbf{r}) \chi_{mn}(\mathbf{r}'), \qquad (5.21)$$

the matrix coefficients satisfy the inhomogeneous version of equation (5.15),

$$\sum_{kl} \left(H_{ij;kl} - E O_{ij;kl} \right) G_{kl;mn} = -\delta_{im}\delta_{jn}. \qquad (5.22)$$

Substituting $G_0(\mathbf{r}, \mathbf{r}')$ and equation (5.17) into equation (5.20) then gives us inhomogeneous linear equations for the reflection and transmission amplitudes r_{pq} and t_{pq}. From these we can find the reflection and transmission *probabilities* in the open channels,

$$R_{pq} = |r_{pq}|^2 \frac{k_q}{k_p}, \qquad T_{pq} = |t_{pq}|^2 \frac{k_q}{k_p}, \qquad (5.23)$$

which satisfy

$$\sum_{q}' \left(R_{pq} + T_{pq} \right) = 1. \qquad (5.24)$$

Here, the prime on the summation indicates open channels, with real k_q. It is also useful to define the total transmission probability for electrons incident in channel p, given by

$$T_p = \sum_{q}' T_{pq} \qquad (5.25)$$

where the single suffix distinguishes the transmission into all open channels q from the transmission probability T_{pq} from channel p to channel q.

Results for the transmission through a kink with $u = 1$ a.u., $v = 2$ a.u., waveguide width $w = 3$ a.u. (figure 5.5) are shown in figure 5.7, using an embedding potential corresponding to a confinement potential of $\mathcal{V} = 10^7$ a.u. The figure shows T_1 for incident electrons in the first channel, $p = 1$, plotted as a function of energy E/E_1, where E_1 is the threshold energy for the first channel. The different curves show results with different sizes of basis set, going from the red curve with a basis set size of 21×20, to the blue curve with 41×40. There is clearly convergence, though as with the eigenvalues (table 5.3) a large basis set is needed.

The transmission is highly structured as a function of energy, and there are energies at which there is almost complete transmission, and others at which T_1 drops practically to zero. We shall discuss this structure in more detail in the following section, when we compare $T_1(E)$ with the density of states of the kink, but we note here that there is structure close to energies at which new exit channels open up, at $E/E_1 = 4, 9, 16\ldots$. The kink has similar transmission features to a right-angled corner, with a generally decreasing trend as the energy increases—after all, the kink is nothing but two right-angled corners. We can compare this with the

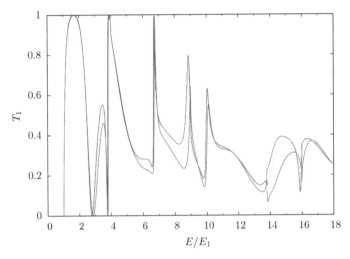

Figure 5.7. Total transmission probability T_1 through the kink for incident electrons in the first channel, as a function of energy E/E_1, where $E_1 = 0.5483$ a.u. is the first threshold energy. The kink is defined by $u = 1$ a.u., $v = 2$ a.u., $w = 3$ a.u. and the box side $a = 5$ a.u. The height of the confining potential barrier is $\mathcal{V} = 10^7$ a.u. Red curve, 21×20 basis functions; green curve, 31×30; blue curve, 41×40.

transmission of electrons round a circular corner, for which T_1 is close to 1 over a wide energy range and relatively featureless.

Knowing the reflection and transmission amplitudes r_{pq} and t_{pq}, hence $\partial\psi(\mathbf{r}_R)/\partial n_S$ and $\partial\psi(\mathbf{r}_L)/\partial n_S$, we can use equation (5.19) to find the wave-function throughout the kink, and what is most interesting, the probability current density $\mathbf{j}(\mathbf{r})$ given by [16]

$$\mathbf{j}(\mathbf{r}) = \Im[\psi^*(\mathbf{r})\,\nabla\psi(\mathbf{r})]. \tag{5.26}$$

This current density is shown in figure 5.8 at different energies, calculated with the 41×40 basis and $\mathcal{V} = 10^7$ a.u. Plot A shows the current at $E/E_1 = 1.82$, where the transmission is close to 1 (figure 5.7), and we see that the flow lines are very smooth through the kink. On the other hand plot B shows the results at $E/E_1 = 2.83$, close to the first zero, and now the flow shows a standing wave pattern (note the vectors in plot B are scaled by a factor of 10 compared with those in A). Moving up in energy to $E/E_1 = 3.78$ in the second deep minimum, plot C, the flow lines again exhibit a standing wave pattern. Plot D gives results at $E/E_1 = 3.92$, again where the transmission is close to 1, and although there is some eddy structure there is also continuous flow through the kink.

As we have already emphasised at the end of section 5.3, this can all be calculated by other methods [12, 15, 17]—and we do have to use large basis sets for this kink example. But think how easily we could put an atom, or other scatterer, into the kink—a matter of adding a potential matrix element to H_{mn} (equation (5.16)).

5.5 Mixed boundary conditions

Implicit in finding the wave-functions (section 5.4.1) and Green function G_0 (section 5.4.2) for the kink, with zero derivative at the open ends, is the fact that we can use embedding to impose *mixed* boundary conditions on the boundary of region I.

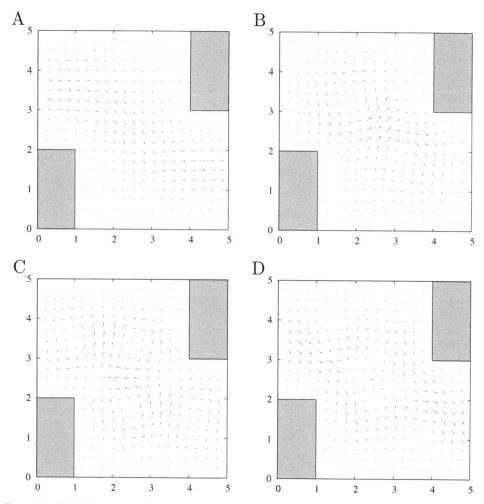

Figure 5.8. Probability current density in the kink, for incident electrons in the first channel with different energies E. A, $E/E_1 = 1.82$, where E_1 is the first threshold energy; B, $E/E_1 = 2.83$ (vectors scaled by 10); C, $E/E_1 = 3.78$; D, $E/E_1 = 3.92$. The currents are calculated with 41×40 basis functions and a confining potential $\mathcal{V} = 10^7$ a.u.

To solve the Schrödinger equation in region I with zero amplitude (the Dirichlet boundary condition) over surfaces S_D and zero derivative (the Neumann boundary condition) over S_N (figure 5.9), we add the confinement embedding potential (equation (5.4)) on S_D, with the embedding potential set to zero on S_N. Slightly altering equation (5.8) (just the range of the surface integral), the variational expression for the energy is given by

$$E = \frac{\int_I \mathrm{d}\mathbf{r} \left\{ \frac{1}{2}\nabla\phi \cdot \nabla\phi + \phi V\phi \right\} + \sqrt{\frac{\mathcal{V}}{2}} \int_{S_D} \mathrm{d}\mathbf{r}_S \, \phi^2}{\int_I \mathrm{d}\mathbf{r} \, \phi^2}. \tag{5.27}$$

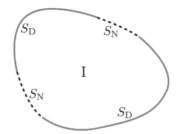

Figure 5.9. We wish to solve the Schrödinger equation in region I with the boundary conditions that $\phi(\mathbf{r}_S) = 0$ over S_D (blue), and $\partial\phi(\mathbf{r}_S)/\partial n_S = 0$ over S_N (dashed).

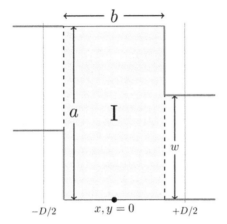

Figure 5.10. Kink treated as a rectangle, with the boundary conditions that $\phi(\mathbf{r}_S) = 0$ over the sides of the kink (blue lines) and $\partial\phi(\mathbf{r}_S)/\partial n_S = 0$ over the ends (dashed lines). Red lines show the straight sections of the waveguide, the grey shaded area is region I and the lines at $x = \pm D/2$ define the basis functions. Comparing with figure 5.5, $b = a - 2u$.

Varying ϕ to minimize E gives wave-functions which satisfy the Schrödinger equation in I subject to the mixed boundary conditions over the enclosing surface.

5.5.1 Kinks revisited

In section 5.4, where we needed to find wave-functions and the Green function satisfying the Dirichlet boundary condition over the walls of the kink and the Neumann condition over the ends, we used basis functions (5.14) which automatically gave zero derivative at $x = 0$ and $x = a$. Here we shall test *really* mixed boundary conditions, by treating the kink as a rectangle, with a mixture of Dirichlet and Neumann boundary conditions along the left- and right-hand sides (figure 5.10). Comparing with figure 5.5, we obtain the same kink if we take the width b of the rectangle as $a - 2u$.

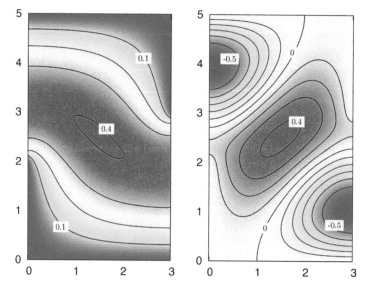

Figure 5.11. Wave-functions for the kink with mixed boundary conditions: $\phi = 0$ on the sides (blue lines in figure 5.10) and $\partial\phi/\partial n_S = 0$ at the open ends (dashed lines in the figure). Left-hand figure, first eigenstate with $E = 0.3819$ a.u.; right-hand figure, fourth eigenstate with $E = 1.7146$ a.u. The extremal contours are labelled, with a contour interval of 0.1. The calculation is for $\mathcal{V} = 10^7$ a.u. with a basis set of 41×40, and $D = 4$ a.u. in the definition of the basis. The geometry of the kink is $a = 5$ a.u., $b = 3$ a.u., and $w = 3$ a.u.

Taking the x-origin in the middle of the rectangle, as basis functions for expanding wave-functions and Green functions we use

$$\chi_{mn}(\mathbf{r}) = \sin(n\pi y/a)\begin{cases} \cos(m\pi x/D), & m = 0, 2, 4..., \\ \sin(m\pi x/D), & m = 1, 3, 5..., \end{cases} \qquad (5.28)$$

where $D > b$ (figure 5.10). These functions have zero derivative at $x = \pm D/2$, giving the range of logarithmic derivatives on the sides of the rectangle at $x = \pm b/2$ which we require.

As a first test of the mixed boundary conditions we calculate the eigenstates of the kink, satisfying the zero-derivative boundary condition at the open ends shown by the dashed lines in figure 5.10. We take $a = 5$ a.u., $b = 3$ a.u., with the width of the open ends $w = 3$ a.u., and in the basis functions we take $D = 4$ a.u.; for the confining embedding potential along the sides of the kink, shown by the blue lines in the figure, we use $\mathcal{V} = 10^7$ a.u. The eigenvalue convergence is much the same as in table 5.3 for the original kink geometry—with a basis set of 31×30 the eigenvalues are accurate to $\sim10^{-2}$ a.u., and with 41×40 the accuracy is $\sim10^{-3}$ a.u.

The wave-functions $\phi(\mathbf{r})$ for the first and fourth eigenvalues are shown in figure 5.11, and they satisfy the boundary conditions extremely well—from the shape of the contours and the shading we see that $\phi(\mathbf{r}_S) = 0$ (as near as we can tell) along the sides of the kink, and the fact that the contours are perpendicular shows that $\partial\phi(\mathbf{r}_S)/\partial n_S = 0$ along the open ends. We should note that these eigenstates are quite different from those shown in figure 5.6, which satisfy a different boundary condition: $\partial\phi(\mathbf{r}_S)/\partial n_S = 0$ a distance u along the waveguide (figure 5.5). A very interesting difference is that the fourth eigenstate shown in figure 5.11 is symmetric with respect to a two-fold rotation about the centre, whereas the fourth eigenstate in figure 5.6 is antisymmetric. With the boundary conditions of figure 5.10, the second and third eigenstates are both antisymmetric with respect to the rotation through 180°.

From the Green function satisfying the zero-derivative boundary condition at the open ends we can calculate the reflection and transmission, using the formulae of section 5.4.2—in fact we should get exactly the same results as in figure 5.7 calculated with different geometry, as R_{pq} and T_{pq} are independent of position along the waveguide. The results for the total transmission $T_1(E)$ (equation (5.25)) for electrons incident in the first channel are shown by the red curve in figure 5.12, calculated with a 41 × 40 basis and $\mathcal{V} = 10^7$ a.u. This is compared with the 41 × 40 results of figure 5.7, shown here by the green curve. Agreement is by no means perfect, but it is reasonable, with all the features present at the same energies. The discrepancies are probably due to lack of convergence in the earlier results, as convergence with the rectangular geometry and mixed boundary conditions is much better than in figure 5.7.

5.5.2 Kink density of states

Transmission is a *dynamic* property of the kink joined on to input and output waveguides. We can calculate the Green function for this system, hence the *static* density of states, by adding embedding potentials at the open ends of the kink (the dashed lines in figure 5.10) to include the effect of the waveguides. In fact, we can mix confinement embedding potentials over surfaces S_D with embedding potentials

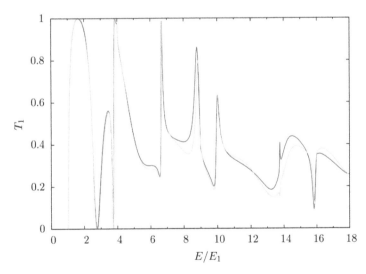

Figure 5.12. Total transmission probability T_1 through the kink for incident electrons in the first channel, as a function of E/E_1, where $E_1 = 0.5483$ a.u. is the first threshold energy. Red curve, calculated with rectangular geometry and mixed boundary conditions (figure 5.10) with $a = 5$ a.u., $b = 3$ a.u., $w = 3$ a.u., a 41 × 40 basis set and $\mathcal{V} = 10^7$ a.u.; green curve, calculated as in section 5.4.2 with a 41 × 40 basis set and $\mathcal{V} = 10^7$ a.u.

over surfaces S_N (figure 5.9) to tackle more general geometries such as a junction of several waveguides or conductors[2].

The embedding potentials which replace the waveguides can be found from equation (2.17). In the left-hand waveguide, the solutions of the Schrödinger equation at energy $E + i\eta$ are given by

$$\psi_p(\mathbf{r}) = \exp(-ik_p\tilde{x})\sin(p\pi\tilde{y}/w), \tag{5.29}$$

where p labels the channel, and $\tilde{x}$ and $\tilde{y}$ are local waveguide coordinates. The wave-vector in the pth channel is given by

$$k_p = \sqrt{2(E + i\eta) - p^2\pi^2/w^2}, \tag{5.30}$$

and we choose the sign of the square root so that k_p is in the upper right quadrant of the complex plane, so that ψ is travelling down the waveguide away from the kink, and decaying. The embedding potential which replaces the left-hand waveguide is then given by a sum over channels,

$$\Sigma(\tilde{y},\tilde{y}') = -\frac{i}{w}\sum_p k_p \sin(p\pi\tilde{y}/w)\sin(p\pi\tilde{y}'/w), \tag{5.31}$$

where $\tilde{y}$, $\tilde{y}'$ vary over the end of the waveguide (the left-hand dashed line in figure 5.10). The embedding potential to replace the right-hand waveguide has the same form.

By including Σ (equation (5.31)) in the Hamiltonian for region I, the Green function $G(\mathbf{r}, \mathbf{r}'; E + i\eta)$ is joined on to the waveguides, and it is forced to go to zero over the sides of the kink by the confinement embedding potential (equation (5.4)). Integrating the corresponding local density of states $-\frac{1}{\pi}\Im G(\mathbf{r}, \mathbf{r}; E + i\eta)$ over region I then gives us the kink density of states $n_I(E)$. As before we take the kink with geometry $a = 5$ a.u., $b = 3$ a.u., $w = 3$ a.u., with $\mathcal{V} = 10^7$ a.u. in the confinement potential and six channels in the waveguide embedding potential. Expanding G in the basis set given by equation (5.28), with $D = 4$ a.u. and 41×40 basis functions, we obtain the results shown in figure 5.13 for $n_I(E)$. In fact, $n_I(E)$ converges very well, much better than the transmission $T_I(E)$, and results with a 21×20 basis lie almost on top of the 41×40 results shown in this figure; the superior convergence is presumably because n_I depends on the diagonal part of the Green function, whereas T_I involves off-diagonal elements connecting the left-hand end of the kink to the right-hand end.

The most striking features of $n_I(E)$ are the peaks, which correspond to localized states in the kink. At $E/E_1 = 0.857$, just below the continuum threshold, there is a discrete, bound state, broadened slightly by $\eta = 0.001$ a.u. In fact it is a general feature that bends and kinks pull off bound states, presumably a result of the one-dimensional nature of waveguides[3]. Just below each threshold at $E = p^2 E_1$, where

---

[2] An interesting project here—think of the applications! Though we might need a very large basis set.
[3] Remember—any attractive potential in 1D has a bound state [16]. But why does a kink or corner behave as an attractive potential?

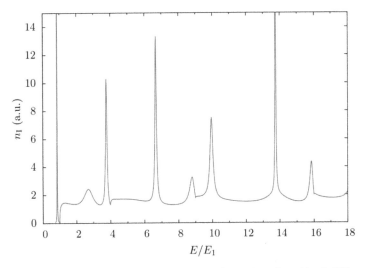

Figure 5.13. Density of states $n_I(E)$ in the kink plotted against E/E_1, where $E_1 = 0.5483$ a.u. is the first threshold energy. The geometry of the kink is $a = 5$ a.u., $b = 3$ a.u., $w = 3$ a.u. The basis set size is 41×40, with $D = 4$ a.u. In the confinement potential $\mathcal{V} = 10^7$ a.u., six channels are used in the waveguide embedding potential, and the Green function is calculated with an imaginary part $\eta = 0.001$ a.u.

the continuum opens up in the pth channel, there is a similar bound state. But above $p = 1$ these are broadened into resonances because they can be scattered into open channels by the kink. However, there are other resonances which are not associated with a particular threshold. If we compare $n_I(E)$ with $T_i(E)$, we find that the peaks in the density of states all correspond to features in the transmission, either peaks or minima: the electrons are getting trapped in the kink. The exception is the bound state as this cannot transmit current.

Apart from the peaks, the other point we notice in figure 5.13 is that $n_I(E)$ is rather constant—similar to the constant density of states of a two-dimensional system. In an infinite two-dimensional system (though the kink is hardly infinite in extent), the density of states is $1/2\pi$ per unit area, and with an area for region I of 15 a.u. we obtain $n_I \approx 2.4$, a little larger than the value between the peaks in figure 5.13. This is presumably due to weight in the density of states being taken up by the peaks, and the dead area near the walls of the kink, where $G(\mathbf{r}, \mathbf{r}) \approx 0$, which reduces the effective area.

5.6 Linear dependence

What has happened in the awful results shown in figure 5.14? Spurious spikes, noise everywhere! It is due to linear dependence in the basis set, a problem which can arise when the basis set is large and region I is considerably smaller than the region over which the basis functions are defined. In this example, the basis is 31×30, and D in the definition of the basis functions (equation (5.28)) is set to 5 a.u., compared with $b = 3$ a.u. for region I (figure 5.10). Fortunately, this is a problem which can be eliminated very easily.

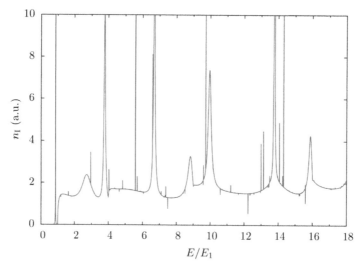

Figure 5.14. Calculation of $n_{\rm I}(E)$ with linear dependence in the basis set. Geometry of the kink, $a = 5$ a.u., $b = 3$ a.u., $w = 3$ a.u. The basis set size is 31×30, with $D = 5$ a.u.

True linear dependence shows up as $\det O = 0$, where the overlap matrix is defined by integrating over region I (equation (2.27)). If we calculate the eigenvectors and eigenvalues of O,

$$\sum_j O_{ij} a_j^{(\mu)} = \lambda^{(\mu)} a_i^{(\mu)}, \tag{5.32}$$

this is equivalent to saying that one or more of the eigenvalues is zero (note that here we are labelling the basis functions by a single index i rather than the two indices m, n of equation (5.28)). In actual calculations, with finite accuracy, the problem arises with *near* linear dependence, when the eigenvalues become very small or even negative: in the example shown in figure 5.14, there are 60 eigenvalues smaller than 10^{-12} including 30 negative eigenvalues of the order of -10^{-14} (O is supposed to be a positive definite matrix).

To eliminate the problem, we transform to a new basis, consisting of the eigenvectors of O with eigenvalues greater than some lower bound ϵ. Suppose we want to remove just two eigenvectors out of N, the remaining $N - 2$ eigenvectors being our new basis. To do this we construct the rectangular matrix E of which the columns are the 'good' eigenvectors,

$$E = \begin{pmatrix} a_1^{(3)} & a_1^{(4)} & a_1^{(5)} & \cdots & a_1^{(N)} \\ a_2^{(3)} & a_2^{(4)} & a_2^{(5)} & \cdots & a_2^{(N)} \\ a_3^{(3)} & a_3^{(4)} & a_3^{(5)} & \cdots & a_3^{(N)} \\ a_4^{(3)} & a_4^{(4)} & a_4^{(5)} & \cdots & a_4^{(N)} \\ \vdots & \vdots & \vdots & \ddots & \vdots \\ a_N^{(3)} & a_N^{(4)} & a_N^{(5)} & \cdots & a_N^{(N)} \end{pmatrix}, \tag{5.33}$$

and then in the new basis the Hamiltonian and overlap matrices become $(N-2) \times (N-2)$ matrices given by

$$\hat{H} = E^{T}HE, \quad \hat{O} = E^{T}OE. \tag{5.34}$$

The overlap matrix in particular is diagonal,

$$\hat{O} = \begin{pmatrix} \lambda^{(3)} & 0 & 0 & \cdots & 0 \\ 0 & \lambda^{(4)} & 0 & \cdots & 0 \\ 0 & 0 & \lambda^{(5)} & \cdots & 0 \\ \vdots & \vdots & \vdots & \ddots & \vdots \\ 0 & 0 & 0 & \cdots & \lambda^{(N)} \end{pmatrix}, \tag{5.35}$$

and explicitly has no eigenvalues smaller than our cut-off ϵ.

We can restore perfect behaviour to figure 5.14 by setting $\epsilon = 10^{-12}$, eliminating the lowest 60 eigenvalues of O. The resulting density of states is almost identical to the results shown in figure 5.13. (Note that eliminating only the very small *negative* eigenvalues is not enough.) In fact we use this bound whenever we are likely to encounter linear dependence—mostly in calculations involving an embedded free-electron Hamiltonian, as the problem does not seem to crop up in embedded surface or interface calculations.

References

[1] Dolmatov V K, Baltenkov A S, Connerade J-P and Manson S T 2004 Structure and photoionization of confined atoms *Radiat. Phys. Chem.* **70** 417–33

[2] Moriarty P 2001 Nanostructured materials *Rep. Prog. Phys.* **64** 297–381

[3] Crampin S, Nekovee M and Inglesfield J E 1995 Embedding method for confined quantum systems *Phys. Rev.* B **51** 7318–20

[4] Brownstein K R 1993 Variational principle for confined quantum systems *Phys. Rev. Lett.* **71** 1427–30

[5] Zhu J-L and Chen X 1994 Spectrum and binding of an off-center donor in a spherical quantum dot *Phys. Rev.* B **50** 4497–502

[6] Gorecki J and Byers Brown W 1989 Variational boundary perturbation theory for enclosed quantum systems *J. Phys. B: At. Mol. Opt. Phys.* **22** 2659–68

[7] Diamond J J, Goodfriend P L and Tsonchev S 1991 General requirements for the use of finite basis sets that do not satisfy the boundary conditions *J. Phys. B: At. Mol. Opt. Phys.* **24** 3669–84

[8] Ting-yun S, Hao-xue Q and Bai-wen L 2000 Variational calculations for a hydrogen atom confined off-centre in an impenetrable spherical box using B-splines *J. Phys. B: At. Mol. Opt. Phys.* **33** 349–55

[9] Li J, Schneider W-D, Berndt R and Crampin S 1998 Electron confinement to nanoscale Ag islands on Ag(111): A quantitative study *Phys. Rev. Lett.* **80** 3332–5

[10] Crampin S, Boon M H and Inglesfield J E 1994 Influence of bulk states on laterally confined surface state electrons *Phys. Rev. Lett.* **73** 1015–8

[11] Ram-Mohan R 2002 *Finite Element and Boundary Element Applications in Quantum Mechanics* (Oxford: Oxford University Press)

[12] Ferry D K, Goodnick S M and Bird J 2009 *Transport in Nanostructures* 2nd edn (Cambridge: Cambridge University Press)

[13] Beenakker C W J and van Houten H 1991 Quantum transport in semiconductor nano-structures In ed H Ehrenreich and D Turnbull *Solid State Physics* vol 44 (New York: Academic Press) p 1

[14] Dix E and Inglesfield J E 1998 An embedding approach to electron waveguides *J. Phys.: Condens. Matter* **10** 5923–41

[15] Yalabik M C 1994 Electronic transport through a kink in an electron waveguide *IEEE T. Electron Dev.* **41** 1843–7

[16] Merzbacher E 1998 *Quantum Mechanics* 3rd edn (New York: John Wiley)

[17] Pendry J B, Prêtre A, Rous P J and Martín-Moreno L 1991 Causal-surface Green's function method *Surf. Sci.* **244** 160–76

IOP Publishing

The Embedding Method for Electronic Structure

John E Inglesfield

Chapter 6

Tight-binding and the embedding self-energy

One of the advantages of the embedding method is that we have complete freedom in how we solve the Schrödinger equation within the embedded region I. All we require is that the solution $\phi(\mathbf{r})$ inside region I should be able to match in amplitude and derivative on to the embedding potential Σ over the embedding surface S. If we have a cluster of atoms within region I embedded into the bulk solid, for example, we could expand ϕ in terms of plane-waves (augmented if we are not using pseudopotentials), muffin-tin orbitals or a linear combination of atomic orbitals (LCAO). But is the configuration space representation of the embedding potential—by which I mean $\Sigma(\mathbf{r}_S, \mathbf{r}_S')$ defined over S—necessarily the best way of proceeding? The problem with using $\Sigma(\mathbf{r}_S, \mathbf{r}_S')$ and an LCAO basis in region I is that we finish off with a mixture of approaches: within region I it is the overlap of atomic orbitals in three dimensions which is important, whereas over S it is the amplitude and derivative of the orbitals over a two-dimensional surface. In this chapter we shall use an LCAO *matrix* formulation for embedding. The matrix representation is more appropriate for local orbital or tight-binding calculations than working with $\Sigma(\mathbf{r}_S, \mathbf{r}_S')$.

What is, in fact, the LCAO matrix representation of the embedding potential is generally called the *self-energy*, and as such it is widely used to replace the leads in calculations of transport through molecules [1]. An early appearance of self-energies in transport theory is in a 1971 paper by Caroli *et al* [2] on tunnelling through a metal–insulator–metal barrier. They introduce the self-energy to include the effects of the metal electrodes on the Green function in the barrier, using an expression that we shall derive in this chapter. Their paper predates the embedding paper, and consequently we shall frequently refer to the LCAO representation of the embedding potential as the self-energy. This also appears in the expression for the density of states of an adsorbate atom in the Grimley–Newns model of chemisorption [3, 4], based on the Anderson model of a magnetic impurity [5]. Following the Inglesfield embedding paper, Baraff and Schlüter [6] showed the link between wave-function embedding in an LCAO representation and Dyson's equation for the Green function.

doi:10.1088/978-0-7503-1042-0ch6

6.1 LCAO embedding

In this section we shall describe several different ways of finding the LCAO representation of the embedding potential, including a variational method analogous to the method we used in chapter 2. All the methods are based on a partitioning of the Hamiltonian and overlap matrices to define regions I and II, and the first approach is a simple Green function way of finding the embedding self-energy.

6.1.1 The embedded Hamiltonian matrix

In the LCAO method, the wave-function for the whole system $\Psi(\mathbf{r})$ is expanded in terms of local orbitals $\chi_i(\mathbf{r})$, which are typically atomic-like wave-functions,

$$\Psi(\mathbf{r}) = \sum_i \alpha_i \chi_i(\mathbf{r}). \tag{6.1}$$

The Schrödinger equation becomes the matrix eigenvalue equation

$$\sum_j H_{ij}\alpha_j = E \sum_j O_{ij}\alpha_j \tag{6.2}$$

with H_{ij} the Hamiltonian and O_{ij} the overlap matrix elements,

$$H_{ij} = \int d\mathbf{r}\, \chi_i(\mathbf{r})H\chi_j(\mathbf{r}), \qquad O_{ij} = \int d\mathbf{r}\, \chi_i(\mathbf{r})\chi_j(\mathbf{r}), \tag{6.3}$$

and without loss of generality we take the χ_is to be real.

In the embedding problem, we split the orbitals into those in region I, represented by the green circles in figure 6.1, and those in region II represented by the brown circles. Unlike in our coordinate treatment of embedding, there is no surface S separating the groups of orbitals, as they overlap spatially: the light green circle represents the orbitals in I which couple to those in region II represented by the light brown circle. The orbitals in region I have coefficients in the wave-function expansion which we shall represent by the vector ϕ_1, and those in region II by ψ_2 (with this notation we keep as close as possible to the notation of section 2.1). The matrix Schrödinger equation for the whole system (equation (6.2)) can then be written in block matrix form as

$$\begin{pmatrix} H_{11} & H_{12} \\ H_{21} & H_{22} \end{pmatrix} \begin{pmatrix} \phi_1 \\ \psi_2 \end{pmatrix} = E \begin{pmatrix} O_{11} & O_{12} \\ O_{21} & O_{22} \end{pmatrix} \begin{pmatrix} \phi_1 \\ \psi_2 \end{pmatrix}, \tag{6.4}$$

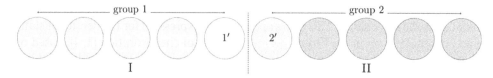

Figure 6.1. Schematic of LCAO embedding. Green circles represent orbitals in region I (group 1), and brown circles orbitals in region II (group 2). Orbitals 1′ in region I are coupled to orbitals 2′.

where the only non-zero elements of H_{12} and O_{12} are those which couple the orbitals across the boundary between the regions, in the sub-blocks $H_{1'2'}$ and $O_{1'2'}$ (figure 6.1).

If we know ϕ_1 we can find ψ_2 by writing the second line of equation (6.4) as

$$(H_{22} - EO_{22})\psi_2 = -(H_{21} - EO_{21})\phi_1. \tag{6.5}$$

This can be solved in terms of the Green function $\mathcal{G}_{22}$ for region II decoupled from I, satisfying

$$(H_{22} - EO_{22})\mathcal{G}_{22} = -I, \tag{6.6}$$

where I is the unit matrix. ψ_2 is then given by

$$\psi_2 = \mathcal{G}_{22}(H_{21} - EO_{21})\phi_1, \tag{6.7}$$

and substituting into the first line of equation (6.4) we obtain

$$(H_{11} - EO_{11})\phi_1 = -(H_{12} - EO_{12})\mathcal{G}_{22}(H_{21} - EO_{21})\phi_1. \tag{6.8}$$

So, ϕ_1 satisfies the following eigenvalue equation within the space of orbitals in region I,

$$[H_{11} + \Sigma_{11}(E)]\phi_1 = EO_{11}\phi_1, \tag{6.9}$$

with

$$\Sigma_{11}(E) = (H_{12} - EO_{12})\mathcal{G}_{22}(H_{21} - EO_{21}). \tag{6.10}$$

This is the embedding potential in matrix form—the self-energy—and the left-hand matrix in equation (6.9) is the embedded LCAO Hamiltonian. This is analogous to equation (2.26) when the embedding potential is evaluated at $\epsilon = E$ so the energy derivative term vanishes. The only non-zero elements of Σ_{11} are for orbitals in group $1'$ (figure 6.1), in other words the sub-block $\Sigma_{1'1'}$, and to construct the self-energy via equation (6.10) we only need the Green function sub-block $\mathcal{G}_{2'2'}$. But we still have to invert the whole matrix $(H_{22} - EO_{22})^{-1}$ to find this sub-block.

6.1.2 Matrix variational method

If we evaluate the embedding potential (equation (6.10)) at a trial energy E_0, the eigenvalues of equation (6.9) are not stationary with respect to variations in E_0 around E. But by adapting the methods of section 2.1 to a matrix representation, we can derive a variational expression for E analogous to equation (2.22).

With the matrix Schrödinger equation given by equation (6.4), the variational expression for the energy is given by

$$E = \frac{\phi_1 H_{11}\phi_1 + 2\phi_1 H_{12}\psi_2 + \psi_2 H_{22}\psi_2}{\phi_1 O_{11}\phi_1 + 2\phi_1 O_{12}\psi_2 + \psi_2 O_{22}\psi_2}. \tag{6.11}$$

We must remember that ϕ_1, ψ_2 are vectors, and a typical term like $\phi_1 H_{11}\phi_1$ implies row vector × block matrix × column vector. As in the configuration space treatment, we assume variational freedom in region I—that is, for vector ϕ_1—and in region II we take the solution of the matrix Schrödinger equation for ψ_2 at energy parameter E_0, which matches on to the trial vector in I. So ψ_2 satisfies

$$(H_{22} - E_0 O_{22})\psi_2 = -(H_{21} - E_0 O_{21})\phi_1, \qquad (6.12)$$

the same as equation (6.5) only with variational parameter E_0 substituted for eigenvalue E.

We can now see how the embedding potential (equation (6.10)) enters the variational expression. From equation (6.12), the last two terms in the numerator of equation (6.11) are given by

$$2\phi_1 H_{12}\psi_2 + \psi_2 H_{22}\psi_2 = E_0\big[2\phi_1 O_{12}\psi_2 + \psi_2 O_{22}\psi_2\big] + \phi_1(H_{12} - E_0 O_{12})\psi_2, \quad (6.13)$$

and substituting equation (6.7) into the last term (with E replaced by E_0) we obtain

$$2\phi_1 H_{12}\psi_2 + \psi_2 H_{22}\psi_2 = E_0\big[2\phi_1 O_{12}\psi_2 + \psi_2 O_{22}\psi_2\big]$$
$$+ \phi_1(H_{12} - E_0 O_{12})\mathcal{G}_{22}(H_{21} - E_0 O_{21})\phi_1. \qquad (6.14)$$

The final term is just the matrix element of the embedding potential.

The terms in square brackets in equation (6.14) are the same as the last two terms in the denominator of equation (6.11), and are the normalization through region II, including the cross-term with ϕ_1. These can be eliminated using the energy derivative of the embedding potential. We rewrite equation (6.12) as

$$(H_{21} - E_0 O_{21})\phi_1 + (H_{22} - E_0 O_{22})\psi_2 = 0, \qquad (6.15)$$

and differentiating with respect to E_0 gives

$$-O_{21}\phi_1 + (H_{22} - E_0 O_{22})\psi_2' - O_{22}\psi_2 = 0, \qquad (6.16)$$

where $\psi_2' := \mathrm{d}\psi_2/\mathrm{d}E_0$. Then, multiplying equation (6.15) by ψ_2', equation (6.16) by ψ_2 and subtracting we obtain

$$\psi_2 O_{21}\phi_1 + \psi_2 O_{22}\psi_2 = -\psi_2'(H_{21} - E_0 O_{21})\phi_1, \qquad (6.17)$$

and consequently

$$2\phi_1 O_{12}\psi_2 + \psi_2 O_{22}\psi_2 = \phi_1 O_{12}\psi_2 - \psi_2'(H_{21} - E_0 O_{21})\phi_1. \qquad (6.18)$$

The left-hand side of equation (6.18) corresponds to the normalization through region II, and using equation (6.7) the right-hand side becomes

$$\phi_1 O_{12}\psi_2 - \psi_2'(H_{21} - E_0 O_{21})\phi_1 = 2\phi_1 O_{12}\mathcal{G}_{22}(H_{21} - E_0 O_{21})\phi_1$$
$$- \phi_1(H_{12} - E_0 O_{12})\mathcal{G}_{22}'(H_{22} - E_0 O_{22})\phi_1$$
$$= -\phi_1 \frac{\mathrm{d}}{\mathrm{d}E_0}\Big[(H_{12} - E_0 O_{12})\mathcal{G}_{22}(H_{21} - E_0 O_{21})\Big]\phi_1. \quad (6.19)$$

This gives the following result for the normalization contribution from region II in terms of the matrix element of the energy derivative of the embedding potential,

$$2\phi_1 O_{12}\psi_2 + \psi_2 O_{22}\psi_2 = -\phi_1\frac{\mathrm{d}\Sigma_{11}}{\mathrm{d}E_0}\phi_1, \tag{6.20}$$

a satisfying generalization of equation (2.21) to the LCAO matrix representation.

Substituting equations (6.14) and (6.20) into equation (6.11) gives us a variational expression for the energy within the space of orbitals in region I,

$$E = \frac{\phi_1 H_{11}\phi_1 + \phi_1\left[\Sigma_{11}(E_0) - E_0\dfrac{\mathrm{d}\Sigma_{11}}{\mathrm{d}E_0}\right]\phi_1}{\phi_1 O_{11}\phi_1 - \phi_1\dfrac{\mathrm{d}\Sigma_{11}}{\mathrm{d}E_0}\phi_1}. \tag{6.21}$$

This has exactly the same structure as equation (2.22). Varying ϕ_1 to find the stationary values of E gives the matrix eigenvalue equation

$$\hat{H}_{11}\phi_1 = E\hat{O}_{11}\phi_1, \tag{6.22}$$

with

$$\hat{H}_{11} = H_{11} + \left[\Sigma_{11}(E_0) - E_0\frac{\mathrm{d}\Sigma_{11}}{\mathrm{d}E_0}\right], \qquad \hat{O}_{11} = O_{11} - \frac{\mathrm{d}\Sigma_{11}}{\mathrm{d}E_0}. \tag{6.23}$$

The eigenvalues evaluated from equation (6.22) are stationary with respect to variations in E_0, behaving like the eigenvalues in figure 2.2.

We could have guessed that the way to modify equation (6.9) to give stationary eigenvalues would be to add the derivative term, as in equation (6.22). But what we have found, in addition to deriving the variational principle, is the correct normalization of the eigenvector ϕ_1 of equation (6.22),

$$\phi_1 O_{11}\phi_1 - \phi_1\frac{\mathrm{d}\Sigma_{11}}{\mathrm{d}E_0}\phi_1 = 1, \tag{6.24}$$

the second term taking care of region II. This brings us back to the concept of *renormalization*. Let us write the eigenvector of equation (6.22) normalized in region I as $\hat{\phi}_1$, with

$$\hat{\phi}_1 O_{11}\hat{\phi}_1 = 1. \tag{6.25}$$

Then, from equation (6.24) the correctly normalized eigenvector is given by

$$\phi_1 = \frac{1}{\sqrt{1 - \hat{\phi}_1\dfrac{\mathrm{d}\Sigma_{11}}{\mathrm{d}E_0}\hat{\phi}_1}}\hat{\phi}_1, \tag{6.26}$$

with the same structure as equation (2.31).

6.1.3 The Green function and Dyson's equation

The embedded Green function satisfies the inhomogeneous version of the eigenvalue equation (6.9) or (6.22),

$$[H_{11} + \Sigma_{11}(E) - EO_{11}]G_{11} = -I, \tag{6.27}$$

with no energy derivative term as in equation (6.23) because the embedding potential is evaluated at the energy E of the Green function.

We can derive this directly, without invoking the eigenvalue equation, using Dyson's equation. The Green function for the whole system I + II satisfies

$$\begin{pmatrix} H_{11} - EO_{11} & H_{12} - EO_{12} \\ H_{21} - EO_{21} & H_{22} - EO_{22} \end{pmatrix} \begin{pmatrix} G_{11} & G_{12} \\ G_{21} & G_{22} \end{pmatrix} = -I. \tag{6.28}$$

If we decouple regions I and II by setting $H_{12}, O_{12} = 0$, the Green function for this system satisfies

$$\begin{pmatrix} H_{11} - EO_{11} & 0 \\ 0 & H_{22} - EO_{22} \end{pmatrix} \begin{pmatrix} \mathcal{G}_{11} & 0 \\ 0 & \mathcal{G}_{22} \end{pmatrix} = -I. \tag{6.29}$$

We see that $\tau := H_{12} - EO_{12}$ is the perturbation which converts $\mathcal{G}$ to G, and iterating Dyson's equation once we have

$$G = \mathcal{G} + \mathcal{G}\tau\mathcal{G} + \mathcal{G}\tau\mathcal{G}\tau G. \tag{6.30}$$

Writing this equation explicitly for the top left-hand block of G we obtain

$$G_{11} = \mathcal{G}_{11} + \mathcal{G}_{11}(H_{12} - EO_{12})\mathcal{G}_{22}(H_{21} - EO_{21})G_{11}. \tag{6.31}$$

But this is just Dyson's equation for the Green function within the space of orbitals in group 1 with the perturbation $(H_{12} - EO_{12})\mathcal{G}_{22}(H_{21} - EO_{21})$ converting $\mathcal{G}_{11}$ to G_{11}, in other words, embedding region I into region II. So we identify the embedding potential or self-energy as

$$\Sigma_{11} = (H_{12} - EO_{12})\mathcal{G}_{22}(H_{21} - EO_{21}), \tag{6.32}$$

recovering equations (6.10) and (6.27). This is the usual derivation of the self-energy.

6.1.4 Baraff–Schlüter approach

In 1986, Baraff and Schlüter [6] presented a particularly neat and concise derivation of the LCAO embedding potential. We take the system shown in the upper part of figure 6.2, consisting of the orbitals in region II (group 2, brown spheres) together with the adjacent orbitals in region I which overlap with region II (orbitals 1′, light green sphere). The Hamiltonian and overlap matrices within the space of orbitals 1′ $H^0_{1'1'}$ and $O^0_{1'1'}$ are arbitrary, though the coupling matrices $H_{1'2}$ and $O_{1'2}$ have to be those appropriate to the combined system I + II. The Green function G^0 for this system, orbitals 1′ + 2, then satisfies

$$\begin{pmatrix} H^0_{1'1'} - EO^0_{1'1'} & H_{1'2} - EO_{1'2} \\ H_{21'} - EO_{21'} & H_{22} - EO_{22} \end{pmatrix} \begin{pmatrix} G^0_{1'1'} & G^0_{1'2} \\ G^0_{21'} & G^0_{22} \end{pmatrix} = -I. \tag{6.33}$$

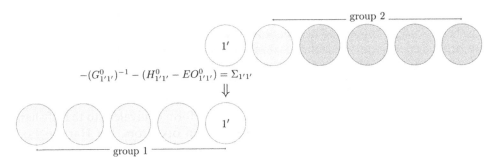

Figure 6.2. Schematic of the Baraff–Schlüter approach. Upper plot: the Green function is calculated for orbitals in region II (group 2, brown circles) + overlapping orbitals from region I (orbitals 1′, light-green circle) from which the embedding potential $\Sigma_{1'1'}$ can be calculated. Lower plot: adding $\Sigma_{1'1'}$ to the Hamiltonian for region I (green circles) embeds this region into region II.

But by definition, the embedding potential $\Sigma_{1'1'}$ which embeds orbitals 1′ into region II satisfies

$$\left(H^0_{1'1'} + \Sigma_{1'1'} - EO^0_{1'1'}\right)G^0_{1'1'} = -I, \tag{6.34}$$

or rearranging,

$$\Sigma_{1'1'} = -\left(G^0_{1'1'}\right)^{-1} - \left(H^0_{1'1'} - EO^0_{1'1'}\right). \tag{6.35}$$

(Note that $(G^0_{1'1'})^{-1}$ is the inverse of the submatrix $G^0_{1'1'}$ only within the space of the 1′ orbitals.) It is easy to show that $\Sigma_{1'1'}$ does not change if we make arbitrary changes to $H^0_{1'1'}$ and $O^0_{1'1'}$ in equations (6.33) and (6.35) [6]. This means that we can include the group 1 orbitals with orbitals 1′, so that $\Sigma_{1'1'}$ given by equation (6.35) is the embedding potential for embedding the whole of region I into region II.

What we have in equation (6.35) is an alternative expression which gives the same embedding potential as equation (6.10). In general it is more complicated to evaluate, as equation (6.35) involves taking the inverse of $G^0_{1'1'}$—so why consider it at all? Well, this formula is the *definition* of the self-energy in contexts such as many-body theory. Moreover, it may lead to elegant ways of calculating the matrix elements of the embedding potential in basis sets other than LCAO.

6.2 The Grimley–Newns chemisorption model

The most common application of the LCAO self-energy is to describe the effect of the metallic leads in calculations of transport through molecules, which we shall discuss in chapter 7. However, an embedding term equivalent to equation (6.10) appeared many years ago in the Anderson model for magnetic impurities in a non-magnetic metal [5], and subsequently in the Grimley–Newns model for chemisorption [3, 4]. These have been very useful for fitting experimental results in a semi-empirical way [7].

In the model for atomic H chemisorbed on a metal substrate, the Hamiltonian in second-quantization notation (as used by Grimley and Newns) is given by

$$\hat{H} = \sum_\sigma \epsilon_a \hat{n}_{a\sigma} + \sum_{\mathbf{k}\sigma} \epsilon_{\mathbf{k}} \hat{n}_{\mathbf{k}\sigma} + \sum_{\mathbf{k}\sigma}\left(V_{a\mathbf{k}}\hat{c}^\dagger_{a\sigma}\hat{c}_{\mathbf{k}\sigma} + V^*_{a\mathbf{k}}\hat{c}^\dagger_{\mathbf{k}\sigma}\hat{c}_{a\sigma}\right) + U\hat{n}_{a\uparrow}\hat{n}_{a\downarrow}, \tag{6.36}$$

where $\hat{c}^\dagger$, $\hat{c}$ are creation and annihilation operators, and $\hat{n} := c^\dagger c$ is the occupation number[1]. So $\hat{n}_{a\sigma}$ denotes the occupation of the adsorbate orbital a, spin σ, with ϵ_a the orbital energy on the isolated atom, and $\hat{n}_{\mathbf{k}\sigma}$ is the occupation of the metal substrate orbital labelled by $\mathbf{k}$, with $\epsilon_\mathbf{k}$ the unperturbed orbital energy. The third term gives the coupling between the adsorbate and substrate orbitals. The final term represents the Coulomb interaction between electrons with spin-up and spin-down on the H-atom. To get the Hamiltonian in a form equivalent to the left-hand side of equation (6.4), only involving one-electron operators, the Hartree–Fock approximation is used,

$$\hat{n}_{a\uparrow}\hat{n}_{a\downarrow} \approx \hat{n}_{a\uparrow}\langle\hat{n}_{a\downarrow}\rangle \qquad \text{or} \qquad \hat{n}_{a\downarrow}\langle\hat{n}_{a\uparrow}\rangle, \tag{6.37}$$

where $\langle\hat{n}_{a\sigma}\rangle$ is the expectation value of $\hat{n}_{a\sigma}$—a number. The Hamiltonian for electrons with spin σ then becomes

$$\hat{H}_\sigma = \epsilon_\sigma\hat{n}_{a\sigma} + \sum_\mathbf{k}\epsilon_\mathbf{k}\hat{n}_{\mathbf{k}\sigma} + \sum_\mathbf{k}\left(V_{a\mathbf{k}}c_{a\sigma}^\dagger c_{\mathbf{k}\sigma} + V_{a\mathbf{k}}^* c_{\mathbf{k}\sigma}^\dagger c_{a\sigma}\right), \tag{6.38}$$

where

$$\epsilon_\sigma = \epsilon_a + U\langle\hat{n}_{a-\sigma}\rangle. \tag{6.39}$$

ϵ_σ is an effective energy level on the adatom, including the Coulomb repulsion with electrons of opposite spin.

The one-electron Hamiltonian given by equation (6.38) can be written in matrix form, in the basis of the adsorbate orbital $|a\rangle$ and substrate orbitals $|\mathbf{k}\rangle$, as

$$H_\sigma = \begin{pmatrix} \epsilon_\sigma & V_{a\mathbf{k}} & V_{a\mathbf{k}'} & V_{a\mathbf{k}''} & \cdots \\ V_{a\mathbf{k}}^* & \epsilon_\mathbf{k} & 0 & 0 & \cdots \\ V_{a\mathbf{k}'}^* & 0 & \epsilon_{\mathbf{k}'} & 0 & \cdots \\ V_{a\mathbf{k}''}^* & 0 & 0 & \epsilon_{\mathbf{k}''} & \cdots \\ \vdots & \vdots & \vdots & \vdots & \ddots \end{pmatrix}. \tag{6.40}$$

The corresponding Green function then satisfies

$$\begin{pmatrix} \epsilon_\sigma - E & V_{a\mathbf{k}} & V_{a\mathbf{k}'} & V_{a\mathbf{k}''} & \cdots \\ V_{a\mathbf{k}}^* & \epsilon_\mathbf{k} - E & 0 & 0 & \cdots \\ V_{a\mathbf{k}'}^* & 0 & \epsilon_{\mathbf{k}'} - E & 0 & \cdots \\ V_{a\mathbf{k}''}^* & 0 & 0 & \epsilon_{\mathbf{k}''} - E & \cdots \\ \vdots & \vdots & \vdots & \vdots & \ddots \end{pmatrix} \begin{pmatrix} G_{aa}^\sigma & G_{a\mathbf{k}}^\sigma & G_{a\mathbf{k}'}^\sigma & G_{a\mathbf{k}''}^\sigma & \cdots \\ G_{\mathbf{k}a}^\sigma & G_{\mathbf{k}\mathbf{k}}^\sigma & G_{\mathbf{k}\mathbf{k}'}^\sigma & G_{\mathbf{k}\mathbf{k}''}^\sigma & \cdots \\ G_{\mathbf{k}'a}^\sigma & G_{\mathbf{k}'\mathbf{k}}^\sigma & G_{\mathbf{k}'\mathbf{k}'}^\sigma & G_{\mathbf{k}'\mathbf{k}''}^\sigma & \cdots \\ G_{\mathbf{k}''a}^\sigma & G_{\mathbf{k}''\mathbf{k}}^\sigma & G_{\mathbf{k}''\mathbf{k}'}^\sigma & G_{\mathbf{k}''\mathbf{k}''}^\sigma & \cdots \\ \vdots & \vdots & \vdots & \vdots & \ddots \end{pmatrix} = -I, \tag{6.41}$$

---

[1] There is notation overload here. I use $\hat{n}_{a\sigma}$ for the occupation number operator (following the usual convention), and $n_a^\sigma(E)$ for the local density of states (equation (6.46)). The energy-dependence of the density of states should make the difference clear.

where we assume that the orbitals are orthonormal, so that $O_{ij} = \delta_{ij}$. We can solve equation (6.41) by splitting the Hamiltonian,

$$H_\sigma = \begin{pmatrix} \epsilon_\sigma & 0 & 0 & 0 & \cdots \\ 0 & \epsilon_\mathbf{k} & 0 & 0 & \cdots \\ 0 & 0 & \epsilon_{\mathbf{k}'} & 0 & \cdots \\ 0 & 0 & 0 & \epsilon_{\mathbf{k}''} & \cdots \\ \vdots & \vdots & \vdots & \vdots & \ddots \end{pmatrix} + \begin{pmatrix} 0 & V_{a\mathbf{k}} & V_{a\mathbf{k}'} & V_{a\mathbf{k}''} & \cdots \\ V_{a\mathbf{k}}^* & 0 & 0 & 0 & \cdots \\ V_{a\mathbf{k}'}^* & 0 & 0 & 0 & \cdots \\ V_{a\mathbf{k}''}^* & 0 & 0 & 0 & \cdots \\ \vdots & \vdots & \vdots & \vdots & \ddots \end{pmatrix}, \tag{6.42}$$

and treating the second term as a perturbation δV on the Green function $\mathcal{G}$ of the first term, with diagonal matrix elements

$$\mathcal{G}_{aa} = \frac{1}{E - \epsilon_\sigma}, \quad \mathcal{G}_{\mathbf{k}\mathbf{k}} = \frac{1}{E - \epsilon_\mathbf{k}}, \quad \mathcal{G}_{\mathbf{k}'\mathbf{k}'} = \frac{1}{E - \epsilon_{\mathbf{k}'}}, \quad \text{etc.} \tag{6.43}$$

We then use Dyson's equation iterated once (compare with equation (6.30)),

$$G_{aa}^\sigma = \mathcal{G}_{aa} + \mathcal{G}_{aa} \sum_\mathbf{k} V_{a\mathbf{k}} \mathcal{G}_{\mathbf{k}\mathbf{k}} V_{a\mathbf{k}}^* G_{aa}^\sigma, \tag{6.44}$$

and substituting from equation (6.43) we obtain for the Green function on the adsorbate

$$G_{aa}^\sigma(E) = \left(E - \epsilon_\sigma - \sum_\mathbf{k} \frac{|V_{a\mathbf{k}}|^2}{E - \epsilon_\mathbf{k}} \right)^{-1}. \tag{6.45}$$

The term $\left(\sum_\mathbf{k} \frac{|V_{a\mathbf{k}}|^2}{E - \epsilon_\mathbf{k}} \right)$ embeds the adsorbate. This term has exactly the same structure as the embedding potential in equation (6.10), namely, a Green function sandwiched between hopping matrix elements.

The adsorbate density of states can be found from this Green function,

$$n_a^\sigma(E) = -\frac{1}{\pi} \Im G_{aa}^\sigma(E + i\eta), \tag{6.46}$$

from which the occupation number for each spin is given by

$$\langle \hat{n}_{a\sigma} \rangle = \int^{E_\mathrm{F}} \mathrm{d}E \, n_a^\sigma(E). \tag{6.47}$$

As the adatom energy level for each spin depends on the occupation number of the opposite spin via equation (6.39), $\langle \hat{n}_{a\sigma} \rangle$ has to be determined self-consistently, with the possibility of $\langle \hat{n}_{a\uparrow} \rangle \neq \langle \hat{n}_{a\downarrow} \rangle$ corresponding to a magnetic solution. This is particularly relevant when the Hamiltonian (6.36) is used in its original application: a magnetic impurity in a non-magnetic host [6].

Using this model the density of states can be written as

$$n_a^\sigma(E) = \frac{1}{2\pi} \frac{\Gamma(E)}{(E - \epsilon_\sigma - \Lambda(E))^2 + [\Gamma(E)/2]^2}, \tag{6.48}$$

where Λ and Γ are the real and $(-2\times)$ imaginary parts of the embedding potential,

$$\Lambda(E) = \Re \sum_{\mathbf{k}} \frac{|V_{a\mathbf{k}}|^2}{(E + i\eta) - \epsilon_{\mathbf{k}}} = \sum_{\mathbf{k}} \frac{|V_{a\mathbf{k}}|^2}{E - \epsilon_{\mathbf{k}}}, \tag{6.49}$$

$$\Gamma(E) = -2\Im \sum_{\mathbf{k}} \frac{|V_{a\mathbf{k}}|^2}{(E + i\eta) - \epsilon_{\mathbf{k}}} = 2\pi \sum_{\mathbf{k}} |V_{a\mathbf{k}}|^2 \delta(E - \epsilon_{\mathbf{k}}). \tag{6.50}$$

In our convention, the embedding potential has a negative imaginary part (equation (2.29)), and to bring it in line with the usual convention for self-energies we take Γ to be positive. Γ is the surface density of states in the region of the adsorbate, modulated by the adsorbate-substrate coupling term $|V_{a\mathbf{k}}|^2$, and from equations (6.49) and (6.50) we see that Λ and Γ are related by the Hilbert transform

$$\Lambda(E) = \frac{1}{2\pi} \mathcal{P} \int_{-\infty}^{+\infty} dE' \frac{\Gamma(E')}{E - E'}, \tag{6.51}$$

where $\mathcal{P}$ denotes the Cauchy principal value inside the integral. Λ and Γ are called the *chemisorption functions*: for small and relatively constant Γ (the wide-band limit) the atomic energy level is broadened into a Lorentzian of width Γ, shifted by Λ.

To see the range of chemisorption behaviour which can be described by this model we follow Newns [4], using a semi-elliptical form for $\Gamma(E)$ shown in figure 6.3 with $\Lambda(E)$ given by equation (6.51). These correspond to an embedding potential with constant $|V_{a\mathbf{k}}|^2$, and a Green function given by

$$\sum_{\mathbf{k}} \frac{1}{E - \epsilon_{\mathbf{k}}} = \frac{E \pm \sqrt{E^2 - W^2/4}}{W^2/8}, \tag{6.52}$$

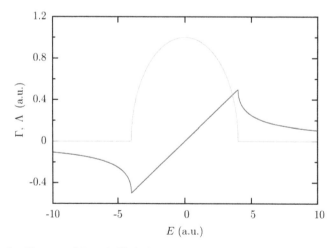

Figure 6.3. Grimley–Newns model: semi-elliptical $\Gamma(E)$ (green curve) with corresponding $\Lambda(E)$ (red curve), for a substrate bandwidth $W = 8$ a.u. and constant $|V_{a\mathbf{k}}|^2 = 1$ a.u.

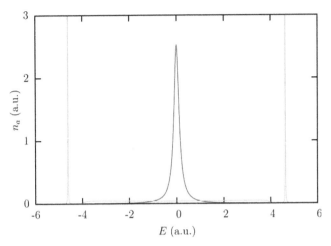

Figure 6.4. Grimley–Newns model: adsorbate density of states $n_a(E)$ for an adsorbate with $\epsilon_a = 0$ and a substrate with semi-elliptical $\Gamma(E)$, bandwidth $W = 8$ a.u. Red curve, weak coupling with $|V_{ak}|^2 = 0.25$ a.u.; green curve, strong coupling with $|V_{ak}|^2 = 16$ a.u.

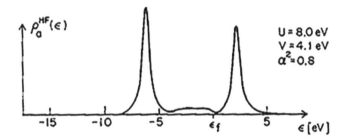

Figure 6.5. Adsorbate density of states for H chemisorbed on Ni. The parameters are fitted to a cluster calculation. (From Schönhammer [9].)

where W is the bandwidth, which we take equal to 8 a.u. The sign in front of the square root is chosen so that Γ is positive. This is, in fact, the Green function at the end of a tight-binding semi-infinite one-dimensional chain of atoms, with hopping equal to $W/4$. The adsorbate density of states $n_a(E)$, given by equation (6.46), is shown in figure 6.4 for an adsorbate with $\epsilon_a = 0$ (we do not consider the Hartree–Fock term), for two values of adsorbate-substrate coupling: the green curve, for $|V_{ak}|^2 = 0.25$ a.u., corresponds to weak coupling with a broadened adsorbate level, while the red curve, for $|V_{ak}|^2 = 16$ a.u., corresponds to strong coupling, with discrete bonding and anti-bonding levels outside the substrate band.

Many different applications of the Grimley–Newns model are given in the book by Davison and Sulston [7], mostly based on the one-dimensional chain model for the substrate for which equation (6.52) is the Green function; an example is the study of H chemisorption on disordered alloys surfaces [8]. However, we must emphasize that the Grimley–Newns model isn't restricted to the one-dimensional chain substrate—$\Gamma(E)$ can be fitted to the density of states of more realistic surfaces. An example of this is given in figure 6.5, which shows the adsorbate density of states for a H-atom chemisorbed on a Ni surface, with parameters fitted to a cluster calculation [9]. This has similar structure to the green curve in figure 6.4, the bonding and antibonding states being broadened by interaction with the Ni bands. And the three-dimensional substrate is inherent, in one of the earliest applications of the model, to the indirect interaction of adsorbed atoms via the substrate electrons [10].

6.3 Finite differences and tight-binding

By discretizing space and using finite-difference methods, the continuum Hamiltonian can be transformed to tight-binding form. In this section, we shall use this to find the relationship between the real space representation of the embedding potential, $\Sigma(\mathbf{r}_S, \mathbf{r}'_S)$, and the LCAO representation Σ_{11}. One of the apparent differences between the expressions for the embedding potential is that in equation (2.17) $\Sigma(\mathbf{r}_S, \mathbf{r}'_S)$ is given by the surface inverse of the Green function in region II, whereas in equation (6.10) the expression for Σ_{11} contains $\mathcal{G}_{22}$ without any inversion. We shall resolve this conundrum using Green function results.

6.3.1 Discretizing the continuum

In finite-difference methods, the Hamiltonian, wave-functions, and Green function are discretized on a grid, which we take to be simple cubic with interval d. The Schrödinger equation then becomes

$$\sum_j H_{ij}\Psi(\mathbf{r}_j) = E\Psi(\mathbf{r}_i), \qquad (6.53)$$

where the discretized Hamiltonian is given in three dimensions by the nearest-neighbour tight-binding form [11],

$$H_{ij} = \begin{cases} -1/2d^2, & \text{for } i \text{ and } j \text{ nearest-neighbour points} \\ 3/d^2 + V(\mathbf{r}_i), & \text{for } i = j \\ 0 & \text{otherwise.} \end{cases} \qquad (6.54)$$

For the overlaps in the tight-binding formalism we take $O_{ij} = \delta_{ij}$.

The discretization points of regions I and II for an interface problem are shown schematically in figure 6.6; here, we *can* define the embedding surface S, with the light green points belonging to group 1 lying on S. The tight-binding form of the Hamiltonian means that all the results in section 6.1 can be applied, so the embedding potential is given in matrix form by substituting equation (6.54) into equation (6.10) to give

$$\Sigma_{1'1''} = \frac{1}{4d^4}\mathcal{G}_{2'2''}, \qquad (6.55)$$

where $1'$ and $1''$ are distinct points in group 1 on the embedding surface, and $2'$ and $2''$ their nearest neighbours in group 2. As before, $\mathcal{G}$ is the Green function for region II decoupled from region I.

This result can be related to the continuum embedding we derived in chapter 2 (we note that the discretized Hamiltonian given by equation (6.54) is already locally

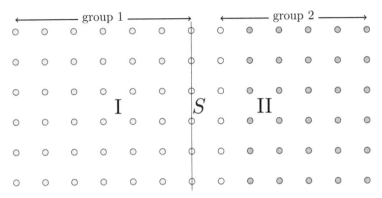

Figure 6.6. Schematic of discretization. Green dots represent points in region I (group 1), and brown dots points in region II (group 2). Light green dots in group 1 which lie on the embedding surface S are coupled to light brown points in group 2.

Hermitian, so combines the first two terms in equation (2.26)). In a continuum picture, $\mathcal{G}$ corresponds to the Green function for region II satisfying a zero-amplitude boundary condition on S. However, $\mathcal{G}$ satisfies the matrix equation (6.6), with a unit vector on the right-hand side, whereas the continuum Green function satisfies the inhomogeneous Schrödinger equation with $\delta(\mathbf{r} - \mathbf{r}')$ on the right. As the Dirac delta function becomes δ_{ij}/d^3 in the discrete representation, the continuum Green function satisfying the zero-amplitude boundary condition is related to $\mathcal{G}$ by

$$\hat{G}_0(\mathbf{r}_i, \mathbf{r}_j) = \frac{1}{d^3}\mathcal{G}_{ij}. \tag{6.56}$$

We use $\hat{G}_0$ for the Green function with zero amplitude on S, and G_0 for the zero-derivative Green function that appears in our original form of the embedding potential (equation (2.17)). So, the matrix embedding potential can be written as

$$\Sigma_{1'1''} = \frac{1}{4d}\hat{G}_0(\mathbf{r}_{2'}, \mathbf{r}_{2''}). \tag{6.57}$$

We now want to convert equation (6.57) into an embedding potential to use in the continuum form of the Schrödinger equation. In the discretized Schrödinger equation the embedding potential appears in a sum over the points in group 1 which lie on S, whereas in the continuum equation we have the integral in equation (2.23). These must be the same, so

$$\delta(n - n_S)\int_S \mathrm{d}\mathbf{r}'_S\, \Sigma_{\mathrm{cont}}(\mathbf{r}_S, \mathbf{r}'_S)\phi(\mathbf{r}'_S) = \sum_{1''}\Sigma_{1'1''}\phi(\mathbf{r}_{1''}). \tag{6.58}$$

(We have added the subscript in Σ_{cont} to emphasise that the continuum embedding potential is not identical to the matrix Σ.) But we can discretize the left-hand side of this equation, the one-dimensional delta function becoming $\frac{1}{d}$ and the two-dimensional integral giving a factor d^2 times the summation, hence

$$\delta(n - n_S) \int_S \mathrm{d}\mathbf{r}'_S \, \Sigma_{\text{cont}}(\mathbf{r}_S, \mathbf{r}'_S)\phi(\mathbf{r}'_S) \approx \left(\frac{1}{d} \times d^2\right) \sum_{1''} \Sigma_{\text{cont}}(\mathbf{r}_{1'}, \mathbf{r}_{1''})\phi(\mathbf{r}_{1''}). \quad (6.59)$$

Comparing with equation (6.58) and substituting equation (6.57) we then obtain

$$\Sigma_{\text{cont}}(\mathbf{r}_{1'}, \mathbf{r}_{1''}) = \frac{1}{d}\Sigma_{1'1''} = \frac{1}{4d^2}\hat{G}_0(\mathbf{r}_{2'}, \mathbf{r}_{2''}). \quad (6.60)$$

Now, because $\hat{G}_0$ is zero on S, $\frac{1}{d^2}\hat{G}_0(\mathbf{r}_{2'}, \mathbf{r}_{2''}) \approx \partial^2\hat{G}_0(\mathbf{r}_S, \mathbf{r}'_S)/\partial n_S \partial n'_S$, so the continuum representation of the embedding potential can be written as

$$\Sigma(\mathbf{r}_S, \mathbf{r}'_S) = \frac{1}{4}\frac{\partial^2\hat{G}_0(\mathbf{r}_S, \mathbf{r}'_S)}{\partial n_S \partial n'_S} \quad (6.61)$$

(we can now drop the subscript on Σ, because it is obvious from its arguments that this is referring to the continuum). We have an expression for the embedding potential which, like the matrix representation, just involves $\hat{G}_0$ *without any surface inverse*.

This result was first obtained by Fisher [12] using Green's theorem results. Substituting $\hat{G}_0$ for G_0 in equation (2.11), and using Green's theorem to replace the integral through region II by a surface integral we obtain

$$\psi(\mathbf{r}) = -\frac{1}{2}\int_S \mathrm{d}\mathbf{r}_S \frac{\partial\hat{G}_0(\mathbf{r}, \mathbf{r}_S)}{\partial n_S}\psi(\mathbf{r}_S). \quad (6.62)$$

Given $\psi(\mathbf{r}_S)$, the surface value of a solution of the Schrödinger equation in II, equation (6.62) gives $\psi(\mathbf{r})$ in the interior of II. (This sort of equation is familiar to us from electrostatics [13].) Letting $\mathbf{r}$ approach S and differentiating gives

$$\frac{\partial\psi(\mathbf{r}_S)}{\partial n_S} = -\frac{1}{2}\int_S \mathrm{d}\mathbf{r}'_S \frac{\partial^2\hat{G}_0(\mathbf{r}_S, \mathbf{r}'_S)}{\partial n_S \partial n'_S}\psi(\mathbf{r}'_S), \quad (6.63)$$

and comparing with equation (2.15) we see that

$$G_0^{-1}(\mathbf{r}_S, \mathbf{r}'_S) = -\frac{1}{4}\frac{\partial^2\hat{G}_0(\mathbf{r}_S, \mathbf{r}'_S)}{\partial n_S \partial n'_S}. \quad (6.64)$$

From equation (2.17) this is equivalent to equation (6.61).

6.3.2 Embedding muffin-tins into a finite-difference grid

Finite-difference methods, with their simple tight-binding structure, offer a powerful way of solving the one-electron Schrödinger equation, especially in systems with awkward geometry for which a plane-wave basis, for example, may not be

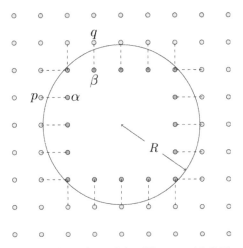

Figure 6.7. Embedding a muffin-tin, radius R, into a finite-difference grid. Grid points just outside the muffin-tin are denoted by p, q, and these are coupled to points α, β just inside the muffin-tin.

appropriate. However, the finite-difference grid must always be much shorter than the wavelength of the solutions of the Schrödinger equation, and this leads to difficulties inside the cores of atoms, for example, where the wavelength is very short. One common approach to this problem is to replace the deep potential inside the atom by a weaker pseudopotential [14]. The embedding solution suggested by Thijssen and Inglesfield [11] is to use standard numerical techniques for solving the Schrödinger equation inside the atomic muffin-tins, and embed these solutions into a finite-difference grid used outside the muffin-tins. (It is outside the muffin-tins that the flexibility of the finite-difference method is most useful.) The grid constitutes region I, and the muffin-tins become region II, replaced by an embedding potential.

Let us consider the solution of the Schrödinger equation at energy E throughout the whole space, $\Psi(\mathbf{r})$. On the finite-difference grid points outside the muffin-tin (the green points in figure 6.7), Ψ is represented by the discrete values ψ_i,

$$\psi_i = \Psi(\mathbf{r}_i). \tag{6.65}$$

Inside the muffin-tin, and up to the first shell of points just outside, we use an expansion in terms of atomic-like solutions $\chi_L(\mathbf{r})$,

$$\Psi(\mathbf{r}) = \sum_L a_L \chi_L(\mathbf{r}). \tag{6.66}$$

L labels the different atomic solutions at fixed energy, and with a spherically symmetric potential inside the muffin-tin it corresponds to the angular momentum quantum numbers, with $L = (l, m)$; the general solution inside the muffin-tin is a linear combination of the χ_L. Over points $\mathbf{r}_p$ on the first shell outside (figure 6.7), we can use both representations, so

$$\psi_p = \sum_L a_L \chi_L(\mathbf{r}_p). \tag{6.67}$$

To find the expansion coefficients a_L in equations (6.66) and (6.67) we first define an inner product across the muffin-tin sphere boundary,

$$\langle \phi | \psi \rangle_{\mathrm{MT}} := \sum_{(\alpha, p)} \phi_\alpha \psi_p, \tag{6.68}$$

where (α, p) denotes a link across the muffin-tin boundary (indicated by the dashed lines in figure 6.7). Then multiplying each side of equation (6.67) by $\chi_{L'}(\mathbf{r}_\alpha)$ and summing over the (α, p) links to give our inner products, we obtain

$$\sum_L \langle \chi_{L'} | \chi_L \rangle_{\mathrm{MT}} \, a_L = \left\langle \chi_{L'} \middle| \psi_p \right\rangle_{\mathrm{MT}}. \tag{6.69}$$

We next define the overlap of the χ_L functions as

$$S_{LL'} := \langle \chi_L | \chi_{L'} \rangle_{\mathrm{MT}}, \tag{6.70}$$

where the inner product is again defined as in equation (6.68). The solution of equation (6.69) is then given by

$$a_L = \sum_{L'} S_{LL'}^{-1} \left\langle \chi_{L'} \middle| \psi_p \right\rangle_{\mathrm{MT}}, \tag{6.71}$$

and the solution of the Schrödinger equation at point $\mathbf{r}_\beta$ (figure 6.7) is

$$\Psi(\mathbf{r}_\beta) = \sum_{L,L'} \chi_L(\mathbf{r}_\beta) S_{LL'}^{-1} \left\langle \chi_{L'} \middle| \psi_p \right\rangle_{\mathrm{MT}}. \tag{6.72}$$

If we now consider the discretized Schrödinger equation (6.53), evaluated at point $\mathbf{r}_q$ (that is, with $\Psi(\mathbf{r}_q)$ on the right-hand side), the contribution to the left-hand side from the (β, q) link becomes

$$H_{q\beta} \Psi(\mathbf{r}_\beta) = -\frac{1}{2d^2} \Psi(\mathbf{r}_\beta) = -\frac{1}{2d^2} \sum_{L,L'} \chi_L(\mathbf{r}_\beta) S_{LL'}^{-1} \left\langle \chi_{L'} \middle| \psi_p \right\rangle_{\mathrm{MT}}$$

$$= -\frac{1}{2d^2} \sum_{(\alpha,p)} \sum_{L,L'} \chi_L(\mathbf{r}_\beta) S_{LL'}^{-1} \chi_{L'}(\mathbf{r}_\alpha) \psi_p. \tag{6.73}$$

This has the form of a non-local term coupling q and p, replacing the coupling of q with its neighbour β inside the muffin-tin—it is an embedding term. Taking into account all the possible links of p and q across the muffin-tin boundary, we have a non-local embedding contribution to the Hamiltonian linking these points given by

$$\Sigma_{pq} = -\frac{1}{2d^2} \sum_{\alpha(p)} \sum_{\beta(q)} \sum_{L,L'} \chi_L(\mathbf{r}_\alpha) S_{LL'}^{-1} \chi_{L'}(\mathbf{r}_\beta), \tag{6.74}$$

where $\alpha(p)$ means all the points α just inside the muffin-tin which are linked to p as nearest neighbours. This is the embedding potential which replaces the muffin-tin.

An undesirable feature of this embedding potential is that $S_{LL'}^{-1}$ is not a symmetric matrix, as we see from the definition of the inner product entering equation (6.70). (We are assuming that the χ_L functions are real—with complex functions we would be dealing with a non-Hermitian $S_{LL'}^{-1}$ matrix.) The non-symmetric part of $S_{LL'}^{-1}$ is small, proportional to d, and in practice we simply symmetrize the matrix.

The energy-dependence of the χ_L means that the embedding potential (equation (6.74)) is (as usual) energy-dependent. So the Schrödinger equation has to be solved self-consistently by evaluating Σ_{pq} at a trial energy E_0, calculating the eigenvalue and iterating. However, we can proceed as in chapter 2 by linearizing the embedding potential about E_0. The linear correction can then be treated as a perturbation in the iteration scheme. An alternative approach, used by Komine and Shiiki [15], is to linearize the energy-dependence of the χ_L functions themselves about an L-dependent pivot energy ϵ_L, writing

$$\chi_L(\mathbf{r}; E) \approx \chi_L(\mathbf{r}; \epsilon_L) + (E - \epsilon_L)\dot{\chi}_L(\mathbf{r}; \epsilon_L). \tag{6.75}$$

This is the linearization technique which is used to construct the LAPW basis functions described in section 3.2.1 (equation (3.19)); it eliminates energy-iteration, though at the expense of using four overlaps like equation (6.70) involving $\dot{\chi}_L$ as well as χ_L.

An important aspect of finite-difference methods is that the Hamiltonian matrix is sparse, so that the eigenvalues and eigenvectors can be found efficiently, even with very large systems with many thousands of grid points. At first it seems that the form of the embedding Hamiltonian (6.74) compromises the sparseness. However, Σ_{pq} is *separable*, and by finding the a_L coefficients explicitly using equation (6.71) an efficient $\mathcal{O}(N)$ scheme for finding the eigenvalues and eigenvectors has been developed [16]. Knowing the a_L coefficients means that we can immediately find the wave-function inside the muffin-tins (equation (6.66)), a necessary step for going to charge density self-consistency in this scheme.

The first tests of the method, to show that it is viable and works in practice, were band-structure calculations using self-consistent potentials taken from KKR calculations [11]. As an example, figure 6.8 shows the band structure for fcc Cu calculated along the symmetry directions in the Brillouin zone using a muffin-tin form of potential (spherically symmetric inside the muffin-tins, flat in the interstitial region), with a maximum value of $l = 3$ in the χ_L expansion inside the muffin-tins. With a grid spacing of $d = 0.03a$ (a is the lattice constant, 6.822 a.u.), excellent agreement is obtained with the KKR band structure using the same potential. The errors are consistent with the error bars shown in the Γ-X segment of the band structure, taken from an empty lattice test with values of d ranging from $0.03a$ to $0.07a$.

In calculations which show the flexibility of the method, Komine and Shiiki [15] calculated the band structure of metallic H in a hypothetical fcc lattice and the electronic structure of the H_2 molecule. Both these calculations were fully self-consistent, with no shape approximations to the potential; it is particularly straightforward to include the full potential in the finite-difference method, as we see from the form of the Hamiltonian (6.54). In the H_2 calculation, they took a grid

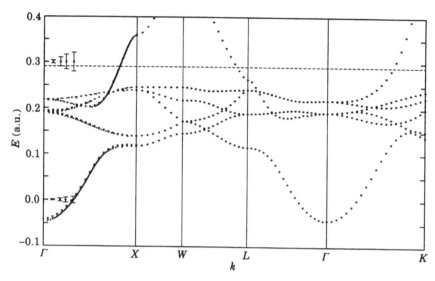

Figure 6.8. Band structure for fcc Cu, lattice constant $a = 6.822$ a.u., calculated using embedded muffin-tins, radius $R = 2.412$ a.u., grid spacing $d = 0.3a$; a maximum $l = 3$ is used in the χ_L expansion. In the Γ-X direction results from different grid spacings are superimposed and the continuous line shows the KKR band structure. Also shown in this segment are estimated errors for d between $0.03a$ and $0.07a$. (From Thijssen and Inglesfield [11].)

spacing $d = 0.25$ a.u. and a muffin-tin sphere radius $R = 0.5$ a.u.; they obtained an equilibrium interatomic spacing of 1.44 a.u., in excellent agreement with other calculations using the LDA in the local-density exchange-correlation scheme (the experimental interatomic distance is 1.40 a.u.).

Joly [17] has independently developed a similar scheme, using muffin-tins embedded into a finite-difference grid to calculate wave-functions for interpreting spectroscopies, in particular x-ray absorption near-edge structure (XANES). In XANES experiments a core-level electron is excited into an unoccupied state with an energy just above the Fermi energy using synchrotron radiation; the probability of this excitation, hence of absorbing the x-ray photon, depends critically on the form of the wave-function of the excited electron—how it is reflected from the atoms around the absorbing atom—and hence contains geometric information on the material being studied. Particularly in open structures, it is important to include the variation of the potential between atoms, and at the same time the region close to the core must be treated accurately. For these reasons, the finite-difference method with an embedded muffin-tin sphere is very appropriate. As an example of these calculations, figure 6.9 shows the Fe K-edge XANES spectra in the protein carbon monoxymyoglobin (MbCO), in which an electron is excited from the innermost core level of the Fe atom, at the centre of the square representing coordinating ligand atoms in the inset. The solid line shows the x-ray absorption spectra, for different polarizations, calculated using Joly's finite-difference method [17], with small muffin-tins embedded into a grid (solid line), compared with experiment (dashed line). The dotted line shows results from a multiple scattering calculation in

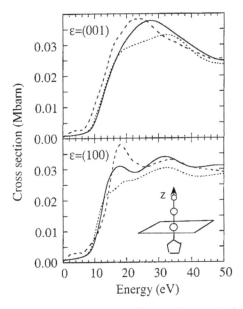

Figure 6.9. Fe K-edge XANES spectra in the protein MbCO, for different polarizations of the exciting x-rays. The full line shows the finite-difference calculation, the dotted line a multiple scattering calculation using a muffin-tin potential (constant between the atoms), and the dashed line experimental results. The inset shows a schematic of the molecule, the Fe atom being in the middle of the square which represents surrounding ligands. (From Joly [17].)

which a muffin-tin potential is used—a spherically symmetric potential inside large muffin-tins with a flat potential in between. Clearly, the finite-difference calculation is superior.

Embedding a muffin-tin into a finite-difference grid is an example of a *finite* region II—using embedding to replace a limited region of space, rather than an extended region or substrate. We shall come across other examples of this in chapter 9, where we use embedding to solve Maxwell's equations in dielectric systems. Of course this method is not restricted to embedding a sphere into the grid, and it could be used in essentially the same form to deal with region II of arbitrary shape or indeed an extended region II; all that we need is the set of solutions of the Schrödinger equation at energy E, analogous to the χ_L (equation (6.66)). One nagging question which might worry the reader is: how is the normal derivative of Ψ in region II, so important as we saw in chapter 2, built into this method? The answer is that the inner product which enters the overlap $S_{LL'}$ (equation (6.70)) involves χ_L at grid points on both sides of the matching surface, and it is the change across the surface which corresponds to the normal derivative.

6.4 LCAO codes for the self-energy

As the LCAO self-energy is so important for studies of transport through molecules [1], replacing the metal contacts, there has been much effort to develop efficient computer programs to calculate Σ. Among the codes for calculating transport through molecules are TranSIESTA [18] and SMEAGOL [19], both based on the SIESTA electronic structure code [20], which uses local orbitals as basis functions. Other codes are based on BAND, the band structure package in the Amsterdam Density Functional programs [21].

The usual model for treating transport is represented in the upper part of figure 6.10, with the molecule (the blue circles bonded by solid lines) connected to semi-infinite metallic leads L and R. To the left and right of planes S_L and S_R, respectively, the leads are treated as perfect bulk, and are replaced by self-energies Σ_L and Σ_R. As shown in the lower part of the figure, the top layers of the leads are usually included with the molecule itself to constitute the 'extended molecule', the orbitals of which we label m. The extended molecule is embedded into the bulk leads by adding Σ_L and Σ_R to its Hamiltonian. The Green function for the whole system is then given by

$$G_{mm} = -(H_{mm} + \Sigma_L + \Sigma_R - EO_{mm})^{-1}, \qquad (6.76)$$

where G_{mm}, H_{mm} and O_{mm} are square matrices within the space of orbitals m. The extended molecule is treated explicitly, going to self-consistency and the electron transport properties evaluated.

We shall study transport in chapter 7, and for the moment we restrict ourselves to a discussion of the self-energies. With the geometry shown in figure 6.10, Σ_L and Σ_R are exactly analogous to the surface embedding potentials described in chapters 3,

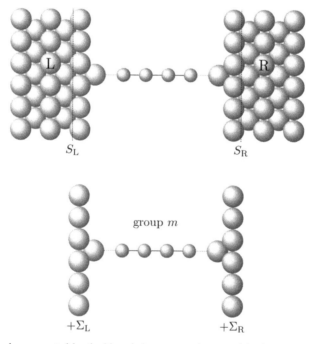

Figure 6.10. Molecule represented by the blue circles connected to metal leads L and R. The top layers of metal, to the left and right of S_L and S_R, respectively, (upper figure) are included with the molecule to constitute an 'extended molecule' (lower figure). The orbitals in the extended molecule constitute group m, and the self-energies Σ_L and Σ_R added to the Hamiltonian in the space of m replace the semi-infinite bulk leads.

but in a local orbital representation. Let us suppose that each layer of atoms to the right of S_R in figure 6.10 only interacts with neighbouring layers (this means that what we call a layer[2] may really consist of several atomic planes). Then, from our work in section 6.1, to construct Σ_R we need $\mathcal{G}_R$, the Green function of the semi-infinite system to the right of S_R decoupled from the layers to the left (equation (6.6)). As all the layers to the right of S_R are in a bulk environment (this is what we assume when we construct the extended molecule), $\mathcal{G}_R$ is a property of the bulk.

6.4.1 Recursion

There are—needless to say—several approaches to constructing $\mathcal{G}_R$ [18]. One approach is a recursion technique [22], starting from the block tri-diagonal structure of the matrix whose inverse gives this Green function

$$\begin{pmatrix} H - EO & h - Eo & 0 & \cdots \\ h - Eo & H - EO & h - Eo & \cdots \\ 0 & h - Eo & H - EO & \cdots \\ \vdots & \vdots & \vdots & \ddots \end{pmatrix} \begin{pmatrix} \mathcal{G}_{00} & \mathcal{G}_{01} & \mathcal{G}_{02} & \cdots \\ \mathcal{G}_{10} & \mathcal{G}_{11} & \mathcal{G}_{12} & \cdots \\ \mathcal{G}_{20} & \mathcal{G}_{21} & \mathcal{G}_{22} & \cdots \\ \vdots & \vdots & \vdots & \ddots \end{pmatrix} = -I. \qquad (6.77)$$

H and O are block matrices for the Hamiltonian and overlap within a principal layer, with h and o coupling neighbouring layers, and E is the complex energy with a positive imaginary part (which can be very small). For simplicity, we assume that the Hamiltonian and overlap matrices are real. The submatrix $\mathcal{G}_{00}$ gives $\mathcal{G}_R$ on principal layer 0 immediately to the right of S_R, with increasing indices corresponding to going deeper into the bulk of the right-hand lead. This matrix equation then gives us a sequence of coupled equations for the first column of $\mathcal{G}_R$,

$$\alpha \mathcal{G}_{00} + b \mathcal{G}_{10} = I, \qquad (6.78)$$

$$b \mathcal{G}_{00} + a \mathcal{G}_{10} + b \mathcal{G}_{20} = 0, \qquad (6.79)$$

where $\alpha = a = EO - H$, $b = Eo - h$. The general equation in this sequence is

$$b \mathcal{G}_{n-1,0} + a \mathcal{G}_{n0} + b \mathcal{G}_{n+1,0} = 0, \quad n \geqslant 1. \qquad (6.80)$$

From equations (6.78) and (6.79) we obtain

$$(\alpha - ba^{-1}b) \mathcal{G}_{00} - ba^{-1}b \mathcal{G}_{20} = I, \qquad (6.81)$$

and substituting for $\mathcal{G}_{n-1,0}$ and $\mathcal{G}_{n+1,0}$ from the equations on either side of equation (6.80) gives

$$-ba^{-1}b \mathcal{G}_{n-2,0} + (a - 2ba^{-1}b) \mathcal{G}_{n0} - ba^{-1}b \mathcal{G}_{n+2,0} = 0. \qquad (6.82)$$

In this way, we have replaced equations (6.78)–(6.80), which couple $\mathcal{G}_{00}, \mathcal{G}_{10}, \mathcal{G}_{20},...,$ by the sequence of equations (6.81) and (6.82) coupling $\mathcal{G}_{00}, \mathcal{G}_{20}, \mathcal{G}_{40}$, etc. Moreover,

---

[2] Sometimes called a *principal layer*.

the structure of equations (6.81) and (6.82) is the same as equations (6.78) and (6.80), with the substitutions

$$\alpha \to (\alpha - ba^{-1}b), \qquad a \to (a - 2ba^{-1}b), \qquad b \to -ba^{-1}b. \qquad (6.83)$$

We can repeat this process, and after n iterations the first equation in the sequence has the form

$$\alpha^{(n)}\mathcal{G}_{00} + b^{(n)}\mathcal{G}_{2^n,0} = I. \qquad (6.84)$$

Now, the imaginary part of E ensures that the Green function decays with distance, so at large n we can neglect the off-diagonal element $\mathcal{G}_{2^n,0}$, and we have

$$\mathcal{G}_{00} \approx (\alpha^{(n)})^{-1}. \qquad (6.85)$$

Then from equation (6.10) the self-energy on the plane immediately to the left of S_R is given by

$$\Sigma_R(E) = (h - Eo)\mathcal{G}_{00}(E)(h - Eo). \qquad (6.86)$$

This procedure for finding the self-energy is just the LCAO version of the layer-doubling method for finding the reflection matrix, hence the embedding potential, which we described in section 3.1.1.

6.4.2 Summing over bulk states

The Green function $\mathcal{G}_R$ for the semi-infinite system can be constructed from the Green function G for the lead extending to infinity in both directions [23]. This can be found from a sum over bulk eigenstates. Fixing the wave-vector $\mathbf{k}_\parallel$ parallel to the layers (which also applies to the recursion technique), eigenvector Ψ^i with eigenvalue ϵ_i satisfies the coupled Schrödinger equations of a linear chain,

$$\begin{aligned}
(h - \epsilon_i o)\Psi^i_{n-1} + (H - \epsilon_i O)\Psi^i_n + (h - \epsilon_i o)\Psi^i_{n+1} &= 0 \\
(h - \epsilon_i o)\Psi^i_n + (H - \epsilon_i O)\Psi^i_{n+1} + (h - \epsilon_i o)\Psi^i_{n+2} &= 0, \text{ etc.}
\end{aligned} \qquad (6.87)$$

Here, Ψ^i_n is a column vector giving the amplitude of the N orbitals in the nth principal layer, and H, h, O and o are $N \times N$ matrices. The periodic structure of equation (6.87) means that the eigenvectors are Bloch functions and can be labelled by the (dimensionless) wave-vector k_z, such that

$$\Psi^i_{n,k_z} = \exp(ik_z)\Psi^i_{n-1,k_z}, \qquad (6.88)$$

where positive k_z gives a wave travelling to the right. Substituting equation (6.88) into equation (6.87) we obtain the generalized eigenvalue equation for $\Psi^i_{k_z}$ (we drop the slab index n) and $\epsilon^i_{k_z}$,

$$(H + 2h\cos k_z)\Psi^i_{k_z} = \epsilon^i_{k_z}(O + 2o\cos k_z)\Psi^i_{k_z}, \qquad (6.89)$$

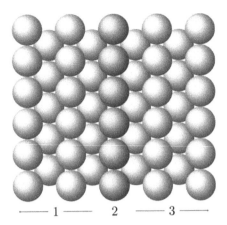

Figure 6.11. The 'ideal' construction. Green and red layers together represent the perfect bulk material, but the red layer (2) is removed by Green function methods to leave separated green layers (groups 1 and 3), each with an 'ideal' surface.

which for each value of k_z has N solutions, labelled by i. The bulk Green function between slabs m and n is then given by

$$G_{mn}(E) = \sum_{k_z} \exp[ik_z(m-n)] \sum_{i=1}^{N} \frac{\Psi_{k_z}^i \left(\Psi_{k_z}^i\right)^\dagger}{E - \epsilon_{k_z}^i}. \tag{6.90}$$

In the numerator the outer product of the column vector $\Psi_{k_z}^i$ and the row vector $(\Psi_{k_z}^i)^\dagger$ gives an $N \times N$ matrix, as we require.

We can now use the so-called 'ideal' construction to find $\mathcal{G}_R$ from G by removing a principal layer [23]. This is shown in figure 6.11: by removing the red principal layer (layer 2) from the bulk, the green layers (groups 1 and 3) are disconnected, each with an 'ideal' surface. The bulk Green function satisfies the equation

$$\begin{pmatrix} H_{11} - EO_{11} & h - Eo & 0 \\ h - Eo & H - EO & h - Eo \\ 0 & h - Eo & H_{33} - EO_{33} \end{pmatrix} \begin{pmatrix} G_{11} & G_{12} & G_{13} \\ G_{21} & G_{22} & G_{23} \\ G_{31} & G_{32} & G_{33} \end{pmatrix} = -I, \tag{6.91}$$

where H_{11}, O_{11}, H_{33} and O_{33} are the Hamiltonian and overlap matrices for the layers in groups 1 and 3, respectively, and H, O, h and o have their previous meaning. This gives

$$(h - Eo)G_{22} + (H_{33} - EO_{33})G_{32} = 0, \tag{6.92}$$

$$(h - Eo)G_{23} + (H_{33} - EO_{33})G_{33} = -I, \tag{6.93}$$

and substituting $(h - Eo)$ from equation (6.92) into equation (6.93) we obtain

$$(H_{33} - EO_{33})[G_{33} - G_{32}(G_{22})^{-1}G_{23}] = -I. \tag{6.94}$$

But $\mathcal{G}_R$ is the inverse of $-(H_{33} - EO_{33})$, so we can write

$$\mathcal{G}_R = G_{33} - G_{32}(G_{22})^{-1}G_{23}, \tag{6.95}$$

from which we can obtain the self-energy. This expression, giving a Green function with one boundary condition in terms of the Green function satisfying a different boundary condition, is reminiscent of the matching Green function method [24] for the continuum case.

6.4.3 Fixing the energy

Instead of using the sum over states (equation (6.90)), the self-energy can be built up from the solutions of the Schrödinger equation at the energy E at which we are working [25, 26]. We have already used this in section 3.6, where the embedding potential was found from the solutions Ψ_i and the normal derivatives Ψ'_i (equation (3.55)), and in section 6.3.2 where the external grid was embedded on to solutions of the atomic Schrödinger equation over the surface of the muffin-tin. Indeed, this idea is inherent in the classic way of solving the Schrödinger equation in a composite system by wave-function matching.

We can construct a matrix eigenvalue equation for the bulk states at energy E, analogous to equation (3.51), in which the Bloch phase factor $\exp(\mathrm{i}k_z)$ is the eigenvalue. From equations (6.87) and (6.88) we have

$$\left[\exp\left(-\mathrm{i}k_z^i\right)(h - Eo) + (H - EO) + \exp\left(\mathrm{i}k_z^i\right)(h - Eo)\right]\Psi^i = 0, \qquad (6.96)$$

where i labels the column vector Ψ^i of length N. This hasn't the required eigenvalue structure, but if we define an auxiliary column vector $\bar{\Psi}^i$,

$$\bar{\Psi}^i := \exp\left(-\mathrm{i}k_z^i\right)\Psi^i, \qquad (6.97)$$

we can rewrite equation (6.96) combined with equation (6.97) as an eigenvalue equation,

$$\begin{pmatrix} -(h - Eo)^{-1}(H - EO) & -1 \\ 1 & 0 \end{pmatrix}\begin{pmatrix} \Psi^i \\ \bar{\Psi}^i \end{pmatrix} = \exp\left(\mathrm{i}k_z^i\right)\begin{pmatrix} \Psi^i \\ \bar{\Psi}^i \end{pmatrix}. \qquad (6.98)$$

This matrix equation, double the size of the original equation (6.96), has $2N$ eigenvalues, which come in pairs: $\exp(\mathrm{i}k_z^i)$ and $\exp(-\mathrm{i}k_z^i)$. As we saw in section 3.6, at real energy in general there are not only eigenvalues with real k_z^i, corresponding to the allowed energy bands, but also complex k_z^i, corresponding to exponentially decaying or increasing solutions, forbidden in the extended solid. In the semi-infinite solid to the right of S_R (figure 6.10), the solutions we need to build up the Green function are the N eigenstates travelling or decaying to the right, with $\Re k_z^i > 0$ and $\Im k_z^i > 0$ (for E with a positive imaginary part, both of these conditions are satisfied simultaneously). This actually gives us the *retarded* Green function and self-energy— that is what we use in our discussion of static properties like the density of states—but we shall have to consider generalizations when we come to transport properties in chapter 7.

The Green function $\mathcal{G}_{00}$ and self-energy Σ_R can now be written in terms of the states allowed in the semi-infinite system. Let's assume that the principal layer immediately to the left of S_R, layer -1 in our numbering convention, is the same as in the bulk material, with the same Hamiltonian and overlap matrix elements (we can always insert an extra layer if necessary). This is the layer on which we define the self-energy Σ_R (figure 6.10). Then a possible solution of the Schrödinger equation starting at layer -1 and proceeding to the right is

$$\text{layer} - 1 \quad \Psi(-1) = \exp\left(-ik_z^i\right)\Psi^i$$

$$\text{layer } 0 \quad \Psi(0) = \Psi^i$$

$$\text{layer } 1 \quad \Psi(1) = \exp\left(ik_z^i\right)\Psi^i, \text{ etc.}$$

As in equation (6.7) we can write $\Psi(0)$ in terms of $\Psi(-1)$ and the (as-yet unknown) $\mathcal{G}_{00}$, giving

$$\Psi^i = \mathcal{G}_{00}(h - Eo)\exp\left(-ik_z^i\right)\Psi^i$$

$$= \mathcal{G}_{00}(h - Eo)\bar{\Psi}^i. \tag{6.99}$$

This is true for all the N solutions Ψ^i, so if we define the matrices Φ_{ij} and $\bar{\Phi}_{ij}$,

$$\Phi_{ij} := \Psi_i^j, \quad \bar{\Phi}_{ij} = \bar{\Psi}_i^j, \tag{6.100}$$

where j runs over the N eigenvectors and i over the N components of each eigenvector, equation (6.99) becomes

$$\Phi = \mathcal{G}_{00}(h - Eo)\bar{\Phi}. \tag{6.101}$$

So the self-energy on the layer to the left of S_R is given by

$$\Sigma_R = (h - Eo)\Phi\bar{\Phi}^{-1}, \tag{6.102}$$

an expression with similar structure to equation (3.55).

6.4.4 Applications

The main applications of self-energy calculations are to transport (chapter 7), but the densities of states of molecules attached to metal contacts have been calculated *en route*, and we shall look at some results in this section.

Figure 6.12 shows density of states results from transport calculations on an H_2 molecule attached to Pt and Pd electrodes [27]. The upper curves show the densities of states on the extended molecule shown in the inset, consisting of H_2 plus a metal atom on each end, and the lower curves show the densities of states inside the electrodes (actually, the results show the *partial* densities of states corresponding to one conduction channel, but this is irrelevant to this discussion). In the case of the Pt/H_2/Pt system, the density of states on the molecule consists of a plateau between about $-1.1eV$ and $+0.8eV$, between two peaks at either end, and a broad peak at $-2.3eV$. The plateau corresponds to a state on H_2 with intermediate coupling to the

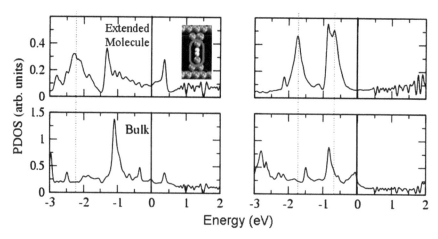

Figure 6.12. Densities of states for an H_2 molecule attached to Pt electrodes (left-hand side) and Pd electrodes (right-hand side). The upper figures show the density of states on the extended molecule, shown in the inset, and the lower figure the density of states in the bulk electrode. (From Khoo *et al* [27].)

substrate, midway between the cases of weak and strong coupling shown in figure 6.4. The peak at -2.3eV is described by Khoo *et al* [27] as a resonance. In the case of Pd/H_2/Pd there is a peak below the Fermi energy ($E_F = 0$ eV), rather than the plateau in Pt/H_2/Pt, presumably corresponding to weak coupling in the Grimley–Newns model (red curve in figure 6.4).

References

[1] Datta S 2005 *Quantum Transport: Atom to Transistor* (Cambridge: Cambridge University Press)

[2] Caroli C, Combescot R, Nozières P and Saint-James D 1971 Direct calculation of the tunneling current *J. Phys. C: Solid St. Phys.* **4** 916–29

[3] Grimley T B and Pisani C 1974 Chemisorption theory in the Hartree-Fock approximation *J. Phys. C: Solid St. Phys.* **7** 2831–48

[4] Newns D M 1969 Self-consistent model of hydrogen chemisorption *Phys. Rev.* **178** 1123–35

[5] Anderson P W 1961 Localized magnetic states in metals *Phys. Rev.* **124** 41–53

[6] Baraff G A and Schlüter M 1986 The LCAO approach to the embedding problem *J. Phys. C: Solid St. Phys.* **19** 4383–91

[7] Davison S G and Sulston K W 2006 *Green-Function Theory of Chemisorption* (Dordrecht: Springer)

[8] Sulston K W, Davison S G and Liu W K 1986 Chemisorption on disordered binary alloys *Phys. Rev.* B **33** 2263–9

[9] Schönhammer K 1977 Theory of hydrogen chemisorption on transition metals *Int. J. Quantum Chem.* **12** 517–27

[10] Grimley T B 1967 The indirect interaction between atoms or molecules adsorbed on metals *Proc. Phys. Soc.* **90** 751–64

[11] Thijssen J M and Inglesfield J E 1994 Embedding muffin-tins into a finite-difference grid *Europhys. Lett.* **27** 65–70

[12] Fisher A J 1990 A modified form for the real-space embedding potential *J. Phys: Condens. Matter* **2** 6079–82

[13] Zangwill A 2013 *Modern Electrodynamics* (Cambridge: Cambridge University Press)

[14] Seitsonen A P, Puska M J and Nieminen R M 1995 Real-space electronic-structure calculations: Combination of the finite-difference and conjugate-gradient methods *Phys. Rev.* B **51** 14057–61

[15] Komine T and Shiiki K 2000 Self-consistent first-principles calculations based on the embedded atomic sphere method *Phys. Rev.* B **61** 7378–82

[16] Benham S P, Thijssen J M and Inglesfield J E 2001 Self-consistent finite-difference electronic structure calculations *Comput. Phys. Commun.* **136** 64–76

[17] Joly Y 2001 X-ray absorption near-edge structure calculations beyond the muffin-tin approximation *Phys. Rev.* B **63** 125120

[18] Brandbyge M, Mozos J, Ordejón P, Taylor J and Stokbro K 2002 Density-functional method for nonequilibrium electron transport *Phys. Rev.* B **65** 165401

[19] Rocha A R, García-Suárez V M, Bailey S, Lambert C, Ferrer J and Sanvito S 2006 Spin and molecular electronics in atomically generated orbital landscapes *Phys. Rev.* B **73** 085414

[20] Soler J, Artacho E, Gale J D, García A, Junquera J, Ordejón P and Sánchez-Portal D 2002 The SIESTA method for ab initio order-N materials simulation *J. Phys.: Condens. Matter* **14** 2745–79

[21] Verzijl C J O, Seldenthuis J S and Thijssen J M 2013 Applicability of the wide-band limit in DFT-based molecular transport calculations *J. Chem. Phys.* **138** 094102

[22] López-Sancho M P, López-Sancho J M and Rubio J 1984 Quick iterative scheme for the calculation of transfer matrices: application to Mo(100) *J. Phys. F: Metal Phys.* **14** 1205–15

[23] Williams A R, Feibelman P J and Lang N D 1982 Green's-function methods for electronic-structure calculations *Phys. Rev.* B **26** 5433–44

[24] Inglesfield J E 1970 Green functions, surfaces, and impurities *J. Phys. C: Solid St. Phys.* **4** 14–8

[25] Sanvito S, Lambert C J, Jefferson J H and Bratkovsky A M 1999 General Green's-function formalism for transport calculations with spd Hamiltonians and giant magnetoresistance in Co- and Ni-based magnetic multilayers *Phys. Rev.* B **59** 11936–48

[26] Rungger I and Sanvito S 2008 Algorithm for the construction of self-energies for electronic transport calculations based on singularity elimination and singular value decomposition *Phys. Rev.* B **78** 035407

[27] Khoo K H, Neaton J B, Choi H J and Louie S G 2008 Contact dependence of the conductance of H_2 molecular junctions from first principles *Phys. Rev.* B **77** 115326

IOP Publishing

The Embedding Method for Electronic Structure

John E Inglesfield

Chapter 7

Electron transport

Much the most important use of embedding potentials/self-energies nowadays is in calculating electron transport through molecules, with a metallic contact at each end [1]. As we discussed in the last chapter, these calculations are usually carried out in a local orbital basis, with the metal contacts and leads replaced by self-energies. Transport calculations can also be carried out within the framework of configuration-space embedding theory (the sort of embedding we developed in chapter 2) [2]. In this chapter we shall use both the configuration-space and local orbital methods, comparing the two approaches, which in fact are formally the same. We shall also find some interesting properties of the embedding potential on the way, which clarify the whole idea of embedding. An alternative approach to these steady-state transport calculations is to solve the time-dependent Schrödinger equation [3]. We shall discuss this in section 10.5.1, in chapter 10 on time-dependent embedding.

7.1 The embedding potential and transport

We have already calculated electron transport in chapter 5, using Green function methods to find the wave-function inside a kink when a wave in a particular input channel was incident on the structure (section 5.4.2). From the wave-function we found the transmission coefficient, giving the transmitted current for unit incident current. As we shall show in this section, the transmission can be found in a much simpler and completely general way in terms of the embedding potentials which embed the structure onto the leads (in the case of the kink, these embedding potentials replace the straight sections of waveguide) [2].

The system we are going to study is shown schematically in figure 7.1, consisting of a molecule or nanostructure connected via left- and right-hand leads to electron reservoirs at Fermi energies $E_{F,L}$ and $E_{F,R}$. Let us assume that the bias across the

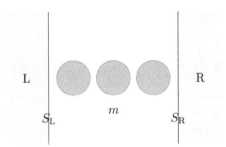

Figure 7.1. Schematic structure for transport calculations: region m contains a molecule or nanostructure, connected to left- and right-hand leads. Embedding potentials are defined on surfaces S_L and S_R.

device is such that $E_{F,L} > E_{F,R}$. Then, from Landauer–Büttiker theory [4, 5], the current through region m is given by

$$I = \frac{2e^2}{h} \int_{E_{F,R}}^{E_{F,L}} dE \, T(E),$$ (7.1)

where T is the total transmission probability across the molecule,

$$T = \sum_{pq}' |t_{pq}|^2.$$ (7.2)

Here, the summation is over the open input and exit channels, p and q respectively (as in section 5.4.2 the prime indicating open channels). t_{pq} is the amplitude of the wave-function in exit channel q with unit incident current in input channel p; the factors of k_p/k_q in equation (5.23) are absorbed in the channel function normalization. This T has no subscripts (compare with equation (5.25)), indicating the transmission probability from all open channels into all open channels.

7.1.1 Channel functions

We begin by considering channel wave-functions, which are central to the Landauer–Büttiker formalism as we see from equation (7.2). It turns out that the embedding potential provides a very convenient way to define the channels [6], irrespective of the form of the leads. For example, they might be semi-infinite metal contacts, or we could have straight electron waveguides as in section 5.4.2. We consider surfaces S_L and S_R which separate the leads from region m (m for molecule or middle), and on these surfaces we evaluate embedding potentials Σ_L and Σ_R from the solutions of the Schrödinger equation in the left- and right-hand leads, respectively.

The current across these surfaces can be written in terms of $\Im\Sigma$ (for the moment let us consider the left-hand lead, and drop the L/R subscripts). Suppose we have a wave-function $\psi(\mathbf{r})$ in this lead; then the current density is given by [7]

$$\mathbf{J} = \frac{1}{2i}(\psi^* \nabla \psi - \psi \nabla \psi^*),$$ (7.3)

and the current across S is

$$I = \frac{1}{2\mathrm{i}} \int_S \mathrm{d}\mathbf{r}_S \left[\psi^*(\mathbf{r}_S)\frac{\partial\psi(\mathbf{r}_S)}{\partial n_S} - \psi(\mathbf{r}_S)\frac{\partial\psi^*(\mathbf{r}_S)}{\partial n_S} \right]. \tag{7.4}$$

Our convention, as always, is that the normal derivative is directed from region I (here m) into region II (here the left-hand contact), so that I is the current across S *into* the lead. But now we can use the definition of the embedding potential (equation (2.16)) to replace the normal derivatives, and equation (7.4) becomes

$$I = \mathrm{i} \int_S \mathrm{d}\mathbf{r}_S \int_S \mathrm{d}\mathbf{r}'_S \left[\psi^*(\mathbf{r}_S)\Sigma(\mathbf{r}_S, \mathbf{r}'_S)\psi(\mathbf{r}_S) - \psi(\mathbf{r}_S)\Sigma^*(\mathbf{r}_S, \mathbf{r}'_S)\psi^*(\mathbf{r}'_S) \right]. \tag{7.5}$$

As the embedding potential has the symmetry of Green functions, $\Sigma(\mathbf{r}_S, \mathbf{r}'_S) = \Sigma(\mathbf{r}'_S, \mathbf{r}_S)$, this simplifies to

$$I = -2 \int_S \mathrm{d}\mathbf{r}_S \int_S \mathrm{d}\mathbf{r}'_S \, \psi^*(\mathbf{r}_S)\Im\Sigma(\mathbf{r}_S, \mathbf{r}'_S)\psi(\mathbf{r}'_S). \tag{7.6}$$

It is because Σ corresponds to outgoing waves from S into region II (see section 2.3, for example), that equation (7.6) gives the current into the lead [6].

The importance of $\Im\Sigma$ suggests that we use its eigenfunctions to define channel functions [6],

$$\int_S \mathrm{d}\mathbf{r}'_S \, \Im\Sigma(\mathbf{r}_S, \mathbf{r}'_S)\psi_p(\mathbf{r}'_S) = \lambda_p \psi_p(\mathbf{r}_S). \tag{7.7}$$

As $\Im\Sigma(\mathbf{r}_S, \mathbf{r}'_S)$ is real and symmetric in $\mathbf{r}_S$ and $\mathbf{r}'_S$, the eigenvalues are real, and the eigenfunctions are real and orthogonal. Of course equation (7.7) only gives the channel function over S—but this defines the full wave-function $\psi^{\mathrm{o}}_p(\mathbf{r})$ outgoing from S back into the lead. From equation (7.6) we see that the current across S associated with this channel function is given by

$$I = -2\lambda_p \int_S \mathrm{d}\mathbf{r}_S \, \psi_p(\mathbf{r}_S)^2. \tag{7.8}$$

This must be positive or zero because of the outgoing boundary condition implicit in Σ, so $\Im\Sigma$ is a negative semi-definite operator with eigenvalues either negative or zero. The eigenfunctions with non-zero eigenvalue correspond to the open channels; there is also an infinite number of closed channel eigenfunctions with zero eigenvalue, corresponding to wave-functions which decay exponentially into the lead.

If we normalize the eigenfunctions to unity over S, so that

$$\int_S \mathrm{d}\mathbf{r}_S \, \psi_p(\mathbf{r}_S)\psi_q(\mathbf{r}_S) = \delta_{pq}, \tag{7.9}$$

we can expand $\Im\Sigma$ in the usual way as

$$\Im\Sigma(\mathbf{r}_S, \mathbf{r}'_S) = \sum_p \lambda_p \psi_p(\mathbf{r}_S)\psi_p(\mathbf{r}'_S). \tag{7.10}$$

This reduces to a sum over the open channel functions, with non-zero eigenvalue,

$$\Im\Sigma(\mathbf{r}_S, \mathbf{r}'_S) = \sum_p{}' \lambda_p \psi_p(\mathbf{r}_S)\psi_p(\mathbf{r}'_S). \tag{7.11}$$

This will prove very useful in the next few pages.

Up to now we have been talking about channel functions which carry current into the lead away from S, but of course we also need functions with current directed from the lead in the opposite direction towards S. As well as the outgoing wave-function, each $\psi_p(\mathbf{r}_S)$ also defines a wave-function in the lead travelling towards S, $\psi_p^i(\mathbf{r})$, and this is the required inward channel function. For any wave $\psi(\mathbf{r})$ travelling towards the surface equation (2.16) is replaced by

$$\frac{\partial\psi(\mathbf{r}_S)}{\partial n_S} = -2\int_S \mathrm{d}\mathbf{r}'_S\, \Sigma^*(\mathbf{r}_S, \mathbf{r}'_S)\psi(\mathbf{r}'_S), \tag{7.12}$$

and the current formula (7.6) is replaced by

$$I = +2\int_S \mathrm{d}\mathbf{r}_S \int_S \mathrm{d}\mathbf{r}'_S\, \psi^*(\mathbf{r}_S)\Im\Sigma(\mathbf{r}_S, \mathbf{r}'_S)\psi(\mathbf{r}'_S). \tag{7.13}$$

The flux over S associated with $\psi_p^i(\mathbf{r})$ is then given by $+2\lambda_p$; note that the amplitude of the function over S is still given by $\psi_p(\mathbf{r}_S)$, and all we must remember is the change in sign of the current.

If the contact is the surface of a semi-infinite piece of metal, a more natural choice of channel functions would seem to be the Bloch functions travelling towards or away from the surface. At fixed E and $\mathbf{k}_\parallel$ there is in fact a unitary transformation between the Bloch functions on S and the channel functions $\psi_p(\mathbf{r}_S)$ defined in [6]. These functions have the advantage of orthonormality over S, a property not satisfied by the Bloch functions. The embedding potential channel functions correspond to the usual choice in the case of a waveguide lead, where the separability of the Schrödinger equation for motion along and across the waveguide leads to an embedding potential which is diagonal in a mode representation. So each embedding potential channel function corresponds to a particular waveguide mode.

7.1.2 Left and right spectral functions

We now make a generalization of the local density of states (equation (2.35)), defining the spectral function by the sum over states,

$$n(\mathbf{r}, \mathbf{r}'; E) = \sum_i \Psi_i(\mathbf{r})\Psi_i^*(\mathbf{r}')\delta(E_i - E) = -\frac{1}{\pi}\Im G(\mathbf{r}, \mathbf{r}'; E + \mathrm{i}\eta). \tag{7.14}$$

In non-equilibrium situations, it is useful to split this up into contributions from waves incident on region m from the left- and right-hand leads, the left and right spectral functions n_L and n_R [1].

To calculate n_L we need the full wave-function $\Psi(\mathbf{r})$ in the molecule when a wave-function $\psi(\mathbf{r})$ is incident on region m from the left-hand lead. We partition the

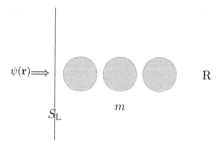

Figure 7.2. Wave-function $\psi(\mathbf{r})$ travels towards S_{L} in the left-hand lead, and is scattered by the potential in region m and the right-hand lead, which together constitute region I.

system slightly differently from figure 7.1, lumping region m, containing the molecule or nanostructure, together with the right-hand lead to constitute region I (figure 7.2). Then from Dyson's equation [7] we have

$$\Psi(\mathbf{r}) = \psi(\mathbf{r}) + \int_{\mathrm{I}} d\mathbf{r}'\, G(\mathbf{r}, \mathbf{r}')\delta V(\mathbf{r}')\psi(\mathbf{r}'),\qquad(7.15)$$

where δV is the potential in region I which scatters the incident wave ψ, and G is the Green function for the combined system, $\mathrm{L} + m + \mathrm{R}$. We now do some Green's theorem manipulations, first of all writing down the Schrödinger equations in region I satisfied by G and ψ,

$$-\frac{1}{2}\nabla^2 G(\mathbf{r}, \mathbf{r}') + [V(\mathbf{r}) + \delta V(\mathbf{r})]G(\mathbf{r}, \mathbf{r}') - EG(\mathbf{r}, \mathbf{r}') = -\delta(\mathbf{r} - \mathbf{r}'),\qquad(7.16)$$

$$-\frac{1}{2}\nabla^2\psi(\mathbf{r}) + V(\mathbf{r})\psi(\mathbf{r}) - E\psi(\mathbf{r}) = 0.\qquad(7.17)$$

Here V is the continuation into I of the potential in the left-hand lead, and $V + \delta V$ is the actual potential in I. Multiplying equation (7.16) by ψ, equation (7.17) by G, subtracting, and integrating over region I we obtain

$$\psi(\mathbf{r}) + \int_{\mathrm{I}} d\mathbf{r}'\, G(\mathbf{r}, \mathbf{r}')\delta V(\mathbf{r}')\psi(\mathbf{r}') = \frac{1}{2}\int_{\mathrm{I}} d\mathbf{r}'[\psi(\mathbf{r}')\nabla^2 G(\mathbf{r}, \mathbf{r}') - G(\mathbf{r}, \mathbf{r}')\nabla^2\psi(\mathbf{r}')].$$

$$(7.18)$$

The left-hand side of this equation is just $\Psi(\mathbf{r})$, and we use Green's theorem to convert the right-hand side to a surface integral over S_{L},

$$\Psi(\mathbf{r}) = \frac{1}{2}\int_{S_{\mathrm{L}}} d\mathbf{r}_S\left[\frac{\partial G(\mathbf{r}, \mathbf{r}_S)}{\partial n_S}\psi(\mathbf{r}_S) - G(\mathbf{r}, \mathbf{r}_S)\frac{\partial\psi(\mathbf{r}_S)}{\partial n_S}\right]\qquad(7.19)$$

(as usual, the other boundaries of region I can be pushed away and make no contribution). We could use this equation as it stands to find the wave-function throughout region I, assuming we know G, but we can obtain a more convenient form using Σ_L, the embedding potential over S_L. Because G is the outgoing Green function and ψ is an incoming wave-function we have

$$\frac{\partial G(\mathbf{r}, \mathbf{r}_S)}{\partial n_S} = -2\int_{S_L} d\mathbf{r}_S'\, G(\mathbf{r}, \mathbf{r}_S')\Sigma_L(\mathbf{r}_S', \mathbf{r}_S), \quad \frac{\partial \psi(\mathbf{r}_S)}{\partial n_S} = -2\int_{S_L} d\mathbf{r}_S'\, \Sigma_L^*(\mathbf{r}_S, \mathbf{r}_S')\psi(\mathbf{r}_S'),$$

$$(7.20)$$

and equation (7.19) becomes [6]

$$\Psi(\mathbf{r}) = -2\mathrm{i}\int_{S_L} d\mathbf{r}_S \int_{S_L} d\mathbf{r}_S'\, G(\mathbf{r}, \mathbf{r}_S)\Im\Sigma_L(\mathbf{r}_S, \mathbf{r}_S')\psi(\mathbf{r}_S'), \quad \text{for } r \text{ in I.} \quad (7.21)$$

We can find G in the central region by embedding region m on to the left and right leads, using Σ_L and Σ_R in the usual way (exactly as in section 5.5.2), and then use equation (7.21) to find the wave-function transmitted into m from the left-hand lead.

Now let us take ψ to be proportional to the incoming (open) channel state ψ_p^i, normalized to unit flux over S_L,

$$\psi(\mathbf{r}) = \frac{1}{\sqrt{2|\lambda_p|}}\psi_p^i(\mathbf{r}). \quad (7.22)$$

Then equation (7.21) becomes

$$\Psi_p(r) = -\mathrm{i}\sqrt{2|\lambda_p|}\int_{S_L} d\mathbf{r}_S\, G(\mathbf{r}, \mathbf{r}_S)\psi_p(\mathbf{r}_S). \quad (7.23)$$

We want something which looks like equation (7.14), and multiplying $\Psi_p(\mathbf{r})$ by $\Psi_p^*(\mathbf{r}')$ and summing over open channels at fixed energy E we obtain

$$\sideset{}{'}\sum_p \Psi_p(\mathbf{r})\Psi_p^*(\mathbf{r}') = -2\int_{S_L} d\mathbf{r}_S \int_{S_L} d\mathbf{r}_S'\, G(\mathbf{r}, \mathbf{r}_S)\Im\Sigma_L(\mathbf{r}_S, \mathbf{r}_S')G^*(\mathbf{r}_S', \mathbf{r}'). \quad (7.24)$$

To go from equation (7.24) to the spectral function, we use energy normalization of the wave-functions [7]: with this normalization, the spectral function is given simply by the sum over states at energy E. In the one-dimensional case (which is what transport essentially is), an energy-normalized wave-function carries a flux of 2π, so the left spectral function is given by dividing equation (7.24) by 2π,

$$n_L(\mathbf{r}, \mathbf{r}'; E) = -\frac{1}{\pi}\int_{S_L} d\mathbf{r}_S \int_{S_L} d\mathbf{r}_S'\, G(\mathbf{r}, \mathbf{r}_S)\Im\Sigma_L(\mathbf{r}_S, \mathbf{r}_S')G^*(\mathbf{r}_S', \mathbf{r}'). \quad (7.25)$$

With the right spectral function given by an equivalent expression, we have

$$n(\mathbf{r}, \mathbf{r}'; E) = n_L(\mathbf{r}, \mathbf{r}'; E) + n_R(\mathbf{r}, \mathbf{r}'; E). \quad (7.26)$$

To illustrate equation (7.25) we go back to the waveguide kink shown in figure 5.10, and calculate the left spectral function in region I when the kink is connected

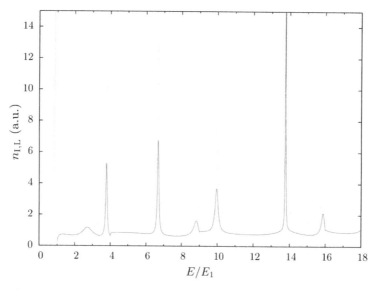

Figure 7.3. Red curve: $n_{I,L}(E)$, left density of states of the kink in figure 5.10 plotted against E/E_1, where E_1 is the first threshold energy. Green curve: $n_I(E)$, the kink density of states. The geometry of the kink is the same as in figure 5.13, and the basis set size is 31 × 30.

to infinitely long straight sections. The embedding potentials at the entrance and exit are given by equation (5.31), and we calculate G in exactly the same way as in section 5.5.2. A convenient quantity to plot is the 'left density of states', which we define as

$$n_{I,L}(E) = \int_I d\mathbf{r}\, n_L(\mathbf{r}, \mathbf{r}; E), \tag{7.27}$$

the contribution to the local density of states from n_L, integrated through region I. This is shown by the red curve in figure 7.3; the green curve shows $n_I(E)$, the kink density of states shown in figure 5.13. We see immediately that $n_{I,L}(E)$ is *exactly* $\frac{1}{2}n_I(E)$, except that the bound state just below the first threshold is missing. This is because the left and right densities of states are the same in kink geometry, but these only include continuum states and exclude discrete, localized states. This can be important in tunnelling at surfaces, where the localized surface states can contribute to the tunnelling current, and Ishida *et al* [8] have shown how this can be included within the embedding framework. Fortunately, the problem does not generally arise when we are dealing with a molecule or nanostructure.

The concept of left and right spectral functions is important in calculating the charge density in the central region m, when there is a finite bias voltage across the leads. Assuming that $E_{F,L} > E_{F,R}$, the charge density contribution from states below $E_{F,R}$ can be calculated from G in the usual way,

$$\rho(\mathbf{r}) = -\frac{2}{\pi} \int^{E_{F,R}} dE\, \Im G(\mathbf{r}, \mathbf{r}; E + i\eta), \tag{7.28}$$

where the factor of two allows for electron spin. This integral can be deformed into a semi-circular contour integral, as in equation (3.26). However, the charge density from states in the energy window between $E_{F,R}$ and $E_{F,L}$ comes only from the electrons that are incident from the left-hand lead, and their contribution to the charge density is given by

$$\rho_L(\mathbf{r}) = 2 \int_{E_{F,R}}^{E_{F,L}} dE \, n_L(\mathbf{r}, \mathbf{r}; E). \tag{7.29}$$

Unfortunately this integral cannot be deformed as the expression (7.25) for n_L involves the non-analytic operations of taking the imaginary part and the complex conjugate. This complication apart, knowing the charge density we can then proceed to self-consistency to find the potential across the conducting structure in region m.

7.1.3 Total transmission

From equation (7.6) the current into the right-hand lead coming from the wave-function Ψ_p, originating in channel p in the left-hand lead, is given by

$$I_p = -2 \int_{S_R} d\mathbf{r}_S \int_{S_R} d\mathbf{r}'_S \, \Psi_p^*(\mathbf{r}_S) \Im \Sigma_R(\mathbf{r}_S, \mathbf{r}'_S) \Psi_p(\mathbf{r}'_S), \tag{7.30}$$

and substituting equation (7.24), the total current from all the open channels p is given by

$$\sideset{}{'}\sum_p I_p = 4 \int_{S_L} d1_L \int_{S_L} d2_L \int_{S_R} d3_R \int_{S_R} d4_R \, G(4_R, 1_L) \Im \Sigma_L(1_L, 2_L) G^*(2_L, 3_R) \Im \Sigma_R(3_R, 4_R). \tag{7.31}$$

To avoid getting bogged down in triple and quadruple dashes, we have simplified the notation for the coordinates over S_L and S_R. Now because each channel p carries unit current (incident on m from the left-hand lead, remember), this corresponds to T in equation (7.2). Simplifying the notation still further, we obtain [6]

$$T = \sideset{}{'}\sum_{pq} |t_{pq}|^2 = 4 \, \mathrm{Tr}[G_{RL} \, \Im \Sigma_L \, G_{LR}^* \, \Im \Sigma_R], \tag{7.32}$$

a result originally obtained by Meir and Wingreen [9] and well-known in the context of self-energies and localized orbital theory. It was derived in this form, using configuration-space embedding potentials, by Wortmann et al [2].

Once again, we can use the waveguide kink (figure 5.10) to illustrate this result. Figure 7.4 shows $T(E)$, the total transmission through the kink calculated from equation (7.32), using the kink Green function and the embedding potentials which replace the straight entrance and exit waveguides. As we would hope, this is the same as the total transmission calculated from equation (5.25), summing over all incident and exit channels. The green curve shows T_1, the total transmission through the kink for incident electrons in the first channel, calculated as in section 5.4.2, and the blue curve T_2 for incident waves in the second channel (we show no more to stop the

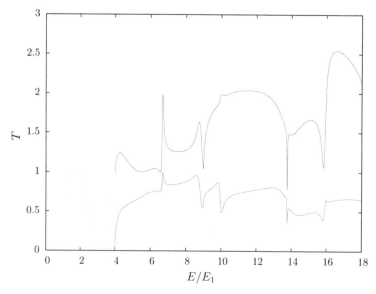

Figure 7.4. Red curve: $T(E)$ total transmission through the kink in figure 5.10 plotted against E/E_1, where E_1 is the first threshold energy; green curve: $T_1(E)$; blue curve: $T_2(E)$. The geometry of the kink is the same as in figure 5.12, and the basis set size is 31×30.

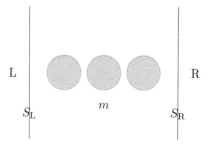

Figure 7.5. Schematic of the computational method: region m consists of the molecule (blue) plus the top layers of the electrodes (light green): the 'extended molecule'. This is embedded on to semi-infinite bulk metallic leads (dark green) over S_L and S_R.

figure getting more and more messy): we see that up to the second threshold, the green and red curves coincide exactly, and if we added on the blue curve they would be the same up to the third threshold—and so on. All the results are calculated using the same kink geometry as in figure 5.12, with a basis of 31×30 and six channels.

7.1.4 Applications

This embedding scheme for calculating transport can be implemented using the LAPW method for calculating the Green function (section 3.2) and the transfer matrix method for finding accurate embedding potentials (section 3.6) [10]. The computational method is shown schematically in figure 7.5. The top layer or two of the electrodes is included with the molecule in region m to give an 'extended molecule' (the same as in figure 6.10), and this is embedded over S_L and S_R on to the semi-infinite bulk leads. Σ_L and Σ_R can be found over S_L and S_R exactly as in section 3.6, and the LAPW method used to find G in region m, embedded on to bulk L and R on each side. The potential in m is found self-consistently, using equations (7.28) and (7.29) to find the charge density if there is a finite bias across the system.

Using this computational scheme Ishida [10] has calculated the conductance of short Au wires and wires incorporating a single O atom, between Au electrodes. As the LAPW codes assume two-dimensional crystal symmetry, it is necessary to employ a supercell technique in the xy-plane parallel to the contact surfaces by repeating the molecule across the contact surfaces. Two supercells were tested, $2\sqrt{2} \times 2\sqrt{2}$ and 2×2, and in fact the smaller proved adequate. With the supercell method, all the functions—embedding potentials, Green function, spectral functions— are calculated for a particular two-dimensional Bloch wave-vector $\mathbf{k}_\parallel$ in the two-dimensional supercell Brillouin zone, and it is necessary to sum over $\mathbf{k}_\parallel$ to obtain integrated properties like charge density and transmission. This calculation was in the small bias limit, so that the charge density and self-consistency in region m could be found using the equilibrium local density of states and equation (7.28).

The charge density for a straight Au_4 wire with the top layer of the Au(001) electrodes is shown in figure 7.6, calculated with embedding on to semi-infinite bulk Au on either side of this extended molecule. The total transmission $T(E)$ for this system, calculated from equation (7.32), is shown in figure 7.7 for the two supercells, and it is clear that there is little difference between the two curves, justifying the use of the smaller 2×2 supercell. A significant feature of these results is that T is close to unity over a wide range of energies starting about -0.5 eV below E_F; the main conduction channel in this range comes from the Au 6s orbitals[1]. These conclusions are in agreement with calculations using localized orbital methods (section 7.2).

When an O atom is inserted into the Au wire the conductance changes considerably. Figure 7.8 shows $T(E)$ calculated for several configurations, and we see that there is a peak in T at E_F, which decreases as more Au atoms are inserted on either side of the O. This can be understood in terms of the equilibrium density of

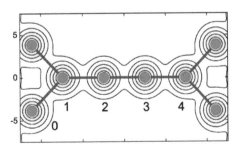

Figure 7.6. Total charge density of Au_4 with the top atomic layers of the Au(001) electrode. This forms region m, the extended molecule, and is embedded on to semi-infinite bulk Au. (From Ishida [10].)

---

[1] The conduction channels describe the ways that the electrons move through the molecule, and are distinct from the channels in the leads. There are various definitions of the conduction channels [11].

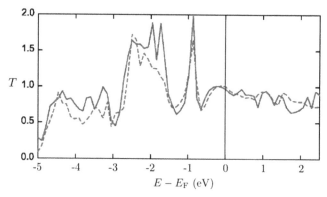

Figure 7.7. $T(E)$, the total transmission of Au_4 between Au(001) electrodes. The red curve is calculated with the $2\sqrt{2} \times 2\sqrt{2}$ supercell and the dashed blue curve with the 2×2 supercell. (From Ishida [10].)

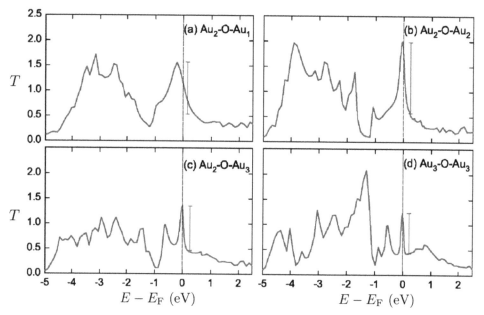

Figure 7.8. $T(E)$, the total transmission of Au wires containing a single O impurity between Au(001) electrodes: (a) $Au_2 - O - Au$, (b) $Au_2 - O - Au_2$, (c) $Au_2 - O - Au_3$, (d) $Au_3 - O - Au_3$. (From Ishida [10].)

states projected on to the O atom, $n_O(E)$, shown in figure 7.9 for the various structures. We see that there is a very prominent peak in n_O at E_F coming from the doubly degenerate O $2p_{x/y}$ orbitals, and the increase in conductance at E_F comes from resonant tunnelling through these orbitals. As the O gets further away from the Au electrodes, the peak in $n_O(E_F)$ sharpens up, whereas the peak in $T(E_F)$ decreases—both due to the decreased coupling of the $2p_{x/y}$ orbitals to the continuum of states in the leads. As well as the increased transmission at E_F, there is a decrease in the plateau

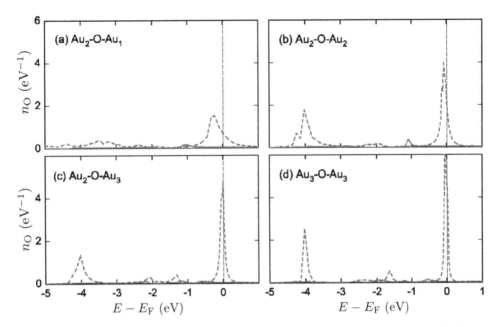

Figure 7.9. $n_O(E)$, the local density of states on the O atom for Au wires containing a single O impurity between Au(001) electrodes: (a) $Au_2 - O - Au$, (b) $Au_2 - O - Au_2$, (c) $Au_2 - O - Au_3$, (d) $Au_3 - O - Au_3$. The red and dashed blue curves correspond to $m = 0$ and $m = \pm 1$ orbital components, respectively. (From Ishida [10].)

for $E > E_F$ to less than 0.5. This is due to increased scattering of electrons in the one-dimensional Au 6s band by the O impurity.

Although the ordinates in figures 7.7 and 7.8 are labelled by the total transmission T, in the original paper Ishida [10] uses the conductance $G(E)$, given by

$$G(E) = \frac{2e^2}{h} T(E),$$ (7.33)

plotting $G(E)/G_0$ with $G_0 = 2e^2/h$, the 'quantum of conductance'. Numerically of course, $G/G_0 = T$. The energy-dependent conductance is a very convenient quantity, as equation (7.1) becomes

$$I = \int_{E_{F,R}}^{E_{F,L}} dE \, G(E),$$ (7.34)

but I avoid using it, reserving G purely for Green functions!

7.2 Transport with localized basis functions

Most computer programs for calculating the electronic structure of molecules invariably use localized basis function, and consequently LCAO methods are ubiquitous for calculating transport through molecules in codes such as TranSIESTA [12]. We have already seen in chapter 6 how the self-energy is the embedding potential in an orbital representation, and in this section we shall compare the LCAO approach to transport, as presented in the books by Datta [1, 13] for example, with the continuum embedding methods of section 7.1.

7.2.1 Current in LCAO

Transport involves currents, so as a starting point we find the expression for current density in a local orbital representation [14], equivalent to the well-known expression (7.3), familiar to us from elementary quantum mechanics [7]. We start off from an LCAO wave-function given by equation (6.1) with the coefficients of the local orbitals satisfying the matrix Schrödinger equation (6.2); here and in the rest of this section, we shall simplify the formalism by assuming real and orthonormal orbitals, with $O_{ij} = \delta_{ij}$. Then the time-dependent version of equation (6.2) is given by

$$\sum_j H_{ij}\alpha_j(t) = i\frac{\partial\alpha_i}{\partial t}, \tag{7.35}$$

with complex conjugate,

$$\sum_j H_{ij}\alpha_j^*(t) = -i\frac{\partial\alpha_i^*}{\partial t}. \tag{7.36}$$

Multiplying equation (7.35) by α_i^*, equation (7.36) by α_i and subtracting gives

$$\sum_j \left(\alpha_i^* H_{ij}\alpha_j - \alpha_i H_{ij}\alpha_j^*\right) = i\frac{\partial\alpha_i^*\alpha_i}{\partial t}, \tag{7.37}$$

an expression for the rate of increase of occupancy of the ith orbital. By analogy with the derivation of equation (7.3), it is then consistent to write the current from orbital j to orbital i as [14]

$$\mathcal{J}_{j\to i} = \frac{1}{i}\left(\alpha_i^* H_{ij}\alpha_j - \alpha_i H_{ij}\alpha_j^*\right). \tag{7.38}$$

We can use the self-energy to write the LCAO current in a form identical to equation (7.6). Referring back to figure 6.1, we represent the orbitals in region I by the vector ϕ_1 and those in region II by ψ_2. Then from equation (7.38) the current from I into II is given by

$$I = i\left(\psi_2 H_{21}\phi_1^* - \psi_2^* H_{21}\phi_1\right), \tag{7.39}$$

and using equations (6.6) and (6.10) this becomes

$$\begin{aligned} I &= i\left(\phi_1^* H_{12}\mathcal{G}_{22}H_{21}\phi_1 - \phi_1 H_{12}\mathcal{G}_{22}^* H_{21}\phi_1^*\right) \\ &= -2\phi_1^*\mathcal{J}\Sigma_{11}\phi_1. \end{aligned} \tag{7.40}$$

If region II represents a lead attached to a molecule (figure 6.10), replaced by self-energy Σ, this expression gives the current from the molecule *into* the lead, because Σ defined by equation (6.10) satisfies outgoing boundary conditions. To bring this into line with the usual notation in transport theory [1, 13], we write this as

$$I = \phi_1^* \Gamma \phi_1, \qquad (7.41)$$

where Γ is $-2\Im\Sigma$ (equation (6.50)).

7.2.2 Spectral functions in LCAO

In section 7.1 the starting point for calculating spectral functions and current was electrons travelling in conduction channels, through the lead towards the molecule. Here we follow Datta's approach [1], and start with the left-hand lead disconnected from the molecule, with standing waves in the lead. Figure 7.10 shows our labelling of the orbitals: orbitals in the left-hand lead L are indexed collectively by 1, and in the extended molecule m + the right-hand lead R (grouped together at this stage) by 2. With lead L disconnected, the amplitudes of the orbitals in group 1 for the ith wave-function are given by vectors Φ^i, energy eigenvalue E_i.[2] Connecting L to the rest of the system, Φ^i evolves into vector Ψ^i with components given by

$$\Psi^i = \begin{cases} \Phi^i + \varphi^i, & \text{orbitals in group 1} \\ \psi^i, & \text{orbitals in group 2,} \end{cases} \qquad (7.42)$$

where φ^i represents a wave reflected back through L, and ψ^i a wave travelling to the right through m + R (figure 7.10).

We now determine ψ^i, which contributes to the left spectral function. The components of Ψ^i satisfy the matrix Schrödinger equation,

$$\begin{pmatrix} H_{11} - E_i & H_{12} \\ H_{21} & H_{22} - E_i \end{pmatrix} \begin{pmatrix} \Phi^i + \varphi^i \\ \psi^i \end{pmatrix} = 0, \qquad (7.43)$$

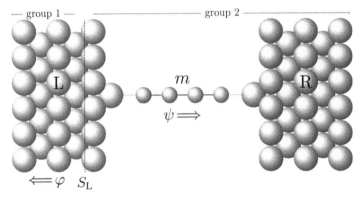

Figure 7.10. Extended molecule m attached to leads L and R. Orbitals in L, to the left of S_L, constitute group 1, and orbitals in m + R, to the right, constitute group 2. Attaching m + R to L sets up wave-functions φ travelling to the left and ψ travelling to the right.

---

[2] The E_i are in the continuum, and discrete (surface) states are not involved.

the first line of which reads

$$(H_{11} - E_i)\Phi^i + (H_{11} - E_i)\phi^i + H_{12}\psi^i = 0. \tag{7.44}$$

But Φ^i satisfies the Schrödinger equation for the disconnected lead,

$$(H_{11} - E_i)\Phi^i = 0, \tag{7.45}$$

so ϕ^i is given by

$$\phi^i = -(H_{11} - E_i)^{-1}H_{12}\,\psi^i = G_{11}(E_i)H_{12}\,\psi^i, \tag{7.46}$$

where G is the Green function for the disconnected lead, exactly as in section 6.1. Now the second line of equation (7.43) is

$$H_{21}\Phi^i + H_{21}\phi^i + (H_{22} - E_i)\psi^i = 0, \tag{7.47}$$

and substituting equation (7.46) gives

$$[H_{22} + H_{21}G_{11}H_{12} - E_i]\psi^i = -H_{21}\Phi^i, \tag{7.48}$$

that is,

$$\psi^i = -[H_{22} + H_{21}G_{11}H_{12} - E_i]^{-1}H_{21}\Phi^i. \tag{7.49}$$

We recognize the second term as Σ_L (equation (6.10)), the self-energy for embedding $m + R$ into L, so the inverse in equation (7.49) is the Green function for the whole system $L + m + R$ in the space of group 2 orbitals, and the final result for ψ^i is simply

$$\psi^i = G_{22}(E_i)H_{21}\Phi^i. \tag{7.50}$$

This is also what we would expect from Dyson's theorem—what we have is basically a scattering problem.

As we are interested in ψ^i in the extended molecule (we remember that this includes contact layers in the electrodes which are perturbed by the molecule—section 6.4) it is convenient to relabel the orbitals at this point. The orbitals in the whole extended molecule now constitute group m, with sub-group l in the principal layer connected across S_L to λ in the left-hand substrate L and sub-group r connected across S_R to ρ in R (figure 7.11). We can then rewrite (7.50) more explicitly as

$$\psi_m^i = \sum_{l\lambda} G_{ml}(E_i)H_{l\lambda}\Phi_\lambda^i. \tag{7.51}$$

Here ψ_m^i is the mth component of vector ψ^i, giving the amplitude of orbital $\chi_m(\mathbf{r})$ in the ith wave-function, and G_{ml} is an element of the full Green function within the space of orbitals m—this can be found from equation (6.76). ψ^i originates in the left-hand lead, and its contribution to the left spectral function (7.25) is then given by

$$\psi_m^i\psi_{m'}^{i*} = \sum_{l\lambda l'\lambda'} G_{ml}H_{l\lambda}\Phi_\lambda^i\Phi_{\lambda'}^{i*}H_{\lambda'l'}G_{l'm'}^*. \tag{7.52}$$

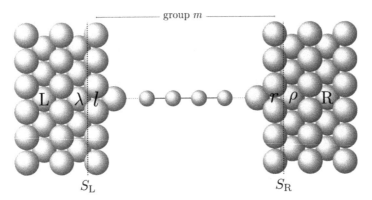

Figure 7.11. Labelling the orbitals for the current calculation in the extended molecule m attached to electrodes L and R. The orbitals in the extended molecule constitute group m, and those in the principal layers connected to the bulk electrodes are labelled l, connected to λ across S_L, and r, connected to ρ across S_R.

With energy normalization of the states Φ^i [7], the left spectral function is simply a sum of equation (7.52) over the states with energy E,

$$n_{L;mm'}(E) = \sum_{E_i=E} \sum_{l\lambda l'\lambda'} G_{ml} H_{l\lambda} \Phi^i_\lambda \Phi^{i\,*}_{\lambda'} H_{\lambda'l'} G^*_{l'm'}, \tag{7.53}$$

where $n_{L;mm'}(E)$ is the orbital form of $n_L(\mathbf{r}, \mathbf{r}'; E)$ (equation (7.25)), and $E_i = E$ is shorthand for states with energy E.

This formula isn't very convenient, involving as it does the states Φ^i in the uncoupled left-hand electrode, but fortunately it can be simplified. With energy normalization the Green function for the uncoupled system can be written as an integral over the states Φ^i,

$$\mathcal{G}_{\lambda\lambda'}(E) = \int dE_i \frac{\Phi^i_\lambda \Phi^{i*}_{\lambda'}}{E - E_i + i\eta}, \tag{7.54}$$

so

$$\Im\mathcal{G}_{\lambda\lambda'}(E) = -\pi \sum_{E_i=E} \Phi^i_\lambda \Phi^{i*}_{\lambda'}. \tag{7.55}$$

Substituting back into equation (7.53) and using the expression for the self-energy (equation (6.10)) we finally obtain

$$n_{L;mm'}(E) = -\frac{1}{\pi} \sum_{ll'} G_{ml} \Im\Sigma_{L;ll'} G^*_{l'm'}, \tag{7.56}$$

exactly the same as equation (7.25)! To bring this into a more conventional form we rewrite this in terms of Γ_L (equation (6.50)) [1],

$$n_{L;mm'}(E) = \frac{1}{2\pi} \sum_{ll'} G_{ml}(E) \Gamma_{L;ll'}(E) G^*_{l'm'}(E). \tag{7.57}$$

Similarly the right spectral function, due to electrons travelling into the molecule from the right-hand lead, is given by

$$n_{R;mm'}(E) = \frac{1}{2\pi} \sum_{rr'} G_{mr}(E)\Gamma_{R;rr'}(E)G^*_{r'm'}(E). \tag{7.58}$$

As in equation (7.26), the total spectral function is given by the sum of n_L and n_R,

$$n_{mm'}(E) = -\frac{1}{\pi}\Im G_{mm'}(E) = n_{L;mm'}(E) + n_{R;mm'}(E), \tag{7.59}$$

with the proviso that $n_L + n_R$ does not include localized states, which *are* included in $\Im G$.

We saw in section 7.1.2 that the left and right spectral functions are important for calculating the charge density across the extended molecule when a finite bias voltage is applied across the leads. Here we consider the density matrix $\rho_{mm'}$, such that the charge density is given by

$$\rho(\mathbf{r}) = 2 \sum_{mm'} \chi_m(\mathbf{r})\rho_{mm'}\chi_{m'}(\mathbf{r}), \tag{7.60}$$

where we assume that the basis functions are real (we make this assumption throughout section 7.2), and the factor of two allows for spin. In the equilibrium case (no bias voltage), $\rho_{mm'}$ is given by

$$\rho_{mm'} = \int dE \, f_0(E - \mu)n_{mm'}(E), \tag{7.61}$$

where we generalize equation (7.28) by using the Fermi distribution f_0, with chemical potential μ, rather than the zero-temperature result with a sharp Fermi cut-off[3]. In the non-equilibrium case, the states coming from the left-hand lead are associated with the distribution function $f_L(E) := f_0(E - \mu_L)$, and similarly for the states coming from the right, so that equation (7.61) becomes

$$\rho_{mm'} = \int dE\left[f_L(E)n_{L;mm'}(E) + f_R(E)n_{R;mm'}(E)\right]. \tag{7.62}$$

The integrand of equation (7.62) is called the correlation function [1], and in many-body language is related to the lesser Green function $G^<$ [15].

7.2.3 Transmission in LCAO

As in section 7.1.3, we now calculate the current due to electrons coming from the left-hand lead and travelling across the molecule into the right-hand lead. Using our LCAO expression for current (equation (7.41)), the wave-function ψ^i, evolved from Φ^i in the disconnected left-hand lead, gives a current into the right-hand lead of

$$I_i = \sum_{rr'} \psi^{i*}_r \Gamma_{R;rr'}\psi^i_r, \tag{7.63}$$

---

[3] This makes our results look more like Green function transport theory, as in Datta [1].

where r, r' label orbitals at the right-hand side of the extended molecule in the principal layer adjoining the right-hand lead (figure 7.11). The current coming from all the states with energy E travelling from left to right is then given (per spin) by

$$I_{L\to R}(E) = \sum_{E_i=E} \sum_{rr'} \psi_r^{i*}\Gamma_{R;rr'}(E)\psi_r^i = \sum_{rr'} n_{L;rr'}(E)\Gamma_{R;r'r}(E). \qquad (7.64)$$

We now substitute from equation (7.57) for the left spectral function, so that our expression for the current (per spin) becomes

$$I_{L\to R}(E) = \frac{1}{2\pi} \sum_{ll'rr'} G_{rl}(E)\Gamma_{L;ll'}(E)G_{l'r}^*(E)\Gamma_{R;r'r}(E)$$

$$= \frac{1}{2\pi} \mathrm{Tr}\left[G_{RL}(E)\Gamma_L(E)G_{LR}^*(E)\Gamma_R(E)\right], \qquad (7.65)$$

where we have used the same subscript notation as in equation (7.32). Because of the symmetry of this expression, we can use equation (7.65) for the current from right to left. The net current, taken as positive from left to right, is then given by

$$I = \frac{1}{\pi} \int \mathrm{d}E \left[f_L(E) - f_R(E)\right]\mathrm{Tr}\left[G_{RL}(E)\Gamma_L(E)G_{LR}^*(E)\Gamma_R(E)\right], \qquad (7.66)$$

where we have multiplied equation (7.65) by a factor of two to take spin into account.

This expression has the same form as equation (7.1) ($2e^2/h = 1/\pi$ in atomic units), if we identify the transmission probability across the molecule as

$$T(E) = \mathrm{Tr}\left[G_{RL}(E)\Gamma_L(E)G_{LR}^*(E)\Gamma_R(E)\right]. \qquad (7.67)$$

We recover Meir and Wingreen's famous expression for transmission [9], the same as equation (7.32) of course. This formula and the correlation function expression for the charge density (equation (7.62)) are the main results of non-equilibrium Green function theory, with the acronym NEGF.

7.3 LCAO transport calculations

LCAO methods for calculating the transmission and conductance are very widely applied, especially to transport through organic molecules, in no small part due to the availability of accurate and efficient programs such as TranSIESTA [12] and SMEAGOL [16]. Central to these codes is the calculation of the self-energies to replace the electrodes, the calculation of the self-consistent potential with a finite bias (section 7.2.2), using density-functional theory for exchange-correlation, and the calculation of the transmission using the Meir–Wingreen Green function expression (7.67) (or something equivalent like equation (7.64)).

7.3.1 Zero-bias calculations

In the limit of zero bias, the self-consistent potential and charge density can be found from the total spectral function $n_{mm'}(E)$ (equation (7.59)), without calculating the left

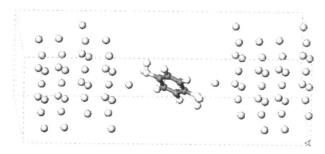

Figure 7.12. Extended molecule for Au(111)/1,4-diaminobenzene/Au(111) conductance calculation, consisting of four Au layers plus a contact atom on each side of the diaminobenzene molecule. Parallel to the surface, a (3 × 3) supercell is used. (From Jing Nang *et al* [17].)

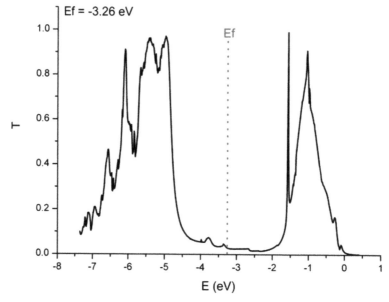

Figure 7.13. Zero-bias transmission $T(E)$ of the Au(111)/1,4-diaminobenzene/Au(111) molecular junction. The Fermi energy E_F is at -3.26 eV, indicated by the dotted line. (From Jing Nang *et al* [17].)

and right spectral functions. As an example, we take the calculation by Jing Nang *et al* [17] of the conductance at zero bias of the 1,4-diaminobenzene molecule between gold electrodes. This molecule consists of amine (NH_2) groups attached at opposite ends of a benzene ring, chosen because the NH_2 group provides an effective contact with Au. The calculations, carried out with the SMEAGOL code [16], used an extended molecule consisting of four layers of Au on each side of the 1,4-diaminobenzene, as shown in figure 7.12.

The calculated transmission $T(E)$ is shown in figure 7.13, with $T(E_F) = 0.028$. This compares with the experimental value of 0.0064 [18], acceptable agreement given the uncertainty of bonding between the molecule and the electrodes [17].

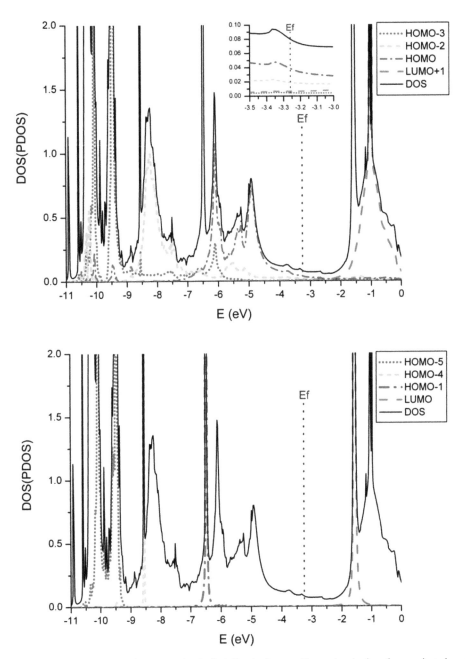

Figure 7.14. Partial densities of states on the Au/1,4-diaminobenzene/Au molecular junction, projected on to different molecular orbitals. The black curve shows the total density of states on the molecule $n(E)$. (From Jing Nang *et al* [17].)

To understand the structure of $T(E)$, Jing Nang *et al* calculated the total density of states $n(E)$ on the 1,4-diaminobenzene molecule in the junction, and the partial density of states (PDOS) projected on to different molecular orbitals (figure 7.14). The first point to notice, from the upper figure, is that $n(E_F)$—hence $T(E_F)$—comes

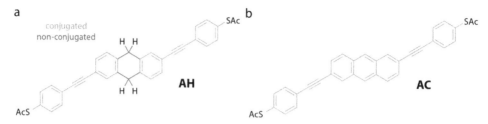

Figure 7.15. (a) Thiolated arylethynylene with 9,10-dihydroanthracene core (AH); (b) thiolated arylethynylene with anthracene core (AC). The conjugated parts of the molecules are in green, and the unconjugated in red. The SAc at each end of the molecules indicate sulphur-acetate groups, the acetate being cut off when the S–Au bond is formed at the molecule–metal contact. (From Perrin *et al* [19].)

mainly from the highest occupied molecular orbital (HOMO) and the HOMO-2 (the orbital two down in energy from the HOMO). The broadened PDOS of these two orbitals shows that they interact strongly with the metal contacts, and a calculation of the orbitals themselves shows p-orbitals on the N-atoms interacting with the benzene π-orbitals and s-orbitals on the gold. The broad peak in T at about $E = -5.5$ eV is clearly associated with broadened peaks in $n(E)$ coming mainly from the HOMO and HOMO-2 PDOS. The PDOS of the the first orbital above the the lowest unoccupied molecular orbital (LUMO + 1) shows a broad peak at $E \approx -1$ eV, about 2 eV above E_F, due to a strong interaction with the electrodes, giving a similar broad peak in $T(E)$. On the other hand the sharp peak in T at $E = -1.6$ eV corresponds to a sharp peak in $n(E)$ coming from the LUMO PDOS (the lower figure), and the shape of this orbital shows that it hardly interacts with the electrode.

7.3.2 Finite-bias calculations

Recently Perrin *et al* [19] have combined transport measurements on thiolated arylethynylene, with a 9,10-dihydroanthracene core, with finite bias calculations based on non-equlibrium Green function theory. This molecule (AH) consists of two conjugated pieces on either side of a non-conjugated core (figure 7.15), and when it is attached to Au contacts via the end S atoms and a voltage applied across the molecule, the current–voltage (I–V) characteristics show negative differential conductance (NDC). The experiments were carried out by attaching the molecule to a break junction in a Au wire, and by varying the curvature of the wire, the separation of the break—the distance between the Au contacts—could be controlled accurately and reproducibly.

A typical I–V characteristic obtained in this way is shown in figure 7.16, for low bias voltages: the NDC features set in at a bias of 100 meV, for both positive and negative bias. The NDC is a result of the weak coupling between the arms of the AH molecule through the non-conjugated core; the fully conjugated molecule with an anthracene core (AC, figure 7.15) shows no NDC effects. In AH, there are in effect separate molecular orbitals on each arm of the molecule, and when a bias voltage is applied the Stark effect splits the energy of these orbitals. As energy is conserved in tunnelling between these energy-broadened orbitals, the conductance drops beyond a certain bias.

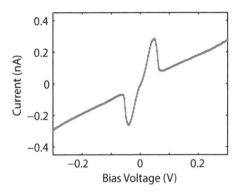

Figure 7.16. A typical I–V characteristic of the thiolated arylethynylene molecule with 9,10-dihydroanthracene core (AH), attached to Au electrodes in a break junction, showing negative differential conductance features symmetric about zero bias. (From Perrin *et al* [19].)

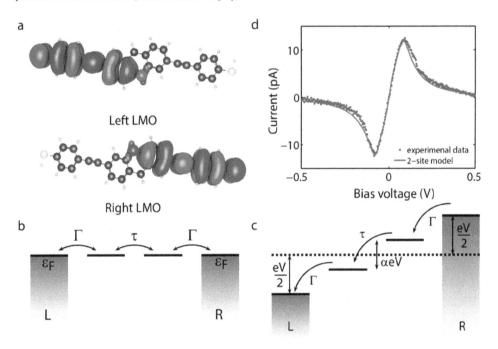

Figure 7.17. Two-site model for conductance of AH molecule. (a) Left- and right-hand effective molecular orbitals, with which the molecule can be modelled, as in (b), by two sites coupled to the contacts; Γ is the imaginary part of the self-energy replacing the contacts, and τ is the coupling between the two sites. (c) Bias voltage V is applied across the molecule, splitting the two sites by αeV, where α is the fraction of the voltage drop across the molecule. (d) I–V characteristic data (blue dots, smooth variation removed), modelled by the two-site model (red line). (From Perrin *et al* [19].)

This two-level model is illustrated in figure 7.17. In AH, the conductance is dominated by the HOMO and HOMO-1 orbitals, and adding and subtracting these gives effective molecular orbitals localized on each arm of the molecule, shown in (a). These localized orbitals can be treated as the states of a two-site model (b), in which

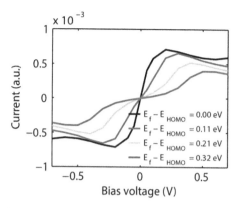

Figure 7.18. I–V characteristics of the AH molecule attached to Au electrodes, calculated self-consistently at each bias voltage using the BAND extension to ADF. The different curves correspond to shifts in E_F relative to the HOMO level. (Figure courtesy of J Thijssen.)

an electrode self-energy, with imaginary part Γ, is added to the Hamiltonian at each site, and the sites are coupled across the unconjugated core by τ. When a bias voltage V is applied between the electrodes, a fraction α of the voltage drop occurs across the molecule, as in (c), splitting the energy of the two sites by αeV—the rest of the voltage drop occurs at the metal–molecule contacts. When αeV is much smaller than τ, the sites are in resonance and the conductance of the molecule is large, but when $\alpha eV \gg \tau$ the resonance is suppressed, and the conductance drops sharply, giving the NDC feature. This can be described by a model Hamiltonian, containing α, τ, and Γ as fitting parameters (constant Γ smaller than τ corresponds to the wide-band limit mentioned in section 6.2). The resulting conductance, compared with experimental data with the smooth background removed, is shown in (d), and the agreement is excellent. Interestingly, the parameter fit gives $\alpha = 0.48$, so that half the voltage drop occurs within the molecule itself.

This model shows that it is essential at finite bias to include the potential variation across the molecule—preferably self-consistently using non-equilibrium Green function theory (section 7.2.2). Having optimized the geometry of AH linked via the end S-atoms to the Au electrodes using first-principles methods, finite bias calculations were carried out using BAND, a sister-program to the Amsterdam Density Functional (ADF) package [20] for periodic systems. This uses localized Slater-type orbitals as basis functions. The top layers in the Au contacts were included in the extended molecule and accurate self-energies were calculated to replace the semi-infinite bulk Au leads (L and R in figures 6.10 and 7.11).

The results from the self-consistent finite bias calculations for AH between Au contacts are shown in figure 7.18, giving the I–V characteristics. The different curves show different values for the energy difference between the Fermi energy and the HOMO energy—what is meant here is the value of E_F in the energy integration of $T(E)$, so that $E_F - E_{HOMO} = 0$ means an energy window in equation (7.66) centred on the HOMO level. The reason for taking different values of $E_F - E_{HOMO}$ is that the HOMO level is very sensitive to such factors as the contact adsorption site, and uncertainties in effects like the image potential. We see that with $E_F = E_{HOMO}$ the I–V characteristic shows very clear NDC features—and remember, this is a first-principles calculation. Moreover, the variation in shape of the I–V curves for different values of $E_F - E_{HOMO}$ corresponds to what is really found in different experiments. Particularly interesting is the red curve, corresponding to $E_F - E_{HOMO} = 0.32$ eV, in which the I–V characteristic is flat between about ± 0.3 eV. This corresponds to an energy gap, which can be understood in terms of the two-site model as a misalignment of the energy levels with E_F. Such an energy gap is found in some of the break junction experiments on AH.

References

[1] Datta S 2005 *Quantum Transport: Atom to Transistor* (Cambridge: Cambridge University Press)

[2] Wortmann D, Ishida H and Blügel S 2002 Embedded Green-function approach to the ballistic electron transport through an interface *Phys. Rev.* B **66** 075113

[3] Kurth S, Stefanucci G, Almbladh C-O, Rubio A and Gross E K U 2005 Time-dependent quantum transport: A practical scheme using density functional theory *Phys. Rev.* B **72** 035308

[4] Landauer R 1970 Electrical resistance of disordered one-dimensional lattice *Phil. Mag.* **21** 863–7

[5] Büttiker M 1986 Four-terminal phase-coherent conductance *Phys. Rev. Lett.* **57** 1761–4

[6] Inglesfield J E, Crampin S and Ishida H 2005 Embedding potential definition of channel functions *Phys. Rev.* B **71** 155120

[7] Merzbacher E 1998 *Quantum Mechanics* 3rd edn (New York: John Wiley)

[8] Ishida H, Wortmann D and Ohwaki T 2004 First-principles calculations of tunneling conductance *Phys. Rev.* B **70** 085409

[9] Meir Y and Wingreen N S 1992 Landauer formula for the current through an interacting electron region *Phys. Rev. Lett.* **68** 2512–5

[10] Ishida H 2007 Embedded Green-function calculation of the conductance of oxygen-incorporated Au and Ag monatomic wires *Phys. Rev.* B **75** 205419

[11] Paulsson M and Brandbyge M 2007 Transmission eigenchannels from nonequilibrium Green's functions *Phys. Rev.* B **76** 115117

[12] Brandbyge M, Mozos J, Ordejón P, Taylor J and Stokbro K 2002 Density-functional method for nonequilibrium electron transport *Phys. Rev.* B **65** 165401

[13] Datta S 1995 *Electronic Transport in Mesoscopic Systems* (Cambridge: Cambridge University Press)

[14] Todorov T N 2002 Tight-binding simulation of current-carrying nanostructures *J. Phys.: Condens. Matter.* **14**(11) 3049–84

[15] Stefanucci G and van Leeuwen R 2013 *Nonequilibrium Many-Body Theory of Quantum Systems* (Cambridge: Cambridge University Press)

[16] Rocha A R, García-Suárez V M, Bailey S, Lambert C, Ferrer J and Sanvito S 2006 Spin and molecular electronics in atomically generated orbital landscapes *Phys. Rev.* B **73** 085414

[17] Ning J, Li R, Shen X, Qian Z, Hou S, Rocha A R and Sanvito S 2007 First-principles calculation on the zero-bias conductance of a gold/1,4-diaminobenzene/gold molecular junction *Nanotechnology* **18**(34) 345203

[18] Venkataraman L, Klare J E, Nuckolls C, Hybertsen M S and Steigerwald M 2006 Dependence of single-molecule junction conductance on molecular conformation *Nature* **442** 904–7

[19] Perrin M L *et al* 2014 Large negative differential conductance in single-molecule break junctions *Nature Nanotech.* **9** 830–4

[20] te Velde G, Bickelhaupt F M, Baerends E J, Fonseca Guerra C, van Gisbergen S J A, Snijders J G and Ziegler T 2001 Chemistry with ADF *J. Comput. Chem.* **22** 931–67

IOP Publishing

The Embedding Method for Electronic Structure

John E Inglesfield

Chapter 8

Relativistic embedding

Relativistic effects, especially spin–orbit coupling, become important in systems involving heavier atoms, and to describe these we must solve the Dirac equation for the electrons, instead of the Schrödinger equation. Most bulk electronic structure methods have been adapted to this end, and in this chapter we shall see how the embedding method can also be generalized to the Dirac equation. As with the Schrödinger equation, embedding for the Dirac equation becomes especially useful for surfaces—spin–orbit coupling leads to important effects like the Rashba splitting of surface states on surfaces like Au(111) [1]. Closely connected with this effect are topological surface states, a topic of tremendous current interest.

Crampin has generalized the embedding method to the Dirac equation [2], based on a variational principle analogous to the method described in chapter 2. This will be described in section 8.1, and we shall see that the method turns out to be remarkably similar in structure to the Schrödinger equation case. This is despite the fact that the Dirac equation is a first-order differential equation rather than second-order like the (time-independent) Schrödinger equation, and that the wave-function is a four-component *two-spinor*. The main difference is that the spinor properties of the wave-function give a (2×2) matrix structure for the embedding potential. We shall see in section 8.2 that Dirac embedding works extremely well when it is used in a real fully self-consistent application—the calculation of the electronic structure of the Au(111) surface, for which even a one-layer calculation describes the Rashba surface state very well [3].

Instead of solving the full Dirac equation, we can use the scalar-relativistic equation, a modified Schrödinger equation which includes the mass-velocity effect due to the changing mass of the electron as it speeds up near the nucleus (section 8.3). Spin–orbit coupling can be added on to this equation [4], which can then be used to treat the same problems as the full Dirac equation, with the advantage that the wave-function is a single spinor with two components—spin-up and spin-down.

doi:10.1088/978-0-7503-1042-0ch8

This equation is widely used for treating relativistic problems, and is the one used in the FLEUR software package, for example[1].

8.1 Embedding the Dirac equation

The Dirac equation[2] which we wish to solve in region I joined on to region II (figure 1.1) is given by [5]

$$(c\boldsymbol{\alpha} \cdot \mathbf{p} + \beta m_e c^2 + V(\mathbf{r}) - W)\Psi(\mathbf{r}) = 0, \tag{8.1}$$

where Ψ is the four-component two-spinor with (mass-)energy W,

$$\Psi(\mathbf{r}) = \begin{pmatrix} \Psi_l(\mathbf{r}) \\ \Psi_s(\mathbf{r}) \end{pmatrix} = \begin{pmatrix} \Psi_1(\mathbf{r}) \\ \Psi_2(\mathbf{r}) \\ \Psi_3(\mathbf{r}) \\ \Psi_4(\mathbf{r}) \end{pmatrix}, \tag{8.2}$$

made up of large and small spinors, Ψ_l and Ψ_s, each with two components (remember that for electron states, as opposed to positron states, Ψ_s is really small [5]). $\boldsymbol{\alpha}$ and β are the (4 × 4) Dirac matrices, where the three components of $\boldsymbol{\alpha}$ can be written in terms of the Pauli spin matrices $\boldsymbol{\sigma} = (\sigma_x, \sigma_y, \sigma_z)$,

$$\alpha_x = \begin{pmatrix} 0 & \sigma_x \\ \sigma_x & 0 \end{pmatrix}, \qquad \alpha_y = \begin{pmatrix} 0 & \sigma_y \\ \sigma_y & 0 \end{pmatrix}, \qquad \alpha_z = \begin{pmatrix} 0 & \sigma_z \\ \sigma_z & 0 \end{pmatrix}, \tag{8.3}$$

and β is given by

$$\beta = \begin{pmatrix} 1 & 0 & 0 & 0 \\ 0 & 1 & 0 & 0 \\ 0 & 0 & -1 & 0 \\ 0 & 0 & 0 & -1 \end{pmatrix}. \tag{8.4}$$

$\mathbf{p}$ in equation (8.1) is the momentum operator, $\mathbf{p} = -i\hbar\nabla$.

8.1.1 Dirac variational principle

The embedding derivation follows the same lines as in section 2.1, the differences arising from the fact that the Dirac equation is a first-order differential equation, and that the wave-function has four components. We find the expectation value of the energy W in equation (8.1) with a trial function $\Phi(\mathbf{r})$ of the same form as equation (2.4), $\Phi = \phi$ in region I and $\Phi = \psi$ in region II, where now Φ, ϕ, and ψ are all four-component two-spinors. ψ satisfies the Dirac equation in region II at a trial energy w,

[1] FLEUR is an electronic structure package based on the FLPAW method, incorporating embedding for semi-infinite systems, developed in Forschungszentrum Jülich, http://www.flapw.de.
[2] Following Crampin [2], we shall write $\hbar$ and m_e explicitly so that any system of units can be used.

such that the large components ψ_l match ϕ_l, but with no constraint on ψ_s. Following Crampin [2], the expectation value of W is then given by

$$W = \frac{\displaystyle\int_{\mathrm{I}} d\mathbf{r}\, \phi^\dagger H\phi + w \int_{\mathrm{II}} d\mathbf{r}\, \psi^\dagger\psi + ic\hbar \int_{S} d\mathbf{r}_S \cdot \phi_l^\dagger \boldsymbol{\sigma}(\phi_s - \psi_s)}{\displaystyle\int_{\mathrm{I}} d\mathbf{r}\, \phi^\dagger\phi + \int_{\mathrm{II}} d\mathbf{r}\, \psi^\dagger\psi}, \tag{8.5}$$

where $H = -ic\hbar\boldsymbol{\alpha} \cdot \nabla + \beta m_e c^2 + V(\mathbf{r})$, and the dagger indicates the Hermitian conjugate, with complex conjugation and column vectors becoming row vectors,

$$\phi^\dagger(\mathbf{r}) = \begin{pmatrix} \phi_1^*(\mathbf{r}) & \phi_2^*(\mathbf{r}) & \phi_3^*(\mathbf{r}) & \phi_4^*(\mathbf{r}) \end{pmatrix}. \tag{8.6}$$

The final term in the numerator, a surface integral over the boundary surface S between regions I and II, comes from the operation of $\boldsymbol{\alpha} \cdot \mathbf{p}$ on the discontinuity between ϕ_s and ψ_s across S. We note the dot product between the surface element $d\mathbf{r}_S$ and $\boldsymbol{\sigma}$, corresponding to taking the spin in the direction of the surface normal, outwards from I to II.

The expectation value of W (equation (8.5)) has similar structure to equation (2.5), with the difference in normal derivatives over S in equation (2.5) replaced by the difference between small components in equation (8.5), and we tackle it in a similar way by eliminating explicit reference to ψ. To do this we introduce the Green function $G(\mathbf{r}, \mathbf{r}'; w)$ satisfying the Dirac equation in region II [5],

$$\left(-ic\hbar\boldsymbol{\alpha} \cdot \nabla + \beta m_e c^2 + V - w\right)G = -I\delta(\mathbf{r} - \mathbf{r}'), \tag{8.7}$$

where G is a (4×4) matrix,

$$G(\mathbf{r}, \mathbf{r}'; w) = \begin{pmatrix} G_{ll}(\mathbf{r}, \mathbf{r}'; w) & G_{ls}(\mathbf{r}, \mathbf{r}'; w) \\ G_{sl}(\mathbf{r}, \mathbf{r}'; w) & G_{ss}(\mathbf{r}, \mathbf{r}'; w) \end{pmatrix}, \tag{8.8}$$

made up of the (2×2) submatrices G_{ll}, G_{ls}, G_{sl} and G_{ss}. I is the (4×4) unit matrix. Now, ψ satisfies the equation,

$$-ic\hbar\boldsymbol{\alpha} \cdot \nabla\psi + \left(\beta m_e c^2 + V - w\right)\psi = 0, \tag{8.9}$$

and taking the Hermitian conjugate of equation (8.7) we have

$$ic\hbar(\nabla G^\dagger) \cdot \boldsymbol{\alpha} + G^\dagger\left(\beta m_e c^2 + V - w\right) = -I\delta(\mathbf{r} - \mathbf{r}'). \tag{8.10}$$

The brackets in $(\nabla G^\dagger)$ indicate that ∇ only operates on $G^\dagger$ and nothing we might put to the right, as we shall see in a moment or two. Multiplying equation (8.9) by $G^\dagger$ to the left, equation (8.10) by ψ to the right, subtracting, and integrating over region II gives

$$\psi = ic\hbar \int_{\mathrm{II}} d\mathbf{r} \left[(\nabla G^\dagger) \cdot \boldsymbol{\alpha}\psi + G^\dagger\boldsymbol{\alpha} \cdot \nabla\psi \right]$$

$$= -ic\hbar \int_{\mathrm{II}} d\mathbf{r}\, \nabla \cdot \left(G^\dagger\boldsymbol{\alpha}\psi\right). \tag{8.11}$$

We now use the divergence theorem [6] to convert the volume integral in equation (8.11) to a surface integral over S,

$$\psi(\mathbf{r}) = -ic\hbar \int_S d\mathbf{r}_S \cdot G(\mathbf{r}, \mathbf{r}_S; \omega)\alpha\psi(\mathbf{r}_S), \qquad (8.12)$$

giving us an expression for $\psi(\mathbf{r})$ within region II in terms of its values over S (ψ satisfies the boundary condition that it vanishes on any outer boundary of region II). Putting $\mathbf{r}$ on S, and substituting for α from equation (8.3) gives us two equations relating $\psi_l(\mathbf{r}_S)$ and $\psi_s(\mathbf{r}_S)$,

$$\psi_l(\mathbf{r}_S) = ic\hbar \int_S d\mathbf{r}'_S \cdot \left[G_{ll}(\mathbf{r}_S, \mathbf{r}'_S; w)\sigma\psi_s(\mathbf{r}'_S) + G_{ls}(\mathbf{r}_S, \mathbf{r}'_S; w)\sigma\psi_l(\mathbf{r}'_S) \right],$$

$$\psi_s(\mathbf{r}_S) = ic\hbar \int_S d\mathbf{r}'_S \cdot \left[G_{sl}(\mathbf{r}_S, \mathbf{r}'_S; w)\sigma\psi_s(\mathbf{r}'_S) + G_{ss}(\mathbf{r}_S, \mathbf{r}'_S; w)\sigma\psi_l(\mathbf{r}'_S) \right], \quad (8.13)$$

from which we obtain

$$\psi_s(\mathbf{r}_S) = -ic\hbar \int_S d\mathbf{r}'_S \cdot \Sigma(\mathbf{r}_S, \mathbf{r}'_S; w)\sigma\psi_l(\mathbf{r}'_S), \qquad (8.14)$$

where the (2×2) matrix Σ satisfies the integral equation

$$\Sigma(\mathbf{r}_S, \mathbf{r}'_S; w) = -G_{ss}(\mathbf{r}_S, \mathbf{r}'_S; w) + ic\hbar \int_S d\mathbf{r}''_S \cdot G_{sl}(\mathbf{r}_S, \mathbf{r}''_S; w)\sigma\Sigma(\mathbf{r}''_S, \mathbf{r}'_S; w). \qquad (8.15)$$

Σ is the Dirac embedding potential[3], a (2×2) matrix, which gives the small component of the wave-function over S in terms of the large component. Using the requirement that $\psi_l(\mathbf{r}_S) = \phi_l(\mathbf{r}_S)$, the last term in the numerator of the expectation value (equation (8.5)) can then be written as

$$\int_S d\mathbf{r}_S \cdot \phi_l^\dagger \sigma(\phi_s - \psi_s) = \int_S d\mathbf{r}_S \cdot \phi_l^\dagger(\mathbf{r}_S)\sigma\left(\phi_s(\mathbf{r}_S) + ic\hbar \int_S d\mathbf{r}'_S \cdot \Sigma(\mathbf{r}_S, \mathbf{r}'_S; w)\sigma\phi_l(\mathbf{r}'_S) \right). \qquad (8.16)$$

We now eliminate the volume integrals over region II in equation (8.5), replacing them by integrals over S involving Σ, very much as we did in section 2.1. First we differentiate equation (8.9) with respect to w, the energy parameter,

$$-ic\hbar\alpha \cdot \nabla\frac{\partial\psi}{\partial w} + \left(\beta m_e c^2 + V - w \right)\frac{\partial\psi}{\partial w} = \psi, \qquad (8.17)$$

and take the Hermitian conjugate of the original equation,

$$ic\hbar\left(\nabla\psi^\dagger\right) \cdot \alpha + \psi^\dagger\left(\beta m_e c^2 + V - w \right) = 0. \qquad (8.18)$$

---

[3] Crampin uses Γ for the embedding potential [2]. Note that $\Sigma = -\Gamma$, as we have chosen the sign so that equation (8.14) looks like equation (2.16).

Multiplying equation (8.17) on the left by $\psi^\dagger$ and equation (8.18) on the right by ψ and subtracting then gives

$$\psi^\dagger \psi = -\mathrm{i}c\hbar\nabla \cdot \left(\psi^\dagger \boldsymbol{\alpha} \frac{\partial \psi}{\partial w}\right), \tag{8.19}$$

which integrating over region II, and once again using the divergence theorem, simplifies to

$$\int_{\mathrm{II}} \psi^\dagger \psi = \mathrm{i}c\hbar \int_S \mathrm{d}\mathbf{r}_S \cdot \psi^\dagger \boldsymbol{\alpha} \frac{\partial \psi}{\partial w}. \tag{8.20}$$

But we know that $\psi_l(\mathbf{r}_S)$, the large component of ψ over S, is fixed and set equal to $\phi_l(\mathbf{r}_S)$, so equation (8.20) reduces to

$$
\begin{aligned}
\int_{\mathrm{II}} \psi^\dagger \psi &= \mathrm{i}c\hbar \int_S \mathrm{d}\mathbf{r}_S \cdot \psi_l^\dagger \boldsymbol{\sigma} \frac{\partial \psi_s}{\partial w} \\
&= \mathrm{i}c\hbar \int_S \mathrm{d}\mathbf{r}_S \cdot \phi_l^\dagger(\mathbf{r}_S)\boldsymbol{\sigma}\left(-\mathrm{i}c\hbar \int_S \mathrm{d}\mathbf{r}_S' \cdot \frac{\partial \Sigma}{\partial w}(\mathbf{r}_S, \mathbf{r}_S'; w)\boldsymbol{\sigma}\phi_l(\mathbf{r}_S')\right),
\end{aligned}
\tag{8.21}
$$

where we have used equation (8.14) to write $\psi_s(\mathbf{r}_S)$ in terms of $\phi_l(\mathbf{r}_S)$. So, we have an expression for the normalization of ψ in region II in terms of $\phi_l(\mathbf{r}_S)$ and the energy derivative of the embedding potential, similar in structure to equation (2.21).

Substituting equations (8.16) and (8.21) into equation (8.5) gives the expectation value for the energy in terms of the trial function ϕ in region I and over S, together with the embedding potential,

$$W = \frac{\displaystyle\int_{\mathrm{I}} \mathrm{d}\mathbf{r}\, \phi^\dagger H\phi + \mathrm{i}c\hbar \int_S \mathrm{d}\mathbf{r}_S \cdot \phi_l^\dagger \boldsymbol{\sigma}\left[\phi_s + \mathrm{i}c\hbar \int_S \mathrm{d}\mathbf{r}_S' \cdot \left(\Sigma - w\frac{\partial \Sigma}{\partial w}\right)\boldsymbol{\sigma}\phi_l\right]}{\displaystyle\int_{\mathrm{I}} \mathrm{d}\mathbf{r}\, \phi^\dagger \phi + c^2\hbar^2 \int_S \mathrm{d}\mathbf{r}_S \cdot \phi_l^\dagger \boldsymbol{\sigma} \int_S \mathrm{d}\mathbf{r}_S' \cdot \frac{\partial \Sigma}{\partial w}\boldsymbol{\sigma}\phi_l}. \tag{8.22}$$

This expression is remarkably similar to the Schrödinger variational principle (equation (2.22)). The main difference is that ϕ_s in equation (8.22) takes the place of $\partial \phi/\partial n_S$ in equation (2.22). In equation (2.22) Σ is a generalized logarithmic derivative, giving the derivative in terms of the amplitude of the wave-function over S; in the Dirac equation case, Σ gives the small component in terms of the large component.

8.1.2 Solving the embedded Dirac equation

Crampin [2] has shown that the embedded expression for W given by equation (8.22) is stationary, with respect to arbitrary variations $\delta\phi$ in ϕ, when

$$H\phi = W\phi, \quad \mathbf{r} \in \text{region I}, \tag{8.23}$$

$$\phi_s(\mathbf{r}_S) = -\mathrm{i}c\hbar \int_S \mathrm{d}\mathbf{r}_S' \cdot \left[\Sigma(\mathbf{r}_S, \mathbf{r}_S'; w) + \left(W - w\right)\frac{\partial \Sigma}{\partial w}(\mathbf{r}_S, \mathbf{r}_S'; w)\right]\boldsymbol{\sigma}\phi_l(\mathbf{r}_S'). \tag{8.24}$$

Equation (8.23) means that ϕ satisfies the Dirac equation in region I, and equation (8.24) means that the relationship between ϕ_l and ϕ_s over S is correct for matching on to the solution of the Dirac equation in region II; the expression in square brackets in equation (8.24) means that the embedding potential, giving this relationship, is evaluated at energy W, to first order in $(W - w)$.

To find ϕ from the variational expression we use a basis set expansion in (almost) the same way that we solved the embedded Schrödinger equation in section 2.2. The difference is that for the Dirac equation we need separate expansions for ϕ_l and ϕ_s,

$$\phi_l(\mathbf{r}) = \sum_{n=1}^{N_l} a_{l,n}\chi_{l,n}(\mathbf{r}), \qquad \phi_s(\mathbf{r}) = \sum_{n=1}^{N_s} a_{s,n}\chi_{s,n}(\mathbf{r}), \tag{8.25}$$

where the $\chi_{l,n}$'s and $\chi_{s,n}$'s are *spinor* basis functions, in general different for the two expansions. Substituting equation (8.25) into equation (8.22), and requiring that W is stationary with respect to variations in the expansion coefficients, we obtain the matrix eigenvalue equation analogous to equation (2.25), namely

$$\begin{pmatrix} H_{ll} & H_{ls} \\ H_{sl} & H_{ss} \end{pmatrix}\begin{pmatrix} a_l \\ a_s \end{pmatrix} = W\begin{pmatrix} O_{ll} & 0 \\ 0 & O_{ss} \end{pmatrix}\begin{pmatrix} a_l \\ a_s \end{pmatrix}. \tag{8.26}$$

Making use of the Dirac matrices (8.3) and (8.4), we find that the matrix elements of the embedded Hamiltonian are given by

$$\left[H_{ll}\right]_{mn} = \int_{I} d\mathbf{r}\, \chi_{l,m}^{\dagger}\left(V + m_e c^2\right)\chi_{l,n} - c^2\hbar^2 \int_{S} d\mathbf{r}_S \cdot \chi_{l,m}^{\dagger}\boldsymbol{\sigma}\int_{S} d\mathbf{r}_S' \cdot \left(\Sigma - w\frac{\partial\Sigma}{\partial w}\right)\boldsymbol{\sigma}\chi_{l,n},$$
$$\tag{8.27}$$

$$[H_{ls}]_{mn} = -ic\hbar\left[\int_{I} d\mathbf{r}\, \chi_{l,m}^{\dagger}\boldsymbol{\sigma}\cdot\nabla\chi_{s,n} - \int_{S} d\mathbf{r}_S \cdot \chi_{l,m}^{\dagger}\boldsymbol{\sigma}\chi_{s,n}\right], \tag{8.28}$$

$$[H_{sl}]_{mn} = -ic\hbar\int_{I} d\mathbf{r}\, \chi_{s,m}^{\dagger}\boldsymbol{\sigma}\cdot\nabla\chi_{l,n}, \tag{8.29}$$

$$[H_{ss}]_{mn} = \int_{I} d\mathbf{r}\, \chi_{s,m}^{\dagger}\left(V - m_e c^2\right)\chi_{s,n}. \tag{8.30}$$

The overlap matrix elements in equation (8.26) are given by

$$[O_{ll}]_{mn} = \int_{I} d\mathbf{r}\, \chi_{l,m}^{\dagger}\chi_{l,n} + c^2\hbar^2 \int_{S} d\mathbf{r}_S \cdot \chi_{l,m}^{\dagger}\boldsymbol{\sigma}\int_{S} d\mathbf{r}_S' \cdot \frac{\partial\Sigma}{\partial w}\chi_{l,n}, \tag{8.31}$$

$$[O_{ss}]_{mn} = \int_{I} d\mathbf{r}\, \chi_{s,m}^{\dagger}\chi_{s,n}. \tag{8.32}$$

These have similar structure to the matrix elements for the embedded Schrödinger equation in equations (2.26) and (2.27). In particular we note that if we transferred the second term in O_{ll} (equation (8.31)) on the right-hand side of equation (8.26) to

H_{ll} on the left-hand side (which we don't do in practice), the matrix element of the embedding potential in equation (8.27) would become

$$-c^2\hbar^2 \int_S \mathrm{d}\mathbf{r}_S \cdot \chi_{l,m}^\dagger \boldsymbol{\sigma} \int_S \mathrm{d}\mathbf{r}'_S \cdot \left(\Sigma + \left(W - w \right) \frac{\partial \Sigma}{\partial w} \right) \boldsymbol{\sigma} \chi_{l,n}, \tag{8.33}$$

showing explicitly that what started off as normalization in region II finishes off by correcting the energy of the embedding potential.

A difficulty arises when using the variational principle for the Dirac equation, because it has an infinite number of negative energy solutions which are unbounded below, corresponding to positron states [7]. This leads to eigenvalues collapsing to unphysical negative energies unless special care is taken; the problem can be avoided by using a so-called kinetically balanced basis [7], in which the small component basis spinors are related to the large component spinors (equation (8.25)) by

$$\chi_{s,n} = \boldsymbol{\sigma} \cdot \mathbf{p} \chi_{l,n}, \tag{8.34}$$

with the numbers of spinors the same, $N_s = N_l = N$.

8.1.3 H-atom in a spherical cavity

As a test of this variational embedding method, Crampin [2] has calculated the lowest bound states of the Dirac equation, with the potential shown in figure 8.1—the H atom potential in a spherical cavity of radius R,

$$V(r) = \begin{cases} -\lambda/r, & r \leqslant R \text{ (region I)}, \\ V_0, & r > R \text{ (region II)}, \end{cases} \tag{8.35}$$

where $\lambda = e^2/(4\pi\epsilon_0)$ and V_0 is positive. Region I consists of the spherical cavity, and the external region where the potential is constant is replaced by an embedding potential over the spherical surface.

In this geometry, with spherical symmetry, we can find Σ most easily from equation (8.14) using the free-electron-like solutions of the Dirac equation in region II in spherical coordinates [2]. The angular momentum quantum numbers are l, j, κ and μ, where l describes the orbital angular momentum $\mathbf{L}$, and j the total angular momentum $\mathbf{J}$, with $\mathbf{J} = \mathbf{L} + \mathbf{S}$ and $\mathbf{S}$ is the electron spin. κ is that very useful quantum number in relativistic theory given by

$$\kappa = \begin{cases} -l - 1 & \text{for } j = l + \dfrac{1}{2} \\ l & \text{for } j = l - \dfrac{1}{2}, \end{cases} \tag{8.36}$$

so that a given κ fixes both l and j. Finally, μ gives the z-component of $\mathbf{J}$. Let us consider the solutions with mass-energy w and real wave-vector k given by

$$c\hbar k = \sqrt{(w - V_0)^2 - m_e^2 c^4}. \tag{8.37}$$

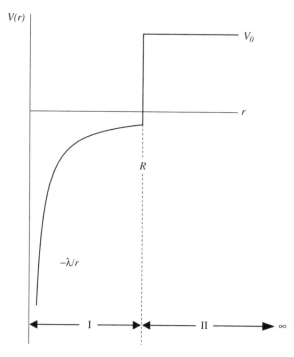

Figure 8.1. H-atom potential confined in a spherical cavity of radius R. The hydrogenic potential with $r < R$ constitutes region I, and the external region II, with constant potential V_0, is replaced by an embedding potential over the spherical surface of region I. (From Crampin [2].)

Then the outgoing wave is given by the two-spinor [5],

$$\psi_{\kappa,\mu}(\mathbf{r}) = \begin{pmatrix} h_l(kr)\Omega_{\kappa,\mu}(\hat{\mathbf{r}}) \\ \dfrac{ikc\hbar S_\kappa}{w - V_0 + m_e c^2} h_{\bar{l}}(kr)\Omega_{-\kappa,\mu}(\hat{\mathbf{r}}) \end{pmatrix}, \tag{8.38}$$

where $\Omega_{\kappa,\mu}(\hat{\mathbf{r}})$ is the spin-angular function (a spinor, sometimes called a spherical spinor) and $\hat{\mathbf{r}}$ is the angle of vector $\mathbf{r}$. $h_l(kr)$ is the Hankel function of the first kind [8], which varies asymptotically as $\sim\exp(iz)/z$. In the small component of equation (8.38), $\bar{l}$ is given by

$$\bar{l} = \begin{cases} l + 1 = -\kappa, & \kappa < 0 \\ l - 1 = \kappa - 1, & \kappa > 0, \end{cases} \tag{8.39}$$

and S_κ in the prefactor of the small component is given by

$$S_\kappa = \begin{cases} -1 & \text{for } j = l + \dfrac{1}{2} \\ +1 & \text{for } j = l - \dfrac{1}{2}. \end{cases} \tag{8.40}$$

With the spherical symmetry of region II the embedding potential can be expanded over S as

$$\Sigma(\mathbf{r}_S, \mathbf{r}'_S; w) = \sum_{\kappa,\mu}\Sigma_{\kappa,\mu}(w)\Omega_{-\kappa,\mu}(\hat{\mathbf{r}}_S)\Omega^{\dagger}_{-\kappa,\mu}(\hat{\mathbf{r}}'_S), \tag{8.41}$$

and substituting this with equation (8.38) into equation (8.14) gives

$$-ic\hbar R^2 h_l(kR)\sum_{\kappa',\mu'}\Sigma_{\kappa',\mu'}\Omega_{-\kappa',\mu'}(\hat{\mathbf{r}}_S)\int d\hat{\mathbf{r}}'_S\Omega^{\dagger}_{-\kappa',\mu'}(\hat{\mathbf{r}}'_S)\hat{\mathbf{r}}'_S \cdot \boldsymbol{\sigma}\,\Omega_{\kappa,\mu}(\hat{\mathbf{r}}'_S)$$

$$= \frac{ikc\hbar S_\kappa}{w - V_0 + m_e c^2}h_{\bar{l}}(kR)\Omega_{-\kappa,\mu}(\hat{\mathbf{r}}). \tag{8.42}$$

But there is an important result in the theory of spin-angular functions [5],

$$\hat{\mathbf{r}}_S \cdot \boldsymbol{\sigma}\,\Omega_{\kappa,\mu}(\hat{\mathbf{r}}_S) = \Omega_{-\kappa,\mu}(\hat{\mathbf{r}}_S), \tag{8.43}$$

and making use of the orthonormality of these functions, equation (8.42) simplifies to give us the required result for the embedding potential with angular quantum numbers (κ, μ),

$$\Sigma_{\kappa,\mu}(w) = -\frac{kS_\kappa}{R^2\left(w - V_0 + m_e c^2\right)}\frac{h_{\bar{l}}(kR)}{h_l(kR)}. \tag{8.44}$$

At energies such that the wave-vector is imaginary we replace k by $i\gamma$, where the decay constant is given by

$$c\hbar\gamma = \sqrt{m_e^2 c^4 - (w - V_0)^2}\,. \tag{8.45}$$

This gives the form of embedding potential needed for calculating the bound states. We note that Crampin [2] has an equivalent expression, but he uses modified spherical Bessel functions with real arguments [8, 9], rather than Hankel functions with imaginary arguments.

With this embedding potential, we can solve the Dirac equation for the confined H atom by solving the matrix eigenvalue equation (8.26). As basis functions for expanding the wave-functions in region I (equation (8.25)), Crampin [2] used the following spinors,

$$\chi_{l,n}^{(\kappa,\mu)} = r^{n-1}\exp(-r)\Omega_{\kappa,\mu}(\hat{r}),$$

$$\chi_{s,n}^{(\kappa,\mu)} = i\hbar\big[(n + \kappa)r^{n-2} - r^{n-1}\big]\exp(-r)\Omega_{-\kappa,\mu}(\hat{r}). \tag{8.46}$$

Here, the large and small component spinors are related by equation (8.34) to ensure kinetic balance. Results for bound states of the confined H-atom, with a confining radius $R = 3$ a.u. and a confining potential $V_0 = 10$ a.u. (figure 8.1), are given in table 8.1: this gives the two lowest (electron-like) eigenvalues, corresponding to the $1s_{1/2}$ and $2s_{1/2}$ states of the free H atom, for different sizes of basis set and different values of the energy w at which the embedding potential is evaluated.

Table 8.1. Confined H-atom [2]: the lowest two electron-like eigenvalues with $s_{1/2}$ symmetry, given as $E = W - m_e c^2$, for different trial energies w and for different sizes of basis set ($N_l = N_s = N$). The confining radius $R = 3$ a.u. and the confining potential $V_0 = 10$ a.u., and 'exact' eigenvalues are found by wave-function matching across R. Energies are in a.u.

N	$w = m_e c^2 - 0.5$	$w = m_e c^2$	$w = W$
2	$-0.411\,1620, 1.698\,0995$	$-0.411\,1527, 1.694\,9300$	$-0.411\,1624, 1.689\,6482$
4	$-0.445\,1482, 0.912\,9418$	$-0.445\,1439, 0.912\,6817$	$-0.439\,6204, 0.971\,4775$
6	$-0.445\,5519, 0.891\,4789$	$-0.445\,5477, 0.891\,2219$	$-0.445\,5520, 0.891\,0268$
8	$-0.445\,5532, 0.891\,2708$	$-0.445\,5488, 0.891\,0141$	$-0.445\,5532, 0.890\,8194$
'exact'	$-0.445\,5532, 0.890\,8194$	$-0.445\,5532, 0.890\,8194$	$-0.445\,5532, 0.890\,8194$

The 'exact' energies were found by matching the external solution (equation (8.38)) to the solution for the internal region which is regular at the origin.

The results are very satisfactory—when w is taken equal to the eigenvalue W (by iteration), convergence to the exact energies is rapid and uniform. Even with a fairly rough value for the trial energy, as in the first two columns of table 8.1, the eigenvalues are reasonably accurate, thanks to the energy-dependence of the embedding potential being included to first order in equation (8.27).

8.1.4 The Dirac Green function and continuum states

As we have seen throughout this book, when we are working in the energy continuum, it is more useful to calculate the Green function of the embedded wave equation than to work with the eigenvalue equation (8.26).

The form of the Dirac Green function is given in equation (8.8), and to calculate this in region I we expand the sub-matrices using the spinor basis functions $\chi_{l,n}$, $\chi_{s,n}$ we used in equation (8.25) [2],

$$G_{ll}(\mathbf{r}, \mathbf{r}'; W) = \sum_{mn} [G_{ll}(W)]_{mn} \chi_{l,m}(\mathbf{r}) \chi_{l,n}^{\dagger}(\mathbf{r}'),$$

$$G_{ls}(\mathbf{r}, \mathbf{r}'; W) = \sum_{mn} [G_{ls}(W)]_{mn} \chi_{l,m}(\mathbf{r}) \chi_{s,n}^{\dagger}(\mathbf{r}'),$$

$$G_{sl}(\mathbf{r}, \mathbf{r}'; W) = \sum_{mn} [G_{sl}(W)]_{mn} \chi_{s,m}(\mathbf{r}) \chi_{l,n}^{\dagger}(\mathbf{r}'),$$

$$G_{ss}(\mathbf{r}, \mathbf{r}'; W) = \sum_{mn} [G_{ss}(W)]_{mn} \chi_{s,m}(\mathbf{r}) \chi_{s,n}^{\dagger}(\mathbf{r}'). \tag{8.47}$$

The full matrix is given by [2]

$$G(W) = \begin{pmatrix} WO_{ll} - H_{ll} & -H_{ls} \\ -H_{sl} & WO_{ss} - H_{ss} \end{pmatrix}^{-1}, \tag{8.48}$$

where H_{ll}, O_{ll}, etc, are given by equations (8.27)–(8.32), but without the energy-derivative terms. From the Green function matrix, the local density of states and the density of states integrated through region I can be calculated [2].

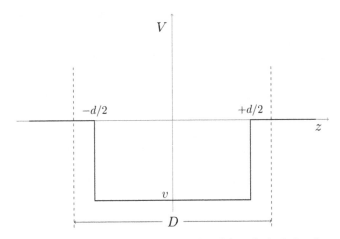

Figure 8.2. One-dimensional potential well. D defines the basis functions.

As an example of Dirac embedding in the continuum, we calculate the density of states inside a one-dimensional square well, potential $V = v$ and width d (figure 8.2). Region I is the well itself, and the free-electron regions on either side, $V = 0$, are replaced by embedding potentials at $z = \pm d/2$. Again we can use equation (8.14) to find the embedding potential—rather simpler here than in spherical geometry, as the outgoing solution of the Dirac equation with mass-energy W, is given for $z > +d/2$ by [9]

$$\psi(z) = \exp(ikz) \begin{pmatrix} 1 \\ 0 \\ \gamma_k \\ 0 \end{pmatrix}, \tag{8.49}$$

where k and γ_k are given by

$$c\hbar k = \sqrt{W^2 - m_e^2 c^4}, \qquad \gamma_k = \frac{c\hbar k}{W + m_e c^2}. \tag{8.50}$$

Note that equation (8.49) is the electron-like solution with spin *parallel* to the z-axis. Then from the definition (equation (8.14)) of the embedding potential, we find for this spin-component of Σ,

$$\Sigma(W) = \frac{ik}{W + m_e c^2}. \tag{8.51}$$

Taking the spin anti-parallel to z gives the same expression for Σ, so the 2×2 embedding potential matrix is given by

$$\Sigma(W) = \begin{pmatrix} \dfrac{ik}{W + m_e c^2} & 0 \\ 0 & \dfrac{ik}{W + m_e c^2} \end{pmatrix}. \tag{8.52}$$

This embedding potential is the same at the left- and right-hand boundaries of region I, at $z = \pm d/2$ (figure 8.2).

As basis functions for expanding the Green function inside the square well we use the relativistic version of equation (2.30), and for spins in the $+z$-direction we have

$$\chi_m(z) \propto \exp(ik_m z)\begin{pmatrix} 1 \\ 0 \\ \gamma_{k_m} \\ 0 \end{pmatrix} \pm \exp(-ik_m z)\begin{pmatrix} 1 \\ 0 \\ -\gamma_{k_m} \\ 0 \end{pmatrix}, \tag{8.53}$$

where $k_m = m\pi/D$; the $+$ sign is taken for m even and the $-$ sign for m odd. D is shown in figure 8.2, and as in chapter 2 it is taken to be a little larger than d. As each spin can be treated separately—they are not mixed by the square-well potential or the embedding potential—we can then write

$$\chi_m(z) = \begin{pmatrix} \chi_{l,m}(z) \\ \chi_{s,m}(z) \end{pmatrix}, \quad \text{where} \begin{cases} \chi_{l,m} = \cos k_m z, \ \chi_{s,m} = i\gamma_{k_m} \sin k_m z, & m \text{ even} \\ \chi_{l,m} = \sin k_m z, \ \chi_{s,m} = -i\gamma_{k_m} \cos k_m z, & m \text{ odd}, \end{cases} \tag{8.54}$$

where the large and small components of χ_m are scalars rather than spinors.

With this form of basis function we have a *single* basis set expansion, rather than separate expansions for the large and small components as in equation (8.25), so that equation (8.47) simplifies to

$$G(z, z'; W) = \sum_{mn} G_{mn}(W)\chi_m(z)\chi_n^\dagger(z')$$

$$= \sum_{mn} \begin{pmatrix} G_{mn}(W)\chi_{l,m}(z)\chi_{l,n}^*(z') & G_{mn}(W)\chi_{l,m}(z)\chi_{s,n}^*(z') \\ G_{mn}(W)\chi_{s,m}(z)\chi_{l,n}^*(z') & G_{mn}(W)\chi_{s,m}(z)\chi_{s,n}^*(z') \end{pmatrix}. \tag{8.55}$$

The matrix G is given by

$$G = (WO - H)^{-1}, \tag{8.56}$$

with the matrix elements of H and O,

$$H_{mn} = [H_{ll}]_{mn} + [H_{ls}]_{mn} + [H_{sl}]_{mn} + [H_{ss}]_{mn}, \tag{8.57}$$

$$O_{mn} = [O_{ll}]_{mn} + [O_{ss}]_{mn}. \tag{8.58}$$

As the charge density in the Dirac equation has contributions from both the large and small components of the wave-function [5], the local density of states inside the well is given by

$$n(\mathbf{r}, W) = -\frac{1}{\pi}\Im\left\{\sum_{mn} G_{mn}(W + i\eta)\left[\chi_{l,m}(z)\chi_{l,n}^*(z) + \chi_{s,m}(z)\chi_{s,n}^*(z)\right]\right\}. \tag{8.59}$$

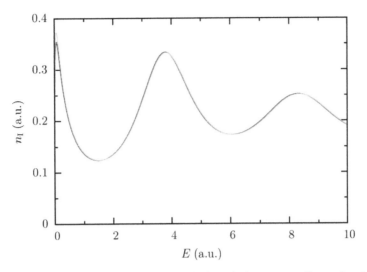

Figure 8.3. $n_I(E)$, the local density of states integrated through the square well as a function of energy $E = W - m_e c^2$ (figure 8.2, $v = -20$ a.u., $d = 5$ a.u., $D = 6$ a.u.). The red curve is calculated using the Dirac equation and the green curve using the Schrödinger equation, with 20 basis functions.

Integrating through region I and using equation (8.58) we obtain for the density of states in the well

$$n_I(W) = -\frac{1}{\pi}\Im \sum_{mn} G_{mn}(W + i\eta)O_{nm}, \qquad (8.60)$$

satisfactorily. This is the same as equation (2.43) [2], though of course with very different matrices.

Using this expression, we have calculated the density of states in a square well of depth 20 a.u. and width 5 a.u., expanding the Green function for the embedded Dirac equation in the basis set given by equation (8.54). This is shown by the red curve in figure 8.3, plotting $n_I(E)$, where $E = W - m_e c^2$, calculated with 20 basis functions and $D = 6$ a.u. The green curve shows $n_I(E)$ calculated using the Schrödinger equation (section 2.4), and we see that the two curves are practically identical except at the narrow peak just above the energy zero—not surprisingly, as $|v| \ll m_e c^2$. In any case, the results give us confidence in using the embedding formalism for the Dirac equation. As we shall see in section 8.2, the standing wave basis functions with spin (equation (8.53)) can be generalized to construct the relativistic linearized augmented plane-waves (RLAPWs) [10], used in relativistic embedded surface calculations.

8.2 Embedded surface calculations with the Dirac equation

A method for calculating the relativistic electronic structure of surfaces and interfaces has been developed by James and Crampin [3] based on the methods described in section 8.1. The structure of this calculation is similar to the surface calculation we described in chapter 3 using the embedded Schrödinger equation: the top few layers of atoms, plus the near-surface region of vacuum constitute region I, and this is embedded on to the semi-infinite substrate on the material side, and on the vacuum side on to a constant potential. The embedding potential is calculated by the Dirac generalization of the transfer matrix method described in section 3.5, and the Green function in region I is expanded using RLAPWs.

8.2.1 The Dirac transfer matrix and embedding potential

To find the embedding potential for the relativistic surface calculation, we use the transfer matrix T_Ω, which relates the Dirac wave-functions in either side of bulk atomic layer Ω (figure 3.19). James and Crampin [3] have shown that the generalization of equation (3.50) to the Dirac equation is

$$\begin{pmatrix} \Psi_{L,l} \\ \Psi_{L,s} \end{pmatrix} = T_\Omega \begin{pmatrix} \Psi_{R,l} \\ \Psi_{R,s} \end{pmatrix}. \tag{8.61}$$

Here $\Psi_{L,l/s}$ consists of the expansion coefficients of the large/small components of two-spinor Ψ over the left-hand surface of Ω, with $\Psi_{R,l/s}$ the expansions over the right-hand surface. Comparing equation (8.61) with equation (3.50), we see that $\Psi_{L/R,s}$ takes the place of $\Psi'_{L/R}$; the length of the vectors in equation (8.61) is 4× the number of two-dimensional reciprocal lattice vectors in the surface expansions.

The Dirac transfer matrix can be found in much the same way as in the non-relativistic case, starting from the Green function relation. Applying equation (8.12) to volume Ω, with S comprising the left- and right-hand surfaces, we obtain

$$\psi(\mathbf{r}) = ic\hbar \left[\int_{S_L} d\mathbf{r}_L \cdot G(\mathbf{r}, \mathbf{r}_L; \omega)\alpha\psi(\mathbf{r}_L) - \int_{S_R} d\mathbf{r}_R \cdot G(\mathbf{r}, \mathbf{r}_R; \omega)\alpha\psi(\mathbf{r}_R) \right], \tag{8.62}$$

where the surface normal is taken into Ω on the left-hand surface S_L and out of the surface on the right-hand surface S_R. Here ψ is any two-spinor satisfying the Dirac equation inside Ω. This equation applies for G satisfying any boundary conditions, and if we choose G so that the submatrix G_{ls} vanishes on S we obtain for the large component of ψ within Ω,

$$\psi_l(\mathbf{r}) = ic\hbar \left[\int_{S_L} d\mathbf{r}_L \cdot G_{ll}(\mathbf{r}, \mathbf{r}_L; \omega)\sigma\psi_s(\mathbf{r}_L) - \int_{S_R} d\mathbf{r}_R \cdot G_{ll}(\mathbf{r}, \mathbf{r}_R; \omega)\sigma\psi_s(\mathbf{r}_R) \right]. \tag{8.63}$$

Putting $\mathbf{r}$ on the two surfaces gives

$$\psi_l(\mathbf{r}_L) = ic\hbar \left[\int_{S_L} d\mathbf{r}'_L \cdot G_{ll}(\mathbf{r}_L, \mathbf{r}'_L; \omega)\sigma\psi_s(\mathbf{r}'_L) - \int_{S_R} d\mathbf{r}'_R \cdot G_{ll}(\mathbf{r}_L, \mathbf{r}'_R; \omega)\sigma\psi_s(\mathbf{r}'_R) \right],$$

$$\psi_l(\mathbf{r}_R) = ic\hbar \left[\int_{S_L} d\mathbf{r}'_L \cdot G_{ll}(\mathbf{r}_R, \mathbf{r}'_L; \omega)\sigma\psi_s(\mathbf{r}'_L) - \int_{S_R} d\mathbf{r}'_R \cdot G_{ll}(\mathbf{r}_R, \mathbf{r}'_R; \omega)\sigma\psi_s(\mathbf{r}'_R) \right],$$

$$\tag{8.64}$$

which we write symbolically as

$$\psi_{L,l} = F_{LL}\psi_{L,s} - F_{LR}\psi_{R,s},$$
$$\psi_{R,l} = F_{RL}\psi_{L,s} - F_{RR}\psi_{R,s}. \tag{8.65}$$

Here, $F_{LL} = ic\hbar G_{ll}(\mathbf{r}_L, \mathbf{r}'_L)\boldsymbol{\sigma} \cdot \mathbf{n}_L(\mathbf{r}'_L)$, etc, where $\mathbf{n}_L(\mathbf{r}'_L)$ is the surface normal at $\mathbf{r}'_L$, with matrix multiplication implying an integration over the surface. This has the same structure as equation (3.64), and it can be rearranged in the same way to give

$$\psi_{R,l} = F_{RR}F_{LR}^{-1}\psi_{L,l} + \left(F_{RL} - F_{RR}F_{LR}^{-1}F_{LL}\right)\psi_{L,s},$$
$$\psi_{R,s} = -F_{LR}^{-1}\psi_{L,l} + F_{LR}^{-1}F_{LL}\psi_{L,s}. \tag{8.66}$$

Comparing with equation (8.61) we see that the Dirac transfer matrix is given by

$$T_\Omega = \begin{pmatrix} F_{RR}F_{LR}^{-1} & \left(F_{RL} - F_{RR}F_{LR}^{-1}F_{LL}\right) \\ -F_{LR}^{-1} & F_{LR}^{-1}F_{LL} \end{pmatrix}. \tag{8.67}$$

As shown by James and Crampin [3], the Green function satisfying the required boundary condition that $G_{ls}(\mathbf{r}, \mathbf{r}_S) = 0$ can be found be embedding Ω with zero embedding potential on the surfaces S_L and S_R: this is exactly the way that G_0 is found for evaluating the Schrödinger transfer matrix (equation (3.67)).

When the atomic layers are stacked together to make an infinite solid, the solutions of the Dirac equation are Bloch waves, and as in equation (3.51) the wavefunctions (at fixed wave-vector parallel to the layers) are multiplied by the Bloch phase-factor in going from layer to layer, so that

$$T_\Omega \begin{pmatrix} \Psi_{l,i} \\ \Psi_{s,i} \end{pmatrix} = \exp(ik_z^i d)\begin{pmatrix} \Psi_{l,i} \\ \Psi_{s,i} \end{pmatrix}. \tag{8.68}$$

The eigenvectors of T_Ω are the large and small components over S_L of the Bloch two-spinor with eigenvalue $\exp(ik_z^i d)$; in the Bloch phase factor, k_z^i is the (possibly complex) wave-vector, and d is the layer spacing. The relevant eigenstates for constructing the embedding potential to replace a semi-infinite solid extending to the left of S_L are those which are travelling or decaying to the left, with $\Re k_z^i < 0$ and $\Im k_z^i < 0$.

In practice, rather than working with the wavy surfaces S_L and S_R wending their way between atoms, it is more convenient to work with planar surfaces, and to do this we follow the same procedure as in section 3.6.1, calculating the transfer matrix $T_{\Omega'}$ for volume Ω', consisting of the original volume Ω extended by buffer regions Δ_L and Δ_R (figure 3.19). We also calculate $T_{\Omega''}$ for volume $\Omega'' = \Delta_L \cup \Delta_R$, and then as we showed in section 3.6.1, transfer matrix $T = T_{\Omega'}T_{\Omega''}^{-1}$ has the same eigenvalues as T_Ω,

$$T \begin{pmatrix} \Upsilon_{l,i} \\ \Upsilon_{s,i} \end{pmatrix} = \exp(ik_z^i d)\begin{pmatrix} \Upsilon_{l,i} \\ \Upsilon_{s,i} \end{pmatrix}. \tag{8.69}$$

The great advantage of calculating T rather than T_Ω is that the transfer matrices $T_{\Omega'}$ and $T_{\Omega''}$ are between flat surfaces, so the Green functions can be easily found.

Figure 3.24 shows the complex band structure of Cu in the ⟨001⟩ direction, calculated in this way, showing the spin–orbit splitting characteristic of the Dirac equation.

We now calculate the embedding potential evaluated on the flat surface P_L (figure 3.19)—again we follow section 3.6.1 almost word for word, equation by equation. From the eigenstates of T, we define new two-spinors, by

$$\begin{pmatrix} \hat{\Psi}_{l,i} \\ \hat{\Psi}_{s,i} \end{pmatrix} = T_{\Omega''}^{-1} \begin{pmatrix} \Upsilon_{l,i} \\ \Upsilon_{s,i} \end{pmatrix}, \tag{8.70}$$

which are related to the eigenstates of T_Ω (equation (8.68)) by [3]

$$\begin{pmatrix} \hat{\Psi}_{l,i} \\ \hat{\Psi}_{s,i} \end{pmatrix} = T_{\Delta_L}^{-1} \begin{pmatrix} \Psi_{l,i} \\ \Psi_{s,i} \end{pmatrix}, \tag{8.71}$$

where T_{Δ_L} is the transfer matrix from the flat surface P_L to the wavy surface S_L (figure 3.19) (the proof of this is given in section 3.6.1). This means that $\hat{\Psi}_i$ is Ψ_i integrated backwards through whatever potential we choose for Δ_L, from S_L to P_L. Next we define two-dimensional plane-wave expansions of the large and small components of $\hat{\Psi}_i$ over P_L, with coefficients

$$\left[\hat{\Phi}_l \right]_{ij} := \hat{\Psi}_{l,i}(\mathbf{g}_j), \quad \left[\hat{\Phi}_s \right]_{ij} := \hat{\Psi}_{s,i}(\mathbf{g}_j). \tag{8.72}$$

For fixed i—that is, for a particular eigenstate of T—we can arrange these coefficients, each of which is a spinor, into vectors of length $2N_g$, where N_g is the number of surface reciprocal lattice vectors we consider in the plane-wave expansion. As there are two eigenstates travelling or decaying to the left for each surface reciprocal lattice vector, $\hat{\Phi}_l$ and $\hat{\Phi}_s$ each constitute a matrix of size $2N_g \times 2N_g$. Then using equation (8.14), and by analogy with equation (3.61), the embedding potential on P_L is given by [11]

$$\Sigma_{\nu\nu'} = \frac{\mathrm{i}}{c\hbar} \sum_{\nu''} \left[\hat{\Phi}_s \hat{\Phi}_l^{-1} \right]_{\nu\nu''} [\sigma_z]_{\nu''\nu'}, \tag{8.73}$$

where ν is an index running from 1 to $2N_g$. The matrix σ_z gives the z-component of spin of each of the elements in $\hat{\Phi}_l$ and $\hat{\Phi}_s$, with z the surface normal to P_L, directed to the right. This embedding potential gives the relationship between the large and small components of any solution of the Dirac equation in the semi-infinite solid to the left of S_L, integrated back through Δ_L to P_L. In other words, it is the embedding potential shifted to P_L, for the semi-infinite solid to the left of S_L (figure 3.20).

8.2.2 Rashba surface states on Au(111)

The embedded surface calculations can now be carried out in much the same way as we described in sections 3.2 and 3.6. The top two or three atomic layers and the near-surface region, plus the buffer region Δ_L, constitute region I; this is embedded on to the bulk embedding potential over P_L to the left, and the vacuum embedding potential on P_R to the right (figure 3.20). The potential between S_L and P_R is determined self-consistently, and in the buffer region it is whatever potential is chosen to find T (as in section 3.6.1). In the surface applications of their method, James and Crampin [3] take a constant vacuum potential to the right of P_R.

The Green function in region I is expanded as in equation (8.55), taking RLAPWs for the basis functions. These functions, introduced by Takeda [10], are the relativistic generalization of the LAPWs defined in section 3.2.1. Adapted to surface calculations, where we are working in the slab between P_L and P_R and we have two-dimensional periodicity in the xy-plane parallel to the surface, the RLAPW in the interstitial region is given by

$$\chi_{gm\sigma}(\mathbf{r}) = \begin{pmatrix} \chi_\sigma \\ \gamma_k \boldsymbol{\sigma} \cdot \hat{\mathbf{k}}^+ \chi_\sigma \end{pmatrix} \exp(i\mathbf{k}^+ \cdot \mathbf{r}) \pm \begin{pmatrix} \chi_\sigma \\ \gamma_k \boldsymbol{\sigma} \cdot \hat{\mathbf{k}}^- \chi_\sigma \end{pmatrix} \exp(i\mathbf{k}^- \cdot \mathbf{r}). \tag{8.74}$$

The basis function now has three subscripts, with $\mathbf{g}$ and m defining the plane-wave,

$$\mathbf{k}^\pm = \left(\mathbf{k}_\parallel + \mathbf{g}, \pm k_m\right). \tag{8.75}$$

As in equation (3.17) we fix the wave-vector $\mathbf{k}_\parallel$ parallel to the surface, $\mathbf{g}$ is a two-dimensional surface reciprocal lattice vector and $k_m = m\pi/D$, where D is a distance parameter somewhat larger than the width of region I. In equation (8.74) the + sign is taken when m is even giving cosine-like functions, and – for m odd giving sine-like states. σ is the spin index denoting the direction of spin, which we can take as spin-up or spin-down in the z-direction, and χ_σ is the corresponding Pauli two-spinor. γ_k is given by equation (8.50), and $\hat{\mathbf{k}}^\pm$ is a unit vector in the direction of $\mathbf{k}^\pm$. Each term in equation (8.74) is a relativistic plane-wave, so that $\chi_{gm\sigma}(\mathbf{r})$ is a relativistic standing wave in the interstitial region, with an overall structure the same as equation (8.53).

Inside the muffin-tins the RLAPW is a linear combination of solutions of the Dirac equation with the spherically symmetric part of the atomic potential, and for $\mathbf{r}$ inside sphere α we have

$$\chi_{gm\sigma}(\mathbf{r}) = \sum_\Lambda \left[A_\Lambda^\alpha(\mathbf{g}, m, \sigma)u_\Lambda^\alpha(\mathbf{r}_\alpha) + B_\Lambda^\alpha(\mathbf{g}, m, \sigma)\dot{u}_\Lambda^\alpha(\mathbf{r}_\alpha)\right], \tag{8.76}$$

where the sum is over the relativistic spherical quantum numbers $\Lambda = (\kappa, \mu)$, and $\mathbf{r}_\alpha$ is the position vector measured from the centre of sphere α. u_Λ^α is the four-component two-spinor solution of the Dirac equation at pivot mass-energy W,

$$u_\Lambda^\alpha(\mathbf{r}) = \begin{pmatrix} g_\kappa^\alpha(r)\Omega_{\kappa,\mu}(\hat{\mathbf{r}}) \\ if_\kappa^\alpha(r)\Omega_{-\kappa,\mu}(\hat{\mathbf{r}}) \end{pmatrix}, \tag{8.77}$$

with $\Omega_{\kappa,\mu}(\hat{\mathbf{r}})$ the spin-angular functions (see equation (8.38)) and $\dot{u}_\Lambda^\alpha = \partial u_\Lambda^\alpha/\partial W$. The coefficients A_Λ^α and B_Λ^α in equation (8.76) are determined by the requirement that the large and small components of the RLAPW are continuous across the boundary of the muffin-tin sphere.

Having set up the RLAPW basis functions, the self-consistent embedded surface calculations can be carried out in much the same way as in non-relativistic calculations. The particular surface which James and Crampin [3] chose to study was Au(111): there is a prominent surface state in the energy gap at the L-point in the bulk Brillouin zone, which is split by the Rashba effect [1]. This effect was first proposed theoretically for the two-dimensional electron gas [12], where a combination of spin–orbit coupling and symmetry-breaking in the direction perpendicular to the layer produces spin-splitting; in the case of the Au(111) surface state, where the region near the atomic nuclei dominates in the spin–orbit interaction, it is the asymmetry of the surface state wave-function near the nuclei which is the principal origin of the effect [13]. In these embedding calculations, James and Crampin used between one and three atomic layers in the surface region, compared with typically ~25 layers needed in slab geometry to reduce the interaction between the surface states on each side of the slab. An LDA exchange-correlation potential was used, with a fairly dense mesh of $\mathbf{k}_\parallel$-points in the surface Brillouin zone. Remarkably similar work-functions were obtained in the different calculations—5.505, 5.507 and 5.508 eV for one, two and three layers, respectively [11], in good agreement with the experimental work-function of 5.31 eV [14].

The dispersion of the surface state is shown in figure 8.4 as a function of surface wave-vector, calculated with just one surface atomic layer in region I [3]. We see very clearly the splitting of the surface state, and at the Fermi energy the wave-vector splitting is $\delta k_F = 0.029$ Å^{-1}; the embedded three-layer calculation gives exactly the same splitting. In their 24-layer pseudopotential calculation Mazzarello *et al* [15] obtained $\delta k_F = 0.032$ Å^{-1}, and a scattering calculation for the semi-infinite crystal [16], using the relativistic layer-Korringa–Kohn–Rostocker method, gave 0.023 Å^{-1}; the experimental value is (typically) $\delta k_F = 0.026$ Å^{-1} [17]. What is remarkable is that the splitting from the embedded one-layer calculation is so good, especially as slab calculations have shown that 40% of the splitting comes from the spin–orbit interaction with sub-surface layers [13]! This shows just how well relativistic embedding works.

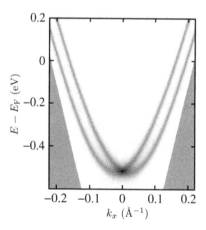

Figure 8.4. Surface state dispersion on Au(111) as a function of wave-vector parallel to the surface, showing Rashba splitting [3]. The shaded area is the continuum of bulk states. The calculation was performed with one surface atomic layer, embedded on to the bulk using the relativistic formalism. (From James and Crampin [3].)

8.3 The scalar-relativistic equation + spin–orbit coupling

The Dirac equation with a radial potential can be written in the form of a modified Schrödinger equation which we can split into two parts, the scalar-relativistic equation [4] and a spin–orbit term. Neglecting spin–orbit, the scalar-relativistic equation still includes some relativistic effects, like the variation in electron mass near the nucleus, and as its name implies, the electron is defined by a one-component wave-function. It is, in fact, generally used inside the muffin-tins in the construction of the LAPWs we described in section 3.2.1, and has been used in this way ever since the first papers on embedding at surfaces [18].

Adding the spin–orbit term on to the scalar-relativistic equation inside the muffin-tins, which is where it is important, we can describe the same physics—such as Rashba splitting of surface states—as by solving the full Dirac equation (section 8.2). The wave-functions become two-component spinors, and we can use standard LAPWs, multiplied by spin-up or spin-down spin states, as basis functions. The embedding potential then has the same form as with the Schrödinger equation, but with additional spin indices. This approach to relativistic electronic structure involves fairly minor modifications to existing embedding calculations compared with Dirac embedding, and important calculations of relativistic surface electronic structure are now being carried out in this way [19].

8.3.1 Spin–orbit derivation

In a spherically symmetric potential $V(r)$, the solution of the Dirac equation is given by the two-spinor (equation (8.77)), where the large and small components $g_\kappa(r)$ and $f_\kappa(r)$ (we drop the muffin-tin label α) satisfy [4, 5]

$$\frac{\mathrm{d}g_\kappa(r)}{\mathrm{d}r} = -\frac{\kappa + 1}{r}g_\kappa(r) + \frac{1}{c\hbar}\Big[W - V(r) + m_e c^2\Big]f_\kappa(r),$$

$$\frac{\mathrm{d}f_\kappa(r)}{\mathrm{d}r} = \frac{\kappa - 1}{r}f_\kappa(r) - \frac{1}{c\hbar}\Big[W - V(r) - m_e c^2\Big]g_\kappa(r). \tag{8.78}$$

We now define energy E as the mass-energy W with the rest-mass contribution subtracted,

$$E := W - m_e c^2, \tag{8.79}$$

and a position-dependent mass $M(r)$ by

$$M(r) := m_e + \frac{1}{2c^2}[E - V(r)]. \tag{8.80}$$

In terms of E and M, equations (8.78) can be written as

$$\frac{\mathrm{d}g_\kappa(r)}{\mathrm{d}r} = -\frac{\kappa + 1}{r}g_\kappa(r) + \frac{2Mc}{\hbar}f_\kappa(r), \tag{8.81}$$

$$\frac{\mathrm{d}f_\kappa(r)}{\mathrm{d}r} = \frac{\kappa - 1}{r}f_\kappa(r) + \frac{1}{c\hbar}[V(r) - E]g_\kappa(r). \tag{8.82}$$

By differentiating equation (8.81) and substituting from both equations we then obtain

$$-\frac{\hbar^2}{2M}\left(\frac{\mathrm{d}^2 g_\kappa}{\mathrm{d}r^2} + \frac{2}{r}\frac{\mathrm{d}g_\kappa}{\mathrm{d}r} - \frac{l(l+1)}{r^2}g_\kappa\right)$$

$$+V(r)g_\kappa - \frac{\hbar^2}{4M^2c^2}\frac{\mathrm{d}V}{\mathrm{d}r}\left(\frac{\mathrm{d}g_\kappa}{\mathrm{d}r} + \frac{\kappa+1}{r}g_\kappa\right) = Eg_\kappa, \qquad (8.83)$$

where using equation (8.36) we have replaced $\kappa(\kappa+1)$ by $l(l+1)$. The Hamiltonian in equation (8.83) can be split into two parts: the scalar-relativistic Hamiltonian H_{sr} and the spin–orbit part H_{so},

$$H_{\mathrm{sr}} = -\frac{\hbar^2}{2M}\left(\frac{\mathrm{d}^2}{\mathrm{d}r^2} + \frac{2}{r}\frac{\mathrm{d}}{\mathrm{d}r} - \frac{l(l+1)}{r^2}\right) + V(r) - \frac{\hbar^2}{4M^2c^2}\frac{\mathrm{d}V}{\mathrm{d}r}\frac{\mathrm{d}}{\mathrm{d}r}, \qquad (8.84)$$

$$H_{\mathrm{so}} = -\frac{\hbar^2}{4M^2c^2}\frac{\kappa+1}{r}\frac{\mathrm{d}V}{\mathrm{d}r}. \qquad (8.85)$$

The main feature of the scalar-relativistic Hamiltonian is that m_e is replaced by M (equation (8.80)), so that relativistic mass-velocity effects are taken into account; the final term in equation (8.84) is the Darwin term, associated with the phenomenon of *Zitterbewegung*[4]. It is straightforward to use the scalar-relativistic Hamiltonian H_{sr} inside the muffin-tins in LAPW calculations [18], and this is important to obtain the correct relative energies of the 5d bands in Au, for example [20].

Our main interest in this section is to describe spin–orbit splitting, which comes from H_{so} (equation (8.85)). Using the result from angular momentum addition [5] that

$$\hbar(\kappa+1)\Omega_{\kappa,\mu}(\hat{\mathbf{r}}) = -\mathbf{L}\cdot\boldsymbol{\sigma}\Omega_{\kappa,\mu}(\hat{\mathbf{r}}), \qquad (8.86)$$

we can rewrite the spin–orbit term as

$$H_{\mathrm{so}} = \frac{\hbar}{4M^2c^2r}\frac{\mathrm{d}V}{\mathrm{d}r}\mathbf{L}\cdot\boldsymbol{\sigma}, \qquad (8.87)$$

where $\mathbf{L}$ is the orbital angular momentum operator, and $\boldsymbol{\sigma}$ is the vector of Pauli spin matrices. This is a more convenient form than equation (8.85), as it can act on the product $Y_L(\hat{\mathbf{r}})\chi_\sigma$, where Y_L is a spherical harmonic and χ_σ is a Pauli two-spinor with spin-index σ—that is, spin-up or spin-down.

In the relativistic electronic structure calculations we consider in this section [19], equation (8.83) is used inside the muffin-tins, while the Schrödinger equation is used in the interstitial region—we see from the form of $M(r)$, the Darwin correction, and the spin–orbit term that all these are only important where the potential is very strong and rapidly varying. In general, the small component of the wave-function is

---

[4] *Zitterbewegung* arises from the oscillatory motion of the electron over a small region of space, giving it an apparent size [5].

neglected (this is the only approximation in what is otherwise a fully relativistic calculation), so the electron wave-functions are two-spinors,

$$\psi(\mathbf{r}) = \begin{pmatrix} \psi_\uparrow(\mathbf{r}) \\ \psi_\downarrow(\mathbf{r}) \end{pmatrix}, \tag{8.88}$$

$\psi_\uparrow(\mathbf{r})$ is the probability amplitude of finding an electron at $\mathbf{r}$ with spin-up, and similarly for $\psi_\downarrow(\mathbf{r})$. LAPW basis functions can be used with little modification: in the interstitial region the basis functions consist of the large component of equation (8.74)—plane-waves multiplied by a spin-up or spin-down two-spinor. Inside the muffin-tins the LAPW is built up of muffin-tin solutions $u_{l,\alpha}$ and $\dot{u}_{l,\alpha}$ satisfying the scalar-relativistic equation with the spherically symmetric part of the potential, *without* spin–orbit coupling. Putting these together, the LAPW is given by

$$\chi_{\mathbf{g}m\sigma}(\mathbf{r}) = \begin{cases} \chi_\sigma[\exp(i\mathbf{k}^+ \cdot \mathbf{r}) \pm \exp(i\mathbf{k}^- \cdot \mathbf{r})], & \mathbf{r} \in \text{interstitial region} \\ \chi_\sigma \displaystyle\sum_L [A_L^\alpha(\mathbf{g}, m)u_{l,\alpha}(r) + B_L^\alpha(\mathbf{g}, m)\dot{u}_{l,\alpha}(r)]Y_L(\theta, \phi), & \mathbf{r} \in \text{muffin tin } \alpha. \end{cases}$$
$$\tag{8.89}$$

The only difference with the LAPWs we discussed in section 3.2.1 is the factor of χ_σ, the Pauli spinor. The spin–orbit term inside the muffin-tins, just like the non-spherical part of the potential, is included when the matrix elements of the Hamiltonian are evaluated between the LAPWs [4, 21].

For embedding applications of this method, we note that the electron wave-function satisfies the Schrödinger equation in the region of the embedding surface. This means that the matching conditions are exactly the same as in the non-relativistic case, the only difference being that the wave-function has two components (equation (8.88)). So the definition of the embedding potential (2.15) becomes [19]

$$\frac{\partial \psi_\sigma(\mathbf{r}_S)}{\partial n_S} = -2 \sum_{\sigma'} \int_S d\mathbf{r}_S' \, \Sigma_{\sigma,\sigma'}(\mathbf{r}_S, \mathbf{r}_S')\psi_\sigma(\mathbf{r}_S'). \tag{8.90}$$

Apart from doubling all matrices to take into account spin-up and spin-down, the formalism for evaluating the embedding potential, for example by using the transfer matrix method, is just the same as we described in section 3.6. This means that we can include (almost) all relativistic effects in embedded surface calculations with the minimum of modifications to the method and computer codes—it's essentially a matter of doubling the size of basis set and including the spin–orbit term inside the muffin-tins.

8.3.2 Application to surface states

Ishida has modified his LAPW code in this way to include relativistic effects [19, 22], and has used this to study surface electronic structure. The embedding geometry is that shown in figure 3.20, with embedding on to the relativistic bulk embedding potential over P_L, and on to a constant vacuum potential over P_R.

To study Rashba splitting on Au(111), Ishida used three atomic layers in the surface region, and obtained a surface state dispersion very similar to figure 8.4, with a wave-vector splitting at the Fermi energy of $\delta k_F = 0.032$ Å^{-1}, in good agreement with the embedded Dirac equation calculation [3], and perfect agreement with the slab calculation of Mazzarello *et al* [15]. An interesting question which this embedding calculation was able to answer is the way that the surface states merge with bulk band edges. Slab calculations cannot handle this satisfactorily, as there is ultimately no distinction between surface and bulk states in slabs of finite thickness; in any case the surface states become more extended as they approach a band edge, leading inevitably to interaction between states on the two surfaces of the slab. Figure 8.5(b) shows the dispersion of the surface states as a function of the wave-vector in the $\bar{\Gamma} - \bar{K}$ direction in the surface Brillouin zone, with the energies taken from the peaks in the surface densities of states shown in figure 8.5(a). These surface densities of states were calculated with a very small energy broadening of 5×10^{-4} a.u. to give very accurate peak positions. We see that the lower surface state merges with the band edge of the magenta-shaded bulk band at about $k_x = 0.28$ a.u., while the upper surface state, after crossing the band gap minimum, finally merges with the band edge of the blue-shaded bulk band at about $k_x = 0.33$ a.u. An interesting feature of figure 8.5(a) is the way that the surface state merges with the bulk band, with a peak developing as the surface state approaches and enters the band edge—this is a very general phenomenon which we see in figure 2.5. The limitations of a slab calculation are shown in figure 8.5(c), where the energy splitting of the surface states $\Delta\epsilon(k_x)$ is plotted for the semi-infinite system (that is, the embedded surface) compared with a 35-layer slab. The behaviour in the slab calculation—even with such a thick slab–is inaccurate beyond $k_x \approx 0.2$ a.u.

Surprisingly, the Rashba splitting of the L-gap surface state is larger for Cu(111) than for Ag(111), though much smaller than for Au(111). High-resolution photoemission experiments have shown a Rashba splitting on Cu(111) of $\delta k_F = 0.0057$ Å^{-1} [23], and in an embedded surface calculation similar to that for Au(111) Ishida obtained $\delta k_F = 0.0057$ Å^{-1} [19]. Splitting on Ag(111) is considerably smaller, and remains unresolved experimentally. It has been suggested that this is due to the 3d bands in Cu being closer to the surface state than the 4d bands in Ag, leading to greater hybridization and a larger spin–orbit matrix element [19].

There is great interest at the moment in 'topologically protected' surface states on the surface of so-called topological insulators—an example of these are the spin–orbit split surface states on Sb(111) [24]. Ishida has used this relativistic surface embedding to calculate the electronic structure of the Sb(111) surface [19] with four atomic layers in the surface region, and the results are shown in figure 8.6. We see clearly from figure 8.6(a) that there are two surface state bands, degenerate at $\bar{\Gamma}$, coming from spin–orbit splitting. In the $\bar{\Gamma} - \bar{K}$ direction one band only crosses the Fermi energy, whereas in the $\bar{\Gamma} - \bar{M}$ direction both bands cross close to $\bar{\Gamma}$; but then further out, towards $\bar{M}$ the bands join the valence and conduction band edges in an interesting way. This is shown in detail in

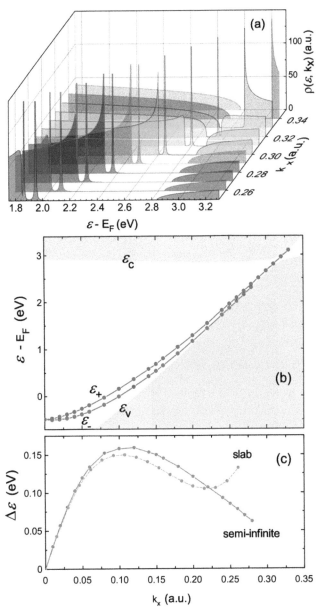

Figure 8.5. Surface electronic structure of Au(111), from an embedded three-layer relativistic calculation. (a) Surface density of states $n_s(E)$ (the local density of states integrated through the surface muffin-tin) plotted for different wave-vectors k_x in the $\bar{\Gamma} - \bar{K}$ direction, calculated with an imaginary part of the energy = 5×10^{-4} a.u. (b) Red and blue lines show the surface state energies $\epsilon_+(k_x)$ and $\epsilon_-(k_x)$ taken from the peaks in (a); the magenta- and blue-shaded areas show the projected bulk energies. (c) The green line shows $\Delta\epsilon(k_x)$, the energy splitting of the surface states, from the embedded surface calculation; the orange line shows $\Delta\epsilon(k_x)$ from a 35-layer slab calculation. (From Ishida [19].)

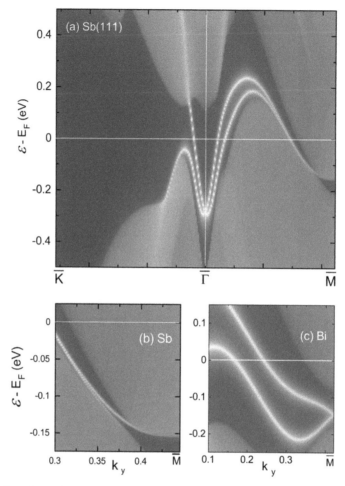

Figure 8.6. Surface electronic structure of Sb(111), from an embedded four-layer relativistic calculation. (a) Surface density of states $n_s(E)$ (the local density of states integrated through the surface muffin-tin) plotted as a function of wave-vector k_y in the $\bar{\Gamma} - \bar{M}$ direction, calculated with an imaginary part of the energy $= 1 \times 10^{-4}$ a.u. The brightness indicates the size of n_s. (b) The same as (a) close to $\bar{M}$ with higher energy resolution, calculated with an imaginary part of the energy $= 2 \times 10^{-5}$ a.u. (c) Surface density of states for Bi (111), $n_s(E)$ plotted as a function of wave-vector k_y in the $\bar{\Gamma} - \bar{M}$ direction. (From Ishida [19].)

figure 8.6(b), calculated with smaller energy broadening: we see that the lower of the spin-split bands joins the top of the valence band at $k_y \approx 0.37$ a.u. while the upper band merges with the bottom of the conduction band at $k_y \approx 0.39$ a.u. There are no surface bands in the gap at $\bar{M}$, in contrast with the behaviour on Bi(111), where as we see from figure 8.6(c), the two bands become degenerate inside the gap at $\bar{M}$. These results are in good agreement with an LCAO calculation of the surface states on Sb(111), in which a self-energy is used to embed the surface into the semi-infinite bulk [25], and with photoemission measurements [26, 27].

References

[1] LaShell S, McDougall B A and Jensen E 1996 Spin splitting of an Au(111) surface state band observed with angle resolved photoelectron spectroscopy *Phys. Rev. Lett.* **77** 3419–22

[2] Crampin S 2004 An embedding scheme for the Dirac equation *J. Phys.: Condens. Matter* **16** 8875–89

[3] James M and Crampin S 2010 Relativistic embedding method: The transfer matrix, complex band structures, transport, and surface calculations *Phys. Rev.* B **81** 155439

[4] Koelling D D and Harmon B N 1977 A technique for relativistic spin-polarised calculations *J. Phys. C: Solid St. Phys.* **10** 3107–14

[5] Strange P 1998 *Relativistic Quantum Mechanics* (Cambridge: Cambridge University Press)

[6] Zangwill A 2013 *Modern Electrodynamics* (Cambridge: Cambridge University Press)

[7] Stanton R E and Havriliak S 1984 Kinetic balance: A partial solution to the problem of variational safety in Dirac calculations *J. Chem. Phys.* **81** 1910–8

[8] Olver Frank W J, Lozier Daniel W, Boisvert Ronald F and Clark Charles W 2010 *NIST Handbook of Mathematical Functions* (Cambridge: Cambridge University Press)

[9] Greiner W 2000 *Relativistic Quantum Mechanics-Wave Equations* 3rd edn (Berlin: Springer-Verlag)

[10] Takeda T 1979 Linear methods for fully relativistic energy-band calculations, *J. Phys. F: Metal Phys.* **9** 815–29

[11] James M 2010 *Relativistic Embedding* PhD thesis (University of Bath)

[12] Bychkov Yu A and Rashba É I 1984 Properties of a 2D electron gas with lifted spectral degeneracy *JETP Lett.* **39** 78–81

[13] Bihlmayer G, Koroteev Yu M, Echenique P M, Chulkov E V and Blügel S 2006 The Rashba-effect at metal surfaces *Surf. Sci.* **600** 3888–91

[14] Hüfner S 2003 *Photoelectron Spectroscopy* 3rd edn (Berlin: Springer-Verlag)

[15] Mazzarello R, Corso A D and Tosatti E 2008 Spin-orbit modifications and splittings of deep surface states on clean Au(111). *Surf. Sci.* **602** 893–905

[16] Henk J, Ernst A and Bruno P 2003 Spin polarization of the l-gap surface states on Au(111) *Phys. Rev.* B **68** 165416

[17] Hoesch M, Muntwiler M, Petrov V, Hengsberger M, Patthey L, Shi M, Falub M, Greber T and Osterwalder J 2004 Spin structure of the Shockley surface state on Au(111) *Phys. Rev.* B **69** 241401

[18] Inglesfield J E and Benesh G A 1988 Surface electronic structure: Embedded self-consistent calculations *Phys. Rev.* B **37** 6682–700

[19] Ishida H 2014 Rashba spin splitting of Shockley surface states on semi-infinite crystals *Phys. Rev.* B **90** 235422

[20] Egede Christensen N 1984 Relativistic band structure calculations *Int. J. Quantum Chem.* **25** 233–61

[21] MacDonald A H, Pickett W E and Koelling D D 1980 A linearised relativistic augmented-plane-wave method utilising approximate pure spin basis functions *J. Phys. C: Solid St. Phys.* **13** 2675–83

[22] Ishida H 2001 Surface-embedded Green-function method: A formulation using a linearized-augmented-plane-wave basis set *Phys. Rev.* B **63** 165409

[23] Tamai A, Meevasana W, King P D C, Nicholson C W, de la Torre A, Rozbicki E and Baumberger F 2013 Spin-orbit splitting of the Shockley surface state on Cu(111) *Phys. Rev.* B **87** 075113

[24] Teo Jeffrey C Y, Fu L and Kane C L 2008 Surface states and topological invariants in three-dimensional topological insulators: Application to $Bi_{1-x}Sb_x$. *Phys. Rev.* B **78** 045426

[25] Narayan A, Rungger I and Sanvito S 2012 Topological surface states scattering in antimony *Phys. Rev.* B **86** 201402

[26] Sugawara K, Sato T, Souma S, Takahashi T, Arai M and Sasaki T 2006 Fermi surface and anisotropic spin-orbit coupling of Sb(111), studied by angle-resolved photoemission spectroscopy *Phys. Rev. Lett.* **96** 046411

[27] Hsieh D *et al* 2009 Observation of unconventional quantum spin textures in topologically ordered materials *Science* **323** 919–22

IOP Publishing

The Embedding Method for Electronic Structure

John E Inglesfield

Chapter 9

Embedding in electromagnetism

Considering the title of the book, a chapter on electromagnetism may seem out of place. However, over the last 20 years or so there has been renewed interest in solving Maxwell's equations with the advent of completely new subjects like photonics and plasmonics [1, 2] and, of course most exciting of all, metamaterials [3]. Here, I hope to convince you that embedding is a useful addition to the armoury of methods for solving these classical equations. The systems which we shall study in this chapter are mostly periodic, with Bloch-type solutions of the electromagnetic wave equations, so there is at least some affinity with electronic structure. Moreover, we shall learn something about embedding on the way.

In this chapter we shall use embedding in the context of solving the Helmholtz wave equation for the electric or magnetic field, periodic in time [4]. The embedding method enters in two ways. Firstly, we often have to solve this equation for a system surrounded by vacuum or some other extended medium, into which electromagnetic waves may decay. In the same way that the semi-infinite substrate in problems of surface electronic structure can be replaced by an embedding potential, an extended medium in the electromagnetic case can also be replaced using embedding. The second application of embedding is to replace a dielectric object, such as a metal sphere in a photonic structure, by an embedding 'potential' (I use inverted commas advisedly) over its surface. This is similar to the way in which we can replace an atom by an embedding potential in the finite difference method of solving the Schrödinger equation (section 6.3.2), and it is useful for similar reasons. The perpendicular components of the electromagnetic fields ($E_\perp$, $H_\perp$) are discontinuous across a dielectric interface, and replacing dielectric objects by embedding 'potentials' over their surfaces means that we only have to expand the fields in a region where they are continuous, outside the objects.

Both the Helmholtz equation [5] and the Schrödinger equation are eigenvalue problems, but a fundamental difference is that the electric and magnetic fields are *vector* fields, whereas the wave-function is a scalar. This means that instead of an embedding potential, we are dealing with an embedding tensor operator $\Sigma(\mathbf{r}_S, \mathbf{r}'_S)$ (hence the inverted commas above), which relates the surface-parallel components of

doi:10.1088/978-0-7503-1042-0ch9

the magnetic and electric fields on the boundary of region II. An additional complication in embedding the electromagnetic problem is that solutions of Laplace's equation [5], the equation for electrostatics, can creep in and upset the method. Fortunately, these modes can be effectively suppressed [6].

9.1 Embedding Maxwell's equations

9.1.1 Variational method for the Helmholtz equation

We start off with the time-varying Maxwell's equations coupling the electric field **E** and magnetic field **H**, which in Gaussian units are [7]

$$\nabla \times \mathbf{H} = \frac{\epsilon}{c}\frac{\partial \mathbf{E}}{\partial t}, \qquad \nabla \times \mathbf{E} = -\frac{\mu}{c}\frac{\partial \mathbf{H}}{\partial t}, \tag{9.1}$$

where c is the velocity of light, and ϵ and μ are the dielectric constant and magnetic permeability, respectively. The dielectric constant is in general a function of position $\epsilon(\mathbf{r})$, but at this stage we assume a constant permeability $\mu = 1$. We look for solutions of equation (9.1) with time-dependence $\exp(-i\,\omega t)$, and substituting for **H** gives the Helmholtz wave equation for **E** [5],

$$\nabla \times \nabla \times \mathbf{E}(\mathbf{r}) = \epsilon(\mathbf{r})\frac{\omega^2}{c^2}\mathbf{E}(\mathbf{r}). \tag{9.2}$$

To avoid carrying c^2 around, we shall set it equal to 1 (in any case, this is true in the reduced units we use for our results). The variational expression for ω^2 equivalent to equation (9.2) is then given by

$$\omega^2 = \frac{\int d\mathbf{r}\, \mathbf{E}^* \cdot (\nabla \times \nabla \times \mathbf{E})}{\int d\mathbf{r}\, \epsilon\, \mathbf{E}^* \cdot \mathbf{E}}, \tag{9.3}$$

which is stationary for variations in $\mathbf{E}(\mathbf{r})$ about the exact solution. Using the vector identity

$$\nabla \cdot (\mathbf{F} \times \nabla \times \mathbf{E}) = (\nabla \times \mathbf{F}) \cdot (\nabla \times \mathbf{E}) - \mathbf{F} \cdot (\nabla \times \nabla \times \mathbf{E}), \tag{9.4}$$

together with the divergence theorem, this becomes

$$\omega^2 = \frac{\int d\mathbf{r}(\nabla \times \mathbf{E}^*) \cdot (\nabla \times \mathbf{E}) - \int_S d\mathbf{r}_S\, \mathbf{n} \cdot (\mathbf{E}^* \times \nabla \times \mathbf{E})}{\int d\mathbf{r}\, \epsilon\, \mathbf{E}^* \cdot \mathbf{E}}, \tag{9.5}$$

where the surface integral in the numerator is over an external surface S. If $\mathbf{E} = 0$ over S this gives

$$\omega^2 = \frac{\int d\mathbf{r}(\nabla \times \mathbf{E}^*) \cdot (\nabla \times \mathbf{E})}{\int d\mathbf{r}\, \epsilon\, \mathbf{E}^* \cdot \mathbf{E}}, \tag{9.6}$$

a form in which the numerator is explicitly Hermitian (compare this with equation (2.28)).

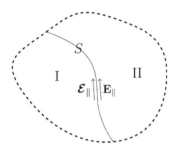

Figure 9.1. $\mathcal{E}(\mathbf{r})$ is the trial field in region I, and in region II **E** is the exact solution of the Helmholtz equation at ω_0^2 such that $\mathbf{E}_\parallel(\mathbf{r}_S) = \mathcal{E}_\parallel(\mathbf{r}_S)$.

We now consider the embedding problem [4], in which we take a trial electric field in equation (9.5) consisting of an arbitrary trial function $\mathcal{E}(\mathbf{r})$ in region I, and in region II the exact solution of equation (9.2) at a trial value ω_0^2, such that the surface-parallel components of the fields match over S (figure 9.1),

$$\mathbf{E}_\parallel(\mathbf{r}_S) = \mathcal{E}_\parallel(\mathbf{r}_S). \tag{9.7}$$

This condition uniquely determines **E** in II. The variational expression (9.5) then becomes

$$\omega^2 = \frac{\int_I d\mathbf{r} (\nabla \times \mathcal{E}^*) \cdot (\nabla \times \mathcal{E}) + \int_{II} d\mathbf{r} (\nabla \times \mathbf{E}^*) \cdot (\nabla \times \mathbf{E})}{\int_I d\mathbf{r}\, \epsilon\, \mathcal{E}^* \cdot \mathcal{E} + \int_{II} d\mathbf{r}\, \epsilon\, \mathbf{E}^* \cdot \mathbf{E}}. \tag{9.8}$$

Again making use of equation (9.4) and the divergence theorem, equation (9.8) can be rewritten as

$$\omega^2 = \frac{\int_I d\mathbf{r} (\nabla \times \mathcal{E}^*) \cdot (\nabla \times \mathcal{E}) + \omega_0^2 \int_{II} d\mathbf{r}\, \epsilon\, \mathbf{E}^* \cdot \mathbf{E} - \int_S d\mathbf{r}_S\, \mathbf{n} \cdot (\mathbf{E}^* \times \nabla \times \mathbf{E})}{\int_I d\mathbf{r}\, \epsilon\, \mathcal{E}^* \cdot \mathcal{E} + \int_{II} d\mathbf{r}\, \epsilon\, \mathbf{E}^* \cdot \mathbf{E}}, \tag{9.9}$$

where the final term in the numerator is an integral over S, the boundary between I and II with **n** the outward surface normal from I to II. This is beginning to look something like equation (2.5), though the last term in the numerator looks complicated. As $\nabla \times \mathbf{E}$ is proportional to the magnetic field **H**, this term is essentially the cross product of $\mathbf{E}_\parallel^*(\mathbf{r}_S)$ with $\mathbf{H}_\parallel(\mathbf{r}_S)$.

The next stage is to eliminate the volume integrals over region II in equation (9.9), rewriting them in terms of surface integrals over S. In region II **E** satisfies the equation

$$\nabla \times \nabla \times \mathbf{E}(\mathbf{r}) = \epsilon(\mathbf{r}) \omega_0^2\, \mathbf{E}(\mathbf{r}), \tag{9.10}$$

and differentiating with respect to the parameter ω_0^2 gives

$$\nabla \times \nabla \times \frac{\partial \mathbf{E}}{\partial \omega_0^2} = \epsilon(\mathbf{r})\mathbf{E}(\mathbf{r}) + \epsilon(\mathbf{r})\omega_0^2 \frac{\partial \mathbf{E}}{\partial \omega_0^2}. \tag{9.11}$$

Following essentially the same procedure as in section 2.1, we now take the dot product of equation (9.11) with $\mathbf{E}^*$, the dot product of equation (9.10) complex conjugated with $\partial \mathbf{E}/\partial \omega_0^2$, subtract, and integrate through region II to give

$$\int_{\mathrm{II}} d\mathbf{r}\, \epsilon \mathbf{E}^* \cdot \mathbf{E} = \int_{\mathrm{II}} d\mathbf{r}\left[\nabla \cdot \left(\frac{\partial \mathbf{E}}{\partial \omega_0^2} \times \nabla \times \mathbf{E}^* \right) - \nabla \cdot \left(\mathbf{E}^* \times \frac{\partial}{\partial \omega_0^2} \nabla \times \mathbf{E} \right) \right]$$

$$= \int_{S} d\mathbf{r}_S\left[\mathbf{n} \cdot \left(\mathbf{E}^* \times \frac{\partial}{\partial \omega_0^2} \nabla \times \mathbf{E} \right) - \mathbf{n} \cdot \left(\frac{\partial \mathbf{E}}{\partial \omega_0^2} \times \nabla \times \mathbf{E}^* \right) \right]. \tag{9.12}$$

We have used the divergence theorem to go from the volume integral through II to the surface integral over S. But $\mathbf{E}_\parallel$ on S is fixed, hence the second term in equation (9.12) vanishes and we are left with the interesting and useful result that

$$\int_{\mathrm{II}} d\mathbf{r}\, \epsilon \mathbf{E}^* \cdot \mathbf{E} = \int_{S} d\mathbf{r}_S\, \mathbf{n} \cdot \left(\mathbf{E}^* \times \frac{\partial}{\partial \omega_0^2} \nabla \times \mathbf{E} \right). \tag{9.13}$$

Substituting equation (9.13) into equation (9.9), and using equation (9.7) gives us

$$\omega^2 = \frac{\displaystyle\int_{\mathrm{I}} d\mathbf{r}(\nabla \times \mathcal{E}^*) \cdot (\nabla \times \mathcal{E}) - \int_{S} d\mathbf{r}_S\, \mathbf{n} \cdot \left[\mathcal{E}^* \times \left(\nabla \times \mathbf{E} - \omega_0^2 \frac{\partial}{\partial \omega_0^2} \nabla \times \mathbf{E} \right) \right]}{\displaystyle\int_{\mathrm{I}} d\mathbf{r}\, \epsilon\, \mathcal{E}^* \cdot \mathcal{E} + \int_{S} d\mathbf{r}_S\, \mathbf{n} \cdot \left(\mathcal{E}^* \times \frac{\partial}{\partial \omega_0^2} \nabla \times \mathbf{E} \right)}, \tag{9.14}$$

an expression in which we have eliminated region II, except for the surface-parallel component of $\nabla \times \mathbf{E}$ on S. This can be written in terms of the surface-parallel component of $\mathcal{E}$ as

$$[(\nabla \times \mathbf{E})_\parallel]_{\mathbf{r}_S} = \int_{S} d\mathbf{r}_S'\, \boldsymbol{\Sigma}(\mathbf{r}_S, \mathbf{r}_S'; \omega_0^2) \cdot [\mathbf{n} \times \mathcal{E}(\mathbf{r}_S')], \tag{9.15}$$

where $\boldsymbol{\Sigma}$ is the embedding tensor operator. Using equation (9.15), and the properties of the scalar triple product, the embedded variational expression takes its final form,

$$\omega^2 = \frac{\displaystyle\int_{\mathrm{I}} d\mathbf{r}(\nabla \times \mathcal{E}^*) \cdot (\nabla \times \mathcal{E}) - \int_{S} d\mathbf{r}_S \int_{S} d\mathbf{r}_S'(\mathbf{n} \times \mathcal{E}^*) \cdot \left(\boldsymbol{\Sigma} - \omega_0^2 \frac{\partial \boldsymbol{\Sigma}}{\partial \omega_0^2} \right) \cdot (\mathbf{n} \times \mathcal{E})}{\displaystyle\int_{\mathrm{I}} d\mathbf{r}\, \epsilon\, \mathcal{E}^* \cdot \mathcal{E} + \int_{S} d\mathbf{r}_S \int_{S} d\mathbf{r}_S'(\mathbf{n} \times \mathcal{E}^*) \cdot \frac{\partial \boldsymbol{\Sigma}}{\partial \omega_0^2} \cdot (\mathbf{n} \times \mathcal{E})}, \tag{9.16}$$

which, like its Schrödinger equation counterpart (equation (2.22)), only involves the trial function in region I and on S.

The simplest way of finding the embedding tensor is to solve Maxwell's equations explicitly in region II, for a given $E_{\parallel}(r_S)$. This boundary condition uniquely defines $E(r)$ in region II, hence we can find $\nabla \times E$ and from equation (9.15) Σ. We shall use examples of this approach throughout the chapter, and it is analogous to the way that we used the logarithmic derivative of the solution of the Schrödinger equation to find the scalar embedding potential (see, for example, equation (2.29)). The embedding tensor can, in fact, be written explicitly in terms of the tensor (or dyadic) Green function for region II, satisfying the inhomogeneous Helmholtz equation [8]

$$\nabla_r \times \nabla_r \times \Gamma(r, r'; \omega^2) - \epsilon(r)\omega^2\Gamma(r, r'; \omega^2) = -1\delta(r - r'), \qquad (9.17)$$

with the homogeneous boundary condition on S that $(n \times \Gamma) = 0$. Then Σ is given by

$$\Sigma(r_S, r'_S) = -[(\nabla_r \times)(\nabla_{r'} \times)\Gamma]_{r_S, r'_S}. \qquad (9.18)$$

This result, which is proved in [4], is analogous to equation (6.61).

We note that Σ defined by equation (9.15) is a non-local form of the impedance boundary condition (actually its inverse), widely used in solving electromagnetic boundary condition problems [9]. Unlike the impedance boundary condition, which is not exact, equation (9.15) is an exact expression giving the surface-parallel component of the magnetic field in terms of the surface-parallel component of the electric field.

9.1.2 First tests, and eliminating Laplace solutions

To find the fields for which equation (9.16) is minimum or stationary, we follow the same procedure as in section 2.2 and expand $\mathcal{E}$ in region I in terms of a set of functions, here vector functions,

$$\mathcal{E}(r) = \sum_n e_n F_n(r). \qquad (9.19)$$

Substituting into equation (9.16), and varying the coefficients e_n, gives the matrix eigenvalue equation,

$$\sum_n A_{mn} e_n = \omega^2 \sum_n B_{mn} e_n, \qquad (9.20)$$

in which the matrices are given by

$$A_{mn} = \int_I dr (\nabla \times F_m^*) \cdot (\nabla \times F_n) - \int_S dr_S \int_S dr'_S (n \times F_m^*) \cdot \left(\Sigma - \omega_0^2 \frac{\partial \Sigma}{\partial \omega_0^2}\right) \cdot (n \times F_n),$$
$$(9.21)$$

and

$$B_{mn} = \int_I dr\, \epsilon F_m^* \cdot F_n + \int_S dr_S \int_S dr'_S (n \times F_m^*) \cdot \frac{\partial \Sigma}{\partial \omega_0^2} \cdot (n \times F_n). \qquad (9.22)$$

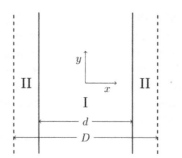

Figure 9.2. Vacuum (region I) between semi-infinite dielectrics (region II, shaded green). The origin $x = 0$ is in the middle, and the z-axis is perpendicular to and directed out of the page to give a right-handed coordinate system. D defines the basis functions.

From the structure of these matrices, we see that the embedding tensor terms grouped together are

$$\int_S dr_S \int_S dr'_S (\mathbf{n} \times \mathbf{F}_m^*) \cdot \left(\mathbf{\Sigma} + (\omega^2 - \omega_0^2) \frac{\partial \mathbf{\Sigma}}{\partial \omega_0^2} \right) \cdot (\mathbf{n} \times \mathbf{F}_n), \qquad (9.23)$$

and, exactly as in equation (2.26), the derivative provides a first-order correction so that the embedding tensor is evaluated at the correct value of ω^2. Having found the eigenvalue ω^2 for an initial trial ω_0^2, ω_0^2 is set equal to ω^2, the embedding tensor is re-evaluated and the procedure iterated. In practice only a few iterations are needed until the output ω^2 is equal to the input ω_0^2 to sufficient accuracy.

As a first application we consider light propagating between semi-infinite dielectrics (figure 9.2) with a dielectric constant $\epsilon < 1$ so that we can have trapped waveguide modes [4, 6]. The vacuum in the middle is region I, and the dielectrics on either side constitute region II, which we shall replace by embedding tensors over the interfaces at $x = \pm d/2$. We consider modes propagating in the y-direction with wave-vector k_y, and the electric field lying in the xy-plane to give transverse magnetic (TM) modes. With fixed k_y, our problem is to find the frequencies ω_i of the guided modes. As the fields vary like $\exp ik_y y$ in the y-direction, we have a one-dimensional problem in the x-direction.

To find the embedding tensor, we use the exact solution of Maxwell's equations in the dielectrics, which for a frequency such that $\omega^2 < \frac{k_y^2}{\epsilon}$ is an evanescent wave; on the right-hand side this has the form

$$\mathbf{E} = E_0 \left(-k_y \hat{\mathbf{x}} + i\gamma \hat{\mathbf{y}} \right) \exp(-\gamma x + ik_y y), \quad x > d/2, \qquad (9.24)$$

where $\gamma = \sqrt{k_y^2 - \epsilon \omega^2}$, and $\hat{\mathbf{x}}$, $\hat{\mathbf{y}}$ are unit vectors in the x- and y-directions. Taking the curl of this field gives

$$\nabla \times \mathbf{E} = E_0 \, i\epsilon \omega^2 \hat{\mathbf{z}} \exp(-\gamma x + ik_y y), \quad x > d/2, \qquad (9.25)$$

so from equation (9.15) the embedding tensor is given by

$$\mathbf{\Sigma} = \frac{\epsilon\omega^2}{\gamma}\hat{\mathbf{z}}\hat{\mathbf{z}}.\tag{9.26}$$

The real value of $\mathbf{\Sigma}$ corresponds to total internal reflection at the vacuum/dielectric interface, but for $\omega^2 > \frac{k_y^2}{\epsilon}$ the modes in region I can leak into the dielectric and we have

$$\mathbf{\Sigma} = \frac{i\epsilon\omega^2}{k_x}\hat{\mathbf{z}}\hat{\mathbf{z}},\tag{9.27}$$

where $k_x = \sqrt{\epsilon\omega^2 - k_y^2}$. When ω^2 is complex, with a positive imaginary part, we take the root with a positive real part—the Fortran convention. The embedding tensor that replaces the left-hand dielectric is identical.

We can now look for solutions of the embedded eigenvalue equation (9.20). As we need transverse solutions, we try expanding $\mathcal{E}$ using transverse plane waves as basis functions (9.19),

$$\mathbf{F}_m^t = \left(k_y\hat{\mathbf{x}} - g_m\hat{\mathbf{y}}\right)\exp i\left(g_m x + k_y y\right), \quad g_m = \frac{2m\pi}{D},\tag{9.28}$$

where m runs from $-M \to +M$. D is chosen to be greater than the separation of the dielectrics (figure 9.2), so that the relationship between the trial electric field and the corresponding magnetic field takes a range of values. Table 9.1 shows the eigenfrequencies for the confined modes, taking the dielectric constant in the semi-infinite media in region II as $\epsilon = 0.5$, and using $D = 3d/2$. We use dimensionless reduced units[1] for wave-vector and frequency, given by

$$\tilde{k}_y = \frac{k_y d}{2\pi}, \qquad \tilde{\omega} = \frac{\omega d}{2\pi c},\tag{9.29}$$

and the results are for $\tilde{k}_y = 1.0$. We see that two of the frequencies converge to the symmetric and antisymmetric waveguide modes, whereas the two other frequencies tend towards zero with increasing basis set size. These correspond to solutions of

Table 9.1. Frequencies of confined modes in vacuum between dielectric media with $\epsilon = 0.5$, at $\tilde{k}_y = 1.0$, for different numbers of transverse waves $(2M + 1)$ in the basis set.

Basis size	$\tilde{\omega}_1$	$\tilde{\omega}_2$	$\tilde{\omega}_3$	$\tilde{\omega}_4$
3	1.08957	1.24163	0.83682	
7	1.07858	1.27214	0.20098	0.96624
11	1.07842	1.26437	0.03102	0.25323
15	1.07842	1.26419	0.00424	0.04766
Exact	1.07842	1.26419	Laplace	Laplace

---

[1] These reflect the scaling properties of Maxwell's equations. The speed of light becomes $\tilde{c} = \tilde{\omega}/\tilde{k} = 1$.

Laplace's equation, appearing with finite frequency at finite basis set size because they are approximated by the expansion.

In general, the approximate solutions to Laplace's equations corrupt the true solutions of Maxwell's equations which we are looking for. In this particular case there are only two Laplace solutions at fixed k_y, corresponding to the two degrees of freedom with two planar surfaces, and increasing the size of basis set ultimately pushes them to frequencies close to zero, out of harm's way. However, in more complicated geometries, such as a lattice of dielectric spheres, the number of Laplace solutions increases without limit as the basis set size increases, and corruption of the true solutions is inevitable. We need a simple method of removing them!

Fortunately we can achieve this quite simply and generally by adding *longitudinal* waves to the basis set [6]. These waves are given by

$$\mathbf{F}_n^l = \left(g_n \hat{\mathbf{x}} + k_y \hat{\mathbf{y}}\right) \exp \mathrm{i}\left(g_n x + k_y y\right), \qquad g_n = \frac{2n\pi}{D}, \tag{9.30}$$

where n runs from $-N \rightarrow +N$. These zero-curl modes soak up the Laplace solutions, and in this one-dimensional case both spurious solutions are removed when the number of longitudinal basis functions is greater than the number of transverse. The frequencies of the confined modes are shown in table 9.2 for different basis set sizes, and we see rapid convergence to the exact values. The convergence behaviour is interesting, with frequencies increasing and converging as N increases at fixed M; these values decrease to the exact eigenfrequencies as M increases.

9.1.3 Leaky electromagnetic waves and the continuum

At frequencies satisfying $\omega^2 > \frac{k_y^2}{\epsilon}$, electromagnetic waves in the dielectric structure of figure 9.2 are no longer confined by total internal reflection, and can leak out of the central vacuum region. This gives a continuum of modes, and we must consider the spectral density, akin to the local density of states for electrons in the continuum above the vacuum level (section 2.5).

Table 9.2. Frequencies of confined modes in vacuum between dielectric media with $\epsilon = 0.5$, at $\tilde{k}_y = 1.0$, for different numbers of transverse waves ($2M + 1$) and longitudinal waves ($2N + 1$) in the basis set.

Trans.	Long.	$\tilde{\omega}_1$	$\tilde{\omega}_2$
3	5	1.07550	1.26737
3	9	1.07850	1.26756
3	13	1.07857	1.26756
7	9	1.07839	1.26420
7	13	1.07842	1.26420
9	11	1.07842	1.26419
	Exact	1.07842	1.26419

We define the (electric) spectral density in the same way as the local density of states (equation (2.34)), by a sum over eigenstates of the Helmholtz equation (9.2),

$$n(\mathbf{r}, \omega) = \sum_i \epsilon(\mathbf{r}) \mathbf{E}_i(\mathbf{r}) \cdot \mathbf{E}_i^*(\mathbf{r}) \delta(\omega - \omega_i). \tag{9.31}$$

It is convenient to include the dielectric constant, because this gives an electrostatic energy density. The spectral density can be written in terms of the tensor Green function (9.17), which in a spectral representation is given by

$$\boldsymbol{\Gamma}(\mathbf{r}, \mathbf{r}'; \omega^2) = \sum_i \frac{\mathbf{E}_i(\mathbf{r}) \mathbf{E}_i^*(\mathbf{r}')}{\omega^2 - \omega_i^2}, \tag{9.32}$$

where the eigenfunctions are normalized by

$$\int d\mathbf{r}\, \epsilon(\mathbf{r}) \mathbf{E}_i(\mathbf{r}) \cdot \mathbf{E}_i^*(\mathbf{r}) = 1, \tag{9.33}$$

including ϵ as a weighting factor. As $d(\omega^2) = 2\omega\, d\omega$ we see that

$$n(\mathbf{r}, \omega) = -\frac{2\omega}{\pi} \epsilon(\mathbf{r}) \Im \boldsymbol{\Gamma}(\mathbf{r}, \mathbf{r}; \omega^2 + i\eta), \tag{9.34}$$

where, as in equation (2.35), η is a positive infinitesimal.

The Green function in region I can be expanded in the same basis as we used for the eigenstates (equation (9.19)),

$$\boldsymbol{\Gamma}(\mathbf{r}, \mathbf{r}'; \omega^2) = \sum_{mn} \Gamma_{mn}(\omega^2) \mathbf{F}_m(\mathbf{r}) \mathbf{F}_n^*(\mathbf{r}'), \tag{9.35}$$

where the Greenian matrix satisfies the inhomogeneous version of equation (9.20),

$$\sum_m \left[A_{km}(\omega^2) - \omega^2 B_{km} \right] \Gamma_{mn}(\omega^2) = -\delta_{kn}. \tag{9.36}$$

Matrices A and B are given by

$$A_{mn}(\omega^2) = \int_{\mathrm{I}} d\mathbf{r} (\nabla \times \mathbf{F}_m^*) \cdot (\nabla \times \mathbf{F}_n) - \int_S d\mathbf{r}_S \int_S d\mathbf{r}_S' (\mathbf{n} \times \mathbf{F}_m^*) \cdot \boldsymbol{\Sigma}(\omega^2) \cdot (\mathbf{n} \times \mathbf{F}_n), \tag{9.37}$$

and

$$B_{mn} = \int_{\mathrm{I}} d\mathbf{r}\, \epsilon \mathbf{F}_m^* \cdot \mathbf{F}_n. \tag{9.38}$$

These are the same as equations (9.21) and (9.22) without the $\partial\boldsymbol{\Sigma}/\partial\omega_0^2$ terms—these cancel because the embedding tensor is evaluated with $\omega_0^2 = \omega^2$, the value at which we evaluate $\boldsymbol{\Gamma}$.

We now evaluate the Green function, hence the spectral density, for the system shown in figure 9.2, a vacuum layer between semi-infinite dielectric regions, using the same parameters that we used for the confined modes presented in table 9.2. The results for $\tilde{n}_{\mathrm{I}}(\tilde{\omega})$, the spectral density integrated through region I as a function of

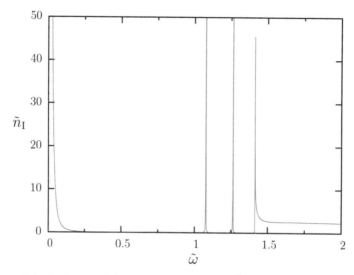

Figure 9.3. Spectral density integrated through region I, $\tilde{n}_I(\tilde{\omega})$, at $\tilde{k}_y = 1.0$ in a vacuum slab between semi-infinite dielectric media with $\epsilon = 0.5$. The basis set consists of nine transverse waves and 11 longitudinal waves, with $D = 3d/2$. The imaginary part of $\tilde{\omega}^2$ is $\tilde{\eta} = 0.0002$.

frequency $\tilde{\omega}$, are shown in figure 9.3 for $\tilde{k}_y = 1.0$ (all in scaled units, remember). A basis set of nine transverse waves and 11 longitudinal waves was used. First, we observe the large peak at $\tilde{\omega} = 0$, broadened by $\tilde{\eta}$, which comes from zero-frequency longitudinal modes; the larger the number of longitudinal basis functions we use, the larger this peak. The physical features are the two peaks at $\tilde{\omega} = 1.078$ and 1.264 corresponding to the confined modes presented in table 9.2, and the continuum of states above $\tilde{\omega} = 1.414$. States in the continuum do not suffer total internal reflection at the vacuum/dielectric interface, and can leak out of region I into the dielectric on either side. But we notice the large peak at the continuum edge—this comes from an *incipient* confined state, exactly analogous to the incipient bound state in figure 2.5 which is pulled off as the surface potential well deepens. In this case, a third confined state appears as $\tilde{k}_y$ increases.

9.1.4 Working with the magnetic field

Instead of working with the Helmholtz equation for the electric field (equation (9.2)), it is sometimes convenient to work with the magnetic field [10, 11]. This is particularly the case in systems which are uniform in one direction, such as systems with cylindrical symmetry, or the waveguide shown in figure 9.2 which is uniform in the z-direction. If we take this last example, the TM polarization we treated in section 9.1.2 with an electric field in the xy-plane corresponds to a magnetic field in the z-direction, which satisfies a *scalar* wave equation. As well as having simpler matrix elements, the scalar problem avoids the problems of longitudinal modes, and we can work with smaller basis sets. Of course this works the other way round, and the polarization with the magnetic field in the xy-plane becomes a scalar problem in E_z.

First, let us derive the embedding principle for the wave equation of the magnetic field. Assuming $\mu = 1$, the Helmholtz equation for $\mathbf{H}$ is given by

$$\nabla \times \frac{1}{\epsilon(\mathbf{r})} \nabla \times \mathbf{H}(\mathbf{r}) = \omega^2 \mathbf{H}(\mathbf{r}). \tag{9.39}$$

Note that the factor of $1/\epsilon(\mathbf{r})$ is inside the outer curl in this equation. Following the general lines of section 9.1.1, the variational expression for ω^2 is given by

$$\omega^2 = \frac{\int d\mathbf{r}\, \mathbf{H}^* \cdot \left(\nabla \times \frac{1}{\epsilon} \nabla \times \mathbf{H} \right)}{\int d\mathbf{r}\, \mathbf{H}^* \cdot \mathbf{H}}, \tag{9.40}$$

and using the vector identity (equation (9.4)) and the divergence theorem we can write this in the combined system I + II as

$$\omega^2 = \frac{\int_{\mathrm{I}} d\mathbf{r}\, \frac{1}{\epsilon_{\mathrm{I}}} (\nabla \times \mathcal{H}^*) \cdot (\nabla \times \mathcal{H}) + \int_{\mathrm{II}} d\mathbf{r}\, \frac{1}{\epsilon_{\mathrm{II}}} (\nabla \times \mathbf{H}^*) \cdot (\nabla \times \mathbf{H})}{\int_{\mathrm{I}} d\mathbf{r}\, \mathcal{H}^* \cdot \mathcal{H} + \int_{\mathrm{II}} d\mathbf{r}\, \mathbf{H}^* \cdot \mathbf{H}}. \tag{9.41}$$

As in equation (9.8), $\mathcal{H}(\mathbf{r})$ is a trial field in region I, and $\mathbf{H}$ is the solution of equation (9.39) at ω_0^2 in region II which matches $\mathcal{H}$ over the surface S. This can be rewritten as

$$\omega^2 = \frac{\int_{\mathrm{I}} d\mathbf{r}\, \frac{1}{\epsilon_{\mathrm{I}}} (\nabla \times \mathcal{H}^*) \cdot (\nabla \times \mathcal{H}) + \omega_0^2 \int_{\mathrm{II}} d\mathbf{r}\, \mathbf{H}^* \cdot \mathbf{H} - \int_{S} d\mathbf{r}_S\, \mathbf{n} \cdot \left(\mathbf{H}^* \times \frac{1}{\epsilon_{\mathrm{II}}} \nabla \times \mathbf{H} \right)}{\int_{\mathrm{I}} d\mathbf{r}\, \mathcal{H}^* \cdot \mathcal{H} + \int_{\mathrm{II}} d\mathbf{r}\, \mathbf{H}^* \cdot \mathbf{H}}. \tag{9.42}$$

We now define the embedding operator to replace region II by the relationship

$$\frac{1}{\epsilon_{\mathrm{II}}(\mathbf{r}_S)} [(\nabla \times \mathbf{H})_{\parallel}]_{\mathbf{r}_S} = \int_{S} d\mathbf{r}'_S\, \mathbf{\Sigma}^{\mathrm{m}}(\mathbf{r}_S, \mathbf{r}'_S; \omega_0^2) \cdot [\mathbf{n} \times \mathbf{H}(\mathbf{r}'_S)], \tag{9.43}$$

where $\mathbf{H}(\mathbf{r})$ is the exact solution of the wave equation in region II at parameter ω_0^2, given the surface-parallel value of $\mathbf{H}$ on the boundary S. Using the same argument as in the case of the electric field, we can show that the normalization of the magnetic field in region II is given by

$$\int_{\mathrm{II}} d\mathbf{r}\, \mathbf{H}^* \cdot \mathbf{H} = \int_{S} d\mathbf{r}_S\, \mathbf{n} \cdot \left(\mathbf{H}^* \times \frac{1}{\epsilon_{\mathrm{II}}} \frac{\partial}{\partial \omega_0^2} \nabla \times \mathbf{H} \right)$$

$$= \int_{S} d\mathbf{r}_S \int_{S} d\mathbf{r}'_S (\mathbf{n} \times \mathbf{H}^*) \cdot \frac{\partial \mathbf{\Sigma}^{\mathrm{m}}}{\partial \omega_0^2} \cdot (\mathbf{n} \times \mathbf{H}), \tag{9.44}$$

and substituting into equation (9.42) we obtain the embedded variational expression,

$$\omega^2 = \frac{\int_I d\mathbf{r} \frac{1}{\epsilon_I}(\nabla \times \mathcal{H}^*) \cdot (\nabla \times \mathcal{H}) - \int_S d\mathbf{r}_S \int_S d\mathbf{r}'_S (\mathbf{n} \times \mathcal{H}^*) \cdot \left(\mathbf{\Sigma}^m - \omega_0^2 \frac{\partial \mathbf{\Sigma}^m}{\partial \omega_0^2}\right) \cdot (\mathbf{n} \times \mathcal{H})}{\int_I d\mathbf{r} \, \mathcal{H}^* \cdot \mathcal{H} + \int_S d\mathbf{r}_S \int_S d\mathbf{r}'_S (\mathbf{n} \times \mathcal{H}^*) \cdot \frac{\partial \mathbf{\Sigma}^m}{\partial \omega_0^2} \cdot (\mathbf{n} \times \mathcal{H})}.$$

(9.45)

This expression is stationary when $\mathcal{H}$ satisfies equation (9.39) in region I, and on the boundary with II the surface-parallel components satisfy

$$\frac{1}{\epsilon_I(\mathbf{r}_S)}[(\nabla \times \mathcal{H})_\parallel]_{\mathbf{r}_S} = \int_S d\mathbf{r}'_S \left[\mathbf{\Sigma}^m(\mathbf{r}_S, \mathbf{r}'_S; \omega_0^2) + (\omega^2 - \omega_0^2)\frac{\partial \mathbf{\Sigma}^m}{\partial \omega_0^2}\right] \cdot [\mathbf{n} \times \mathcal{H}(\mathbf{r}'_S)].$$

(9.46)

Comparing the right-hand side with equation (9.43), we see that the surface-parallel components of $\frac{1}{\epsilon}\nabla \times \mathbf{H}$, corresponding to the surface-parallel components of $\mathbf{E}$, match on either side of the boundary. The surface-parallel components of $\mathbf{H}$ match by construction.

This all simplifies in a system uniform in the z-direction, for a polarization in which $\mathbf{H} = (0, 0, H_z)$. The vector Helmholtz equation (9.39) becomes the scalar wave equation,

$$-\nabla \cdot \left(\frac{1}{\epsilon(\mathbf{r})}\nabla H_z(\mathbf{r})\right) = \omega^2 H_z(\mathbf{r}),$$

(9.47)

and the corresponding variational expression is

$$\omega^2 = \frac{\int_I d\mathbf{r} \frac{1}{\epsilon_I}\nabla \mathcal{H}_z^* \cdot \nabla \mathcal{H}_z - \int_S d\mathbf{r}_S \int_S d\mathbf{r}'_S \, \mathcal{H}_z^*\left(\mathbf{\Sigma}^m - \omega_0^2 \frac{\partial \mathbf{\Sigma}^m}{\partial \omega_0^2}\right)\mathcal{H}_z}{\int_I d\mathbf{r} \, |\mathcal{H}_z|^2 + \int_S d\mathbf{r}_S \int_S d\mathbf{r}'_S \, \mathcal{H}_z^* \frac{\partial \mathbf{\Sigma}^m}{\partial \omega_0^2}\mathcal{H}_z}.$$

(9.48)

Here, the embedding potential satisfies an equation similar to equation (2.15),

$$\frac{1}{\epsilon_{II}(\mathbf{r}_S)}\frac{\partial H_z(\mathbf{r}_S)}{\partial n_S} = \int_S d\mathbf{r}'_S \, \Sigma^m(\mathbf{r}_S, \mathbf{r}'_S; \omega_0^2)H_z(\mathbf{r}'_S),$$

(9.49)

where H_z satisfies the wave equation in II with the specified boundary value $H_z(\mathbf{r}_S)$. In this case equation (9.48) is stationary when $\mathcal{H}_z(\mathbf{r})$ satisfies the scalar Helmholtz equation (9.47) in region I, together with the condition on S that

$$\frac{1}{\epsilon_I(\mathbf{r}_S)}\frac{\partial \mathcal{H}_z(\mathbf{r}_S)}{\partial n_S} = \int_S d\mathbf{r}'_S \left[\Sigma^m(\mathbf{r}_S, \mathbf{r}'_S; \omega_0^2) + (\omega^2 - \omega_0^2)\frac{\partial \mathbf{\Sigma}^m}{\partial \omega_0^2}\right]\mathcal{H}_z(\mathbf{r}'_S).$$

(9.50)

This ensures that $\frac{1}{\epsilon}\frac{\partial H_z}{\partial n_s}$ as well as H_z are continuous across the dielectric boundary. In the systems we are dealing with here, uniform in the z-direction, these conditions correspond to $\mathbf{E}_\parallel$ and $\mathbf{H}_\parallel$ continuous.

Let us now apply the scalar magnetic variational principle to the vacuum slab sandwiched between semi-infinite dielectrics shown in figure 9.2. We shall calculate the magnetic Green function, with a spectral representation given by

$$\Gamma^m(\mathbf{r}, \mathbf{r}'; \omega^2) = \sum_i \frac{H_{z,i}(\mathbf{r})H_{z,i}^*(\mathbf{r}')}{\omega^2 - \omega_i^2}, \tag{9.51}$$

where $H_{z,i}$ is an eigenfunction of equation (9.47) with eigenvalue ω_i^2. The embedded Green function in region I can be expanded in terms of scalar basis functions,

$$\Gamma^m(\mathbf{r}, \mathbf{r}'; \omega^2) = \sum_{m,n}\Gamma_{mn}^m(\omega^2)F_m(\mathbf{r})F_n(\mathbf{r}'), \tag{9.52}$$

where the Γ^m matrix satisfies equation (9.36), but with A and B matrices given by

$$A_{mn}(\omega^2) = \int_I d\mathbf{r}\, \frac{1}{\epsilon_I}\nabla F_m^* \cdot \nabla F_n - \int_S d\mathbf{r}_S \int_S d\mathbf{r}_S'\, F_m^*(\mathbf{r}_S)\Sigma^m(\mathbf{r}_S, \mathbf{r}_S'; \omega^2)F_n(\mathbf{r}_S'), \tag{9.53}$$

and

$$B_{mn} = \int_I d\mathbf{r}\, F_m^*(\mathbf{r})F_n(\mathbf{r}). \tag{9.54}$$

These matrix elements are very easy to evaluate for the vacuum/dielectric sandwich: we use plane-wave basis functions like equation (9.28), but without the vector parts

$$F_m(x, y) = \exp i(g_m x + k_y y), \qquad g_m = \frac{2m\pi}{D}. \tag{9.55}$$

The embedding potential to replace the dielectrics on both sides is given by

$$\Sigma^m = \frac{ik_x}{\epsilon} \tag{9.56}$$

with $k_x = \sqrt{\epsilon\omega^2 - k_y^2}$ (it is interesting that k_x appears in the numerator of equation (9.56), but in the denominator of equation (9.27)).

For comparison with the calculations of section 9.1.3 we should calculate the *electric* spectral density, rather than the magnetic density given by

$$n^m(\mathbf{r}, \omega) = \sum_i \mathbf{H}_i(\mathbf{r}) \cdot \mathbf{H}_i^*(\mathbf{r})\delta(\omega - \omega_i) = -\frac{2\omega}{\pi}\Im\Gamma^m(\mathbf{r}, \mathbf{r}, \omega^2 + i\eta). \tag{9.57}$$

Comparing this equation with (9.31) we see that we can obtain the electric spectral density by applying Maxwell's equations, so that

$$n(\mathbf{r}, \omega) = -\frac{2}{\pi\epsilon(\mathbf{r})\omega}\, \nabla_{\mathbf{r}} \times \nabla_{\mathbf{r}'} \times \Im\Gamma^m(\mathbf{r}, \mathbf{r}'; \omega^2 + i\eta)\Big|_{\mathbf{r}'=\mathbf{r}}. \tag{9.58}$$

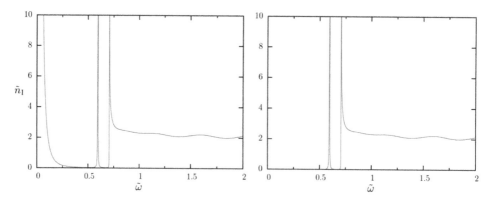

Figure 9.4. Electric spectral density integrated through region I, $\tilde{n}_I(\tilde{\omega})$, at $\tilde{k}_y = 0.5$ in a vacuum slab between semi-infinite dielectric media with $\epsilon = 0.5$. Left-hand figure: calculated from the electric wave equation with a basis set of 17 transverse waves and 21 longitudinal waves. Right-hand figure: calculated from the magnetic wave equation with a basis of 17 plane-waves. The basis is defined by $D = 3d/2$. The imaginary part of $\tilde{\omega}^2$ is $\tilde{\eta} = 0.0002$.

Results for $\tilde{n}_I$, the electric spectral density integrated through region I, calculated in this way are shown in figure 9.4, together with results calculated à la section 9.1.3. We take $\tilde{k}_y = 0.5$ this time, but otherwise the parameters are the same as before, with $\epsilon = 0.5$ in the semi-infinite dielectrics on either side of the vacuum and the basis defined by $D = 3d/2$. A basis of 17 plane-waves is used in the magnetic calculation, and 17 transverse with 21 longitudinal waves in the electric calculation. Apart from the large zero-frequency peak in the electric calculation (left-hand figure), the results for $\tilde{n}_I(\tilde{\omega})$ are identical in the two calculations—this is why we plot them side-by-side rather than on the same graph, where we couldn't distinguish them.

9.2 Embedding dielectric spheres

Electromagnetic problems are generally more complicated than planar dielectric structures, and in this section we shall show how dielectric spheres can conveniently be replaced by embedding operators [6]. We shall use this to calculate the photonic band structure of a lattice of spheres, and because we calculate the electric field only in region I, the uniform region between the spheres, a smaller basis set is needed than in conventional plane-wave methods. The cloud in this particular silver lining is that we must use rather complicated (and unfamiliar) vector spherical waves [12] to evaluate the matrix elements of the embedding operator.

9.2.1 Matrix elements of the embedding operator

Let us calculate the embedding tensor operator to replace a dielecric sphere of radius ρ with a uniform dielectric constant ϵ. Instead of finding the operator explicitly, we find its matrix elements between basis functions $\mathbf{F}_m$ and $\mathbf{F}_n$ from the definition in equation (9.15),

$$\int_S \mathrm{d}\mathbf{r}_S \int_S \mathrm{d}\mathbf{r}'_S (\mathbf{n} \times \mathbf{F}_m^*) \cdot \mathbf{\Sigma}(\mathbf{r}_S, \mathbf{r}'_S; \omega_0^2) \cdot (\mathbf{n} \times \mathbf{F}_n) = -\int_S \mathrm{d}\mathbf{r}_S \, \mathbf{F}_m^* \cdot (\mathbf{n} \times \nabla \times \mathbf{E}_{\mathrm{II},n}).$$

$$(9.59)$$

In this section we work with electric fields, and $\mathbf{E}_{\text{II},n}$ is the electric field which satisfies the Helmholtz equation inside the dielectric sphere—region II—at parameter ω_0^2, with surface-parallel components matching the surface-parallel components of $\mathbf{F}_n$ over the surface of the sphere. As usual, $\mathbf{n}$ is directed from region I into the sphere. (This approach for calculating the matrix elements of the embedding potential without finding $\Sigma(\mathbf{r}_S, \mathbf{r}_S')$ explicitly can be used quite generally.)

The first stage in calculating the matrix elements is to expand the plane-wave basis functions in terms of spherical waves. This is well-known when we are dealing with scalar functions—something we learn early-on in scattering theory—but is more difficult in the present case, with vector waves. Following Zangwill [5] we can show that a solution of the Helmholtz equation in free space is given by

$$\mathbf{E}(\mathbf{r}) = \nabla \times [\mathbf{r}u(\mathbf{r})], \qquad (9.60)$$

where $u(\mathbf{r})$ satisfies the scalar wave equation[2],

$$-\nabla^2 u = k^2 u. \qquad (9.61)$$

In particular, u can be chosen to be a spherical wave, and this gives us a transverse vector spherical wave,

$$\mathbf{M}_{lm}(r, \theta, \phi; k) = \nabla \times \left(\mathbf{r}\, j_l(kr) P_l^{|m|}(\cos\theta) \begin{cases} \cos m\phi \ (m \geqslant 0) \\ \sin m\phi \ (m < 0) \end{cases} \right), \qquad (9.62)$$

where we follow Stratton's convention for spherical harmonics [12]. Applying curl to both sides of the vector Helmholtz equation we obtain a second solution,

$$\mathbf{N}_{lm}(r, \theta, \phi; k) = \frac{1}{k} \nabla \times \mathbf{M}_{lm}, \qquad (9.63)$$

a transverse vector spherical wave with its polarization perpendicular to $\mathbf{M}_{lm}$. Quoting Stratton [12], the expansions of the two transverse plane waves travelling along the z-axis are then given by

$$\hat{\mathbf{x}} \exp ikz = \sum_{l=1}^{\infty} i^l \frac{2l+1}{l(l+1)} (\mathbf{M}_{l,-1} - i\mathbf{N}_{l1}), \qquad (9.64)$$

$$\hat{\mathbf{y}} \exp ikz = -\sum_{l=1}^{\infty} i^l \frac{2l+1}{l(l+1)} (\mathbf{M}_{l1} - i\mathbf{N}_{l,-1}). \qquad (9.65)$$

The vector spherical wave expansion of a transverse plane wave travelling in direction $\hat{\mathbf{k}}$ with the electric field directed along $\hat{e}$ has the form

$$\mathbf{F}^t(\mathbf{r}) = \hat{e} \exp i\mathbf{k} \cdot \mathbf{r} = \sum_{l=1}^{\infty} \sum_{m=-l}^{l} \left[a_{lm}(\hat{\mathbf{k}}, \hat{e}) \mathbf{M}_{lm} + b_{lm}(\hat{\mathbf{k}}, \hat{e}) \mathbf{N}_{lm} \right], \qquad (9.66)$$

---

[2] We are deliberately using k here, rather than ω, as we are expanding plane-waves with wave-vector $\mathbf{k}$.

where the coefficients a_{lm} and b_{lm} can be found from equations (9.64) and (9.65) using the rotation matrices for spherical harmonics.

By comparison with the transverse waves, the longitudinal waves are easy, as all we have to do is to take the grad. The longitudinal spherical waves (L) regular at the origin are given by

$$\mathbf{L}_{lm}(r, \theta, \phi; k) = \nabla \left(j_l(kr) P_l^{|m|}(\cos\theta) \begin{cases} \cos m\phi \ (m \geq 0) \\ \sin m\phi \ (m < 0) \end{cases} \right). \tag{9.67}$$

By taking the grad of both sides of the scalar plane-wave expansion,

$$\exp ikz = \sum_{l=0}^{\infty} i^l (2l + 1) P_l(\cos\theta), \tag{9.68}$$

we obtain for a longitudinal plane-wave travelling in the z-direction,

$$\hat{\mathbf{z}} \exp ikz = \frac{1}{k} \sum_{l=0}^{\infty} i^{l-1}(2l + 1)\mathbf{L}_{l0}, \tag{9.69}$$

with the general form,

$$\mathbf{F}^l(\mathbf{r}) = \hat{\mathbf{k}} \exp i\mathbf{k} \cdot \mathbf{r} = \sum_{l=0}^{\infty}\sum_{m=-l}^{l} c_{lm}(\hat{\mathbf{k}})\mathbf{L}_{lm}. \tag{9.70}$$

To find the matrix elements using equation (9.59), we need the electric field $\mathbf{E}_{\mathrm{II}}$ inside the sphere which matches the plane-wave basis functions of the form of equations (9.66) and (9.70). For the transverse plane-wave given by equation (9.66), matching the surface-parallel components of the field gives [6, 12]

$$\mathbf{E}_{\mathrm{II}} = \sum_{l=1}^{\infty}\sum_{m=-l}^{l} \left(\frac{j_l(k\rho)}{j_l(k'\rho)} a_{lm}(\hat{\mathbf{k}}, \hat{\epsilon})\mathbf{M}'_{lm} + \frac{I_l(k\rho)}{I_l(k'\rho)} b_{lm}(\hat{\mathbf{k}}, \hat{\epsilon})\mathbf{N}'_{lm} \right). \tag{9.71}$$

The wave-vector inside the sphere is given by $k' = \sqrt{\epsilon}\,\omega_0$, where ω_0^2 is the frequency parameter. $\mathbf{M}'_{lm}$ and $\mathbf{N}'_{lm}$ are the spherical waves evaluated at this wave-vector, and the function I_l is given in terms of the spherical Bessel function j_l by

$$I_l(kr) = \frac{1}{kr}\frac{\partial}{\partial r}\left[r j_l(kr) \right]. \tag{9.72}$$

The electric field inside the sphere which matches on to the longitudinal wave is given by [6]

$$\mathbf{E}_{\mathrm{II}} = \frac{1}{kr} \sum_{l=1}^{\infty}\sum_{m=-l}^{l} \frac{j_l(kr)}{I_l(k'r)} c_{lm}(\hat{\mathbf{k}})\mathbf{N}'_{lm}. \tag{9.73}$$

Interestingly, this contains only $\mathbf{N}_{lm}$ spherical waves. It is easy to evaluate $\nabla \times \mathbf{E}_{\mathrm{II}}$ in equation (9.59) using [6]

$$\nabla \times \mathbf{M}'_{lm} = k'\mathbf{N}'_{lm} \qquad \text{and} \qquad \nabla \times \mathbf{N}'_{lm} = k'\mathbf{M}'_{lm}. \tag{9.74}$$

However, evaluating the surface integrals in equation (9.59) is tedious, involving the orthogonality relations between spherical waves given in Stratton [12]; for the results we refer the (interested) reader to [6].

9.2.2 Photonic band structure of dielectric spheres

We now calculate the frequency of electromagnetic modes in a three-dimensional lattice of dielectric spheres, using the embedding matrix elements outlined in section 9.2.1 [6]. The spheres constitute region II, and we expand the electric field in the space in between, region I, in a basis consisting of both transverse and longitudinal plane waves, with coefficients e^t and e^l, respectively,

$$\mathcal{E}(\mathbf{r}) = \sum_{m=1}^{M} \sum_{p=1}^{2} e^t_{mp}\, \hat{\epsilon}_p \exp i(\mathbf{k} + \mathbf{g}_m) \cdot \mathbf{r} + \sum_{n=1}^{N} e^l_n \widehat{(\mathbf{k} + \mathbf{g}_n)} \exp i(\mathbf{k} + \mathbf{g}_n) \cdot \mathbf{r}. \quad (9.75)$$

Here, $\mathbf{k}$ is the Bloch wave-vector at which we are working, and $\mathbf{g}_m$, $\mathbf{g}_n$ are reciprocal lattice vectors of the 'crystal'; the summations are over $2M$ transverse waves, including the two polarization directions $\hat{\epsilon}_p$, and N longitudinal waves with polarization vector $\widehat{\mathbf{k} + \mathbf{g}_n}$. The A and B matrix elements in the eigenvalue equation (9.20) involve integrals over one unit cell, and knowing the matrix elements of the embedding operator these are straightforward to evaluate. As before, we first evaluate the embedding operator at a trial frequency parameter ω_0^2 and iterate until the eigenvalue $\omega^2 = \omega_0^2$. This can proceed more-or-less automatically.

The first example we shall look at is a simple cubic lattice of touching dielectric spheres with dielectric constant $\epsilon = 8$ in vacuum. This study was carried out by Kemp and Inglesfield [6], who found that in the expansions of the plane-wave basis functions (9.66) and (9.70) a maximum value of $l = 8$ gave better than 1% accuracy in frequencies. As in the simple vacuum/dielectric slab example studied in section 9.1.2, it was found that a large number of longitudinal waves is necessary to suppress the spurious Laplace solutions.

Convergence results are shown in table 9.3, giving frequencies in reduced units for different basis set sizes at the X-point in the Brillouin zone (note that the number of transverse waves includes both polarizations, in other words it corresponds to $2M$). We see that the convergence behaviour is consistent, with frequencies decreasing as the number of transverse waves increases at a fixed number of longitudinal waves, but increasing as the number of longitudinal waves increases. The convergence is most satisfactory, and the results are in very good agreement with calculations using a layer-KKR method developed by Stefanou et al [13, 14]. This compares with the very poor convergence with a direct plane-wave expansion of the field [15], without using embedding to eliminate the difficulties associated with the discontinuity across the dielectric interface.

A second example of the embedding method in action is a calculation for a face-centred cubic lattice of touching *vacuum* spheres in a background dielectric of $\epsilon = 9$ [6]. The approach is just the same as before, with the spheres replaced by the embedding operator—the fact that we have vacuum spheres immersed in the

Table 9.3. Frequencies in reduced units at X in a simple cubic lattice of touching spheres with dielectric constant $\epsilon = 8$ in vacuum, calculated using embedding. $\tilde{\omega}_1$ and $\tilde{\omega}_2$ are doubly degenerate, and $\tilde{\omega}_3$ is singly degenerate. Results are shown for different numbers of transverse and longitudinal plane waves in the basis set. The bottom results are calculated using a layer-KKR method [13, 14].

Trans.	Long.	$\tilde{\omega}_1$	$\tilde{\omega}_2$	$\tilde{\omega}_3$
14	7	0.20995	0.26863	0.42838
54	27	0.23459	0.28014	0.42879
162	27	0.23390	0.27959	0.42795
162	81	0.24171	0.28385	0.43097
162	123	0.24283	0.28405	0.43135
246	81	0.24170	0.28384	0.43095
246	123	0.24282	0.28404	0.43132
246	251	0.24514	0.28450	0.43180
246	436	0.24526	0.28453	0.43183
	Layer-KKR	0.246	0.284	0.433

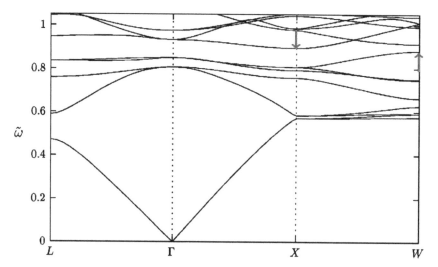

Figure 9.5. Band structure of the face-centred cubic lattice of touching vacuum spheres in a background dielectric of $\epsilon = 9$. The reduced frequency is plotted as a function of wave-vector in the symmetry directions of the Brillouin zone. The arrows indicate the top of the eighth band at W and the bottom of the ninth band at X, between which there is an absolute band gap. (Adapted from Kemp and Inglesfield [6].)

dielectric makes no difference to the method, and all that we have to do is move ϵ from one part of the calculation to another. This is a particularly interesting system because it shows an absolute photonic band gap. The band structure, calculated with a basis of 178 transverse (including both polarizations) and 89 longitudinal waves is shown in figure 9.5, and the results are in very good agreement with Moroz and Sommers [16], who calculated the band structure using a KKR scattering method. The absolute band gap between the top of the eighth band at W and the bottom of the ninth band at X, indicated by red arrows in the figure, is clearly visible. With this size of basis the gap is $\Delta\tilde{\omega} = 0.0123$, in good agreement with the KKR results which give $\Delta\tilde{\omega} = 0.014$. Our converged result (338 transverse, 259 longitudinal) is $\Delta\tilde{\omega} = 0.0138$. Interestingly enough, the inverse problem of a face-centred cubic lattice of spheres with $\epsilon = 9$ in a background of vacuum gives no such band gap.

9.3 Plasmonics of metal cylinders

Metallic objects can support surface plasmons of one sort or another, and the embedding method provides a very effective way of solving Maxwell's equations in such systems, bringing us nicely into the realm of plasmonics. The dielectric constant now becomes $\epsilon(\omega)$, a function of frequency, and in principle we could use the measured dielectric function. However, for a general understanding of metallic behaviour, a convenient form to use is the dielectric function of a free-electron gas, the Drude dielectric function given by [17]

$$\epsilon(\omega) = 1 - \frac{\omega_p^2}{\omega(\omega + i/\tau)},$$ (9.76)

where ω_p is the bulk plasmon frequency and τ is the lifetime of the plasmons. The imaginary part of ϵ broadens all the modes, so we usually calculate the Green function and spectral density in metallic systems, which in any case has the advantage that we do not have to worry about iterating on the frequency at which the embedding operator is calculated.

The metal systems which have been so far studied using the methods described in this chapter are lattices of spheres [18] and cylinders [10], and a line of cylinders [11]. These all show surface plasmons, modes due to an oscillating charge distribution on the surface of the sphere or cylinder, and surface plasmon-polaritons, a mixed mode which is light-like at small wave-vectors and surface plasmon-like at larger wave-vectors. In this section we use embedding to study the electromagnetic modes in a line of metal cylinders.

9.3.1 Embedding with $\epsilon(\omega)$

Before we can begin the embedding calculation, there is an one question we must dispose of—are the embedding methods derived in sections 9.1.1 and 9.1.4 still valid when region II has a *frequency-dependent* dielectric function $\epsilon_{II}(\omega)$? The problem is that in the derivation of the variational principles (equations (9.16) and (9.45)), we differentiate the Helmholtz equation in region II with respect to ω_0^2 to obtain the normalization of the field in region II, that is, in deriving equations (9.12) and (9.44). Any frequency variation in ϵ_{II} is neglected in going from equation (9.10) to equation (9.11) in the case of the electric field method, for example.

In fact the final variational results for the eigenvalue ω^2 are still valid, as long as $\partial\Sigma/\partial\omega_0^2$ includes the frequency dependence of ϵ_{II}. The reason for this is that equation (9.16) is stationary with respect to variations $\delta\mathcal{E}(\mathbf{r})$ when the trial field $\mathcal{E}(\mathbf{r})$ satisfies the Helmholtz equation in region I at ω^2, and the boundary conditions on S given by $\Sigma(\omega^2)$. But we know from equation (9.23) that the derivative term $\partial\Sigma/\partial\omega_0^2$, which appears in the numerator and denominator of equation (9.16), corrects $\Sigma(\omega_0^2)$ to give $\Sigma(\omega^2)$ to first order in $(\omega^2 - \omega_0^2)$. The derivative clearly includes *all* the frequency-dependence of Σ, including that of $\epsilon_{II}(\omega)$. The same argument holds for equation (9.45) with respect to the magnetic trial field $\mathcal{H}(\mathbf{r})$.

The problem does not arise when we calculate the Green function, for example the electric Green function $\Gamma(\mathbf{r}, \mathbf{r}'; \omega^2)$ (equation (9.32)) with matrix elements given by equation (9.36). In calculating Γ we fix ω^2 so we know $\Sigma(\omega^2)$ at the right frequency to fix the boundary conditions, and $\partial\Sigma/\partial\omega_0^2$ doesn't enter. All this is true also for the magnetic Green function.

9.3.2 Embedding a line of cylinders

With this background we can now study an infinitely extended line of infinitely long Ag cylinders in free space (figure 9.6), the system treated by Giannakis *et al* [11]. There is considerable interest in structures like this, as they offer the possibility of guiding electromagnetic waves by the coupling of surface plasmons. We shall take advantage of the cylindrical geometry, and calculate modes propagating in the y-direction in which the electric field is in the xy-plane perpendicular to the cylinder axis, with the magnetic field parallel to the z-axis (called transverse electric (TE) modes[3]). This gives us the scalar problem for H_z which we discussed in section 9.1.4.

To use the embedding method for this problem, we take region I to be the free space between the cylinders, and between planes on either side—the uncoloured region in figure 9.7. Region II then consists of several disconnected regions, the cylinders themselves (blue in figure 9.7) and the semi-infinite regions of free space beyond the plane boundaries at $x = \pm d/2$ (green in the figure). Because of the

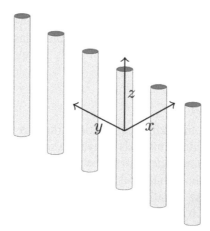

Figure 9.6. The line of metal cylinders: each cylinder is infinitely long and the line is infinitely extended. The modes considered here propagate in the y-direction with Bloch wave-vector k_y, with **E** in the xy-plane, and **H** in the z-direction.

---

[3] This refers to the orientation of the field relative to the cylinder axis. The bound states of the problem are waveguide modes propagating along the line of cylinders, for which the TE description seems inconsistent.

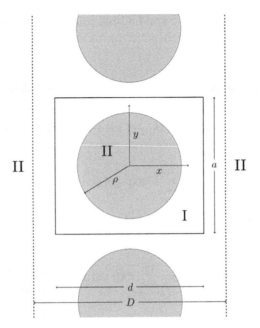

Figure 9.7. Cross-section of a portion of the line of cylinders, radius ρ, separation a. The Helmholtz equation for H_z is solved explicitly in region I (uncoloured), and region II, consisting of the cylinders (blue) and the semi-infinite external regions (green), is replaced by embedding potentials. The solid line outlines the unit cell. D defines the basis functions in the x-direction.

one-dimensional periodicity along the line, we can define a Bloch wave-vector k_y, and we need only consider one unit cell, as shown in figure 9.7.

In region I we expand $H_z(x, y)$ in terms of the basis functions

$$F_j(x, y) = \exp i \left[g_{j,x} x + \left(k_y + g_{j,y} \right) y \right]. \tag{9.77}$$

Here, the reciprocal lattice vectors are given by

$$g_{j,x} = \frac{2\pi p(j)}{D}, \qquad g_{j,y} = \frac{2\pi q(j)}{a}, \tag{9.78}$$

where $p(j)$ and $q(j)$ are integers associated with the jth basis function, D is a parameter somewhat wider than the width of slab d (as in section 9.1.2), and a is the separation of the cylinders.

We now calculate the matrix elements of the embedding potential for replacing a cylinder of radius ρ with dielectric constant ϵ, between these basis functions. Over the surface of the cylinder, the plane wave (equation (9.77)) can be expanded as a sum over cylindrical functions,

$$F_j(\mathbf{r}_S) = \sum_{m=-\infty}^{\infty} i^m \exp im\left(\phi - \phi_j \right) J_m\left(k_j \rho \right), \tag{9.79}$$

where point $\mathbf{r}_S$ on the surface is given in cylindrical coordinates by (ρ, ϕ), and the plane-wave wave-vector by $\mathbf{k}_j = (k_j, \phi_j)$. The J_m are Bessel functions. We follow the same procedure as in section 9.2.1 and find the solution of the wave equation inside the cylinder at frequency ω_0 which matches on to equation (9.79) over $\mathbf{r}_S$. This is given by

$$H_{\text{II},j}(r,\,\phi) = \sum_{m=-\infty}^{\infty} \text{i}^m \exp \text{i}m\big(\phi - \phi_j\big)\frac{J_m(k'r)}{J_m(k'\rho)}J_m\big(k_{j}\rho\big), \qquad (9.80)$$

where $k' = \sqrt{\epsilon}\,\omega$. The derivative of $H_{\text{II},j}$ at the surface of the cylinder with respect to the surface normal (*into* the cylinder from region I into region II) is then given by

$$\frac{\partial H_{\text{II},j}}{\partial n_S}(\rho,\,\phi) = -k' \sum_{m=-\infty}^{\infty} \text{i}^m \exp \text{i}m\big(\phi - \phi_j\big)\frac{J_m'(k'\rho)}{J_m(k'\rho)}J_m\big(k_{j}\rho\big), \qquad (9.81)$$

so from the definition of the matrix element of the embedding potential (equation (9.49)) we have

$$\int_S \text{d}\mathbf{r}_S \int_S \text{d}\mathbf{r}_S'\, F_i^* \Sigma^m F_j = -\frac{2\pi k'\rho}{\epsilon} \sum_{m=-\infty}^{\infty} \exp \text{i}m\big(\phi_i - \phi_j\big)\frac{J_m'(k'\rho)}{J_m(k'\rho)}J_m(k_{i}\rho)J_m(k_{j}\rho). \qquad (9.82)$$

In actual calculations the limits of the summation over the azimuthal quantum number m are taken as $\pm m_{\max}$. We shall see that $m_{\max}$ enters the results for metal cylinders in a very interesting way.

We use the same approach to find the matrix elements of the embedding potentials to replace the semi-infinite regions of free space on either side of region I. The solution of the free-space wave equation in II, which matches the basis function (9.77) over the embedding plane at $x = d/2$, is given by

$$H_{\text{II},j}(x,\,y) = \exp \text{i}\big(g_{j,x} - k''\big)d/2 \exp \text{i}\Big[k''x + \big(k_y + g_{j,y}\big)y\Big], \quad x \geqslant d/2, \quad (9.83)$$

with $k'' = \sqrt{\omega^2 - (k_y + g_{j,y})^2}$. (There is the usual convention for the sign of the square root when the argument is negative or complex.) The outward normal derivative of equation (9.83) over the embedding plane is then

$$\frac{\partial H_{\text{II},j}}{\partial n_S}(y) = \text{i}k'' \exp \text{i}g_{j,x}d/2 \exp \text{i}\big(k_y + g_{j,y}\big)y, \qquad (9.84)$$

so the matrix element of the embedding potential at $x = +d/2$ is

$$\int_S \text{d}\mathbf{r}_S \int_S \text{d}\mathbf{r}_S'\, F_i^* \Sigma^m F_j = \text{i}k''a\delta_{q(i),q(j)} \exp \text{i}(g_{j,x} - g_{i,x})d/2. \qquad (9.85)$$

Combining the matrix elements at $x = \pm d/2$ we have

$$\int_S \text{d}\mathbf{r}_S \int_S \text{d}\mathbf{r}_S'\, F_i^* \Sigma^m F_j = 2\text{i}k''a\delta_{q(i),q(j)} \cos(g_{i,x} - g_{j,x})d/2. \qquad (9.86)$$

This embedding potential, like those in sections 9.1.2 and 9.1.4, fully describes the effect of a semi-infinite region of space beyond region I where we calculate the fields explicitly. In many calculations of electromagnetic waves, something similar is done by using an absorbing boundary condition at the edge of the computational regions, a procedure which—unlike embedding—inevitably involves approximations.

9.3.3 Spectral density and photonic band structure

For the line of Ag cylinders we work with reduced units defined by the lattice constant a, so that the cylinder radius, wave-vector, and frequency become

$$\tilde{\rho} = \frac{2\pi\rho}{a}, \qquad \tilde{k}_y = \frac{k_y a}{2\pi}, \qquad \tilde{\omega} = \frac{\omega a}{2\pi c}. \qquad (9.87)$$

We use a bulk plasmon frequency of $\hbar\omega_p = 6.18$ eV in the Drude expression for the dielectric function (9.76), and with a lattice spacing of 75 nm (typical of a nanostructure) this becomes $\tilde{\omega}_p = 0.3738$; the plasmon lifetime is taken to be $\tilde{\tau} = 1000$. We shall take region I with a width $d = a$,[4] so that $\tilde{d} = 2\pi$.

As we are dealing with an open system in which modes are broadened by leaking into free space when the frequency is above the light-line ($\omega > ck_y$), and in any case are all broadened by the plasmon lifetime entering ϵ, it is convenient to calculate the spectral density. We calculate $\tilde{n}_{\mathrm{I}}(\tilde{\omega})$, the electric spectral density integrated through region I, using equation (9.58) to find this from the magnetic Green function Γ^{m}. The A and B matrices which we need to find Γ^{m} (equation (9.36)) are easy to calculate (as in equations (9.52) and (9.53)), and involve integrals over the unit cell shown in figure 9.7. We take $D = 3d/2$ in the definition of the basis functions (9.77), and typically use 400 basis functions, though this depends on the value of the cylinder radius $\tilde{\rho}$ and $m_{\max}$ in the expansion of the cylinder embedding potential (9.82).

Figure 9.8 shows $\tilde{n}_{\mathrm{I}}(\tilde{\omega})$ for cylinders with a radius $\tilde{\rho} = 2.8$ (touching cylinders have $\tilde{\rho} = \pi$) at $k_y = 0.4$, for different values of $m_{\max}$—for the range of frequencies shown in the figure, all the modes are below the light-line, with the broadening due to $\tilde{\tau}$. The main feature is the very large central peak at $\tilde{\omega}_p/\sqrt{2} = 0.264$, which increases as $m_{\max}$ increases. This is due to the fact that on an isolated metal cylinder with the Drude dielectric function, there is a surface plasmon at this frequency independent of m; in the line of cylinders, the surface plasmons with large m barely interact and the mode will be at the surface plasmon frequency of an isolated cylinder. So, as $m_{\max}$ increases, this peak will increase without limit. (There is in fact a physical limit to $m_{\max}$, coming from Landau damping, but this is *much* greater than the values we use.) Apart from the central peak, however, the other features (all plasmons) in $\tilde{n}_{\mathrm{I}}(\tilde{\omega})$ have converged as a function of frequency by $m_{\max} \approx 14$.

We can obtain the photonic band structure by mapping $\tilde{n}_{\mathrm{I}}$ as a function of $\tilde{\omega}$ and $\tilde{k}_y$, the magnitude of $\tilde{n}_{\mathrm{I}}$ being indicated by the brightness ($\tilde{n}_{\mathrm{I}}^{1/6}$ is plotted, to compress the brightness range). Figure 9.9 shows the band structure obtained in this way for $\tilde{\rho} = 2.094$ with $m_{\max} = 8$ (left-hand figure), and for $\tilde{\rho} = 2.8$ with $m_{\max} = 14$ (right-hand figure). These values of $m_{\max}$ are adequate for convergence—except of course

---

[4] But remember that because embedding exactly describes the semi-infinite regions of free space, the electromagnetic modes are independent of d.

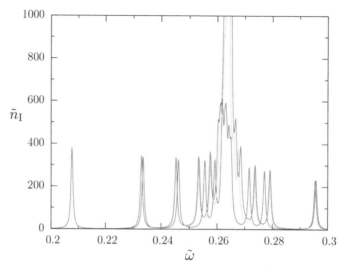

Figure 9.8. Electric spectral density integrated through region I, $\tilde{n}_I(\tilde{\omega})$, at $\tilde{k}_y = 0.4$, cylinder radius $\tilde{\rho} = 2.8$. Red curve, $m_{\max} = 8$; green curve, $m_{\max} = 12$; blue curve, $m_{\max} = 16$. 338 basis functions are used for $m_{\max} = 8$ and 400 for $m_{\max} = 12$ and 16.

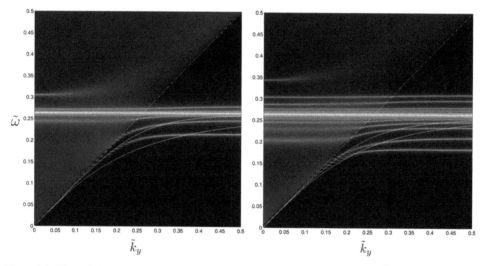

Figure 9.9. Photonic band structure: spectral density mapped as a function of $\tilde{\omega}$ and $\tilde{k}_y$, with the brightness indicating $\tilde{n}_I$. Left-hand figure, cylinder radius $\tilde{\rho} = 2.094$, right-hand figure, $\tilde{\rho} = 2.8$. The dashed white line shows the light-line $\tilde{\omega} = \tilde{k}_y$. The two thin white lines below the light-line show modes on a metal slab of thickness equal to the cylinder diameter. (Figure courtesy of N Giannakis.)

for the central peak. As well as a uniform background above the light-line, there are clear surface plasmon modes, showing little dispersion, with the accumulation of modes at $\tilde{\omega}_p/\sqrt{2}$. Above the light-line the plasmons are resonances, as they can decay into propagating electromagnetic waves, and especially below $\tilde{\omega}_p/\sqrt{2}$ they are considerably broadened. If we follow a single plasmon peak, it sharpens up when it crosses the light-line, the residual width being due to the imaginary part of the dielectric function. In addition to the weakly dispersing surface plasmon bands, the two bands that peel off below the light-line are plasmon-polariton bands. These become almost non-dispersing beyond $\bar{k}_y \approx 0.25$, probably because they are repelled by the flat plasmon bands above. Their behaviour is roughly similar to the two surface plasmon-polaritons on a thin metal slab, shown by the thin white lines in figure 9.9.

9.3.4 Eigenmodes

A resonance peak in the spectral density has a Lorentzian shape given by the Breit–Wigner formula [19]

$$n(\omega) \propto \frac{1}{(\Re\omega_i - \omega)^2 + (\Im\omega_i)^2},\qquad(9.88)$$

where $\Re\omega_i$ and $\Im\omega_i$ can be identified as the real and imaginary part of the complex frequency ω_i: the peak is at $\Re\omega_i$ and the half-width at half-maximum is given by $\Im\omega_i$. These can be found from the complex eigenvalues of equation (9.20), or its magnetic field equivalent for the case of cylindrical geometry in TE polarization. We discussed the origin of complex eigenvalues in section 2.6: from equation (9.51) we see that the Green function as a function of the complex variable ω^2 has poles along the real axis, which are discrete if the whole system I + II is finite; when region II is extended or lossy, so that modes can leak from I, the poles merge to give a branch cut along part of the real axis. Γ^m is analytic in the upper half of the complex ω^2-plane, above the real axis, and this is where we evaluate properties like the spectral density (we add the positive infinitesimal η to move above the real axis, as in equation (9.34)). Analyticity in the upper half-plane ensures causality. A resonance peak in the spectral density like equation (9.88) corresponds to a pole at ω_i^2 in the analytic continuation of Γ^m across the branch cut, into the unphysical lower half-plane. The embedding potentials involve a square root, for example, in equation (9.83) we have $k'' = \sqrt{\omega^2 - (k_y + g_{j,y})^2}$, and to obtain the analytic continuation of Σ^m we stay on the same branch for the square root as $\Im\omega^2$ goes from positive to negative (c.f. Figure 2.8).

For TE modes we expand the trial magnetic field in region I in the same form as equation (9.19), with the basis functions of equation (9.77). The A and B matrices in the eigenvalue equation (9.20) are then given by

$$A_{mn} = \int_I \mathrm{dr}\, \nabla F_m^* \cdot \nabla F_n - \sum_{i=1}^{3} \int_{S_i} \mathrm{dr}_S \int_{S_i} \mathrm{dr}_S' \, F_m^* \left(\Sigma_i^m - \omega_0^2 \frac{\partial \Sigma_i^m}{\partial \omega_0^2} \right) F_n(\mathbf{r}_S') \qquad(9.89)$$

and

$$B_{mn} = \int_I \mathrm{dr}\, F_m^* F_n + \sum_{i=1}^{3} \int_{S_i} \mathrm{dr}_S \int_{S_i} \mathrm{dr}_S' \, F_m^* \frac{\partial \Sigma_i^m}{\partial \omega_0^2} F_n, \qquad(9.90)$$

where the summations are over the three embedding surfaces namely, the surface of the cylinder and the planes at $x = \pm d/2$, with corresponding embedding potentials Σ_i^m (remember that $\epsilon_I = 1$ in comparing these equations with equation (9.53)). Following the discussion in section 9.3.1 we must include the frequency-dependence of the Drude dielectric function in evaluating $\partial\Sigma^m/\partial\omega_0^2$ for the metal cylinder. To find the eigenvalues and eigenvectors, we follow the same procedure as in section 2.3, guessing ω_0^2 in Σ^m and iterating. As we are dealing with resonances, the eigenvalues have a negative imaginary part, but because the imaginary part is small for the narrow resonances we study here, it is accurate—because of the linearization of the embedding potential—to evaluate Σ^m and $\partial\Sigma^m/\partial\omega_0^2$ at $\Re\omega_0^2$. This is in fact a real advantage of using embedding to find the complex eigenvalues—we can evaluate Σ^m and $\partial\Sigma^m/\partial\omega_0^2$ in the usual way at real frequencies and still obtain accurate values for the resonance frequencies and lifetimes.

As an example of complex eigenvalues we consider cylinders with radius $\tilde{\rho} = 2.094$, calculating the modes with wave-vector $\tilde{k}_y = 0.4$, and taking $m_{max} = 8$. As the modes all lie under the light-line, their width comes from the plasmon lifetime τ in the dielectric function (9.76). The results are shown in table 9.4 [11], together with the frequencies of modes calculated using the finite-difference time-domain (FDTD) method [20], which solves the spatially discretized wave equation in discretized time steps. The FDTD calculation used absorbing boundaries on each side of the line of cylinders, roughly corresponding to free-space embedding potentials. The agreement is very good, though the FDTD method, with its spatial

Table 9.4. Frequencies of modes at $\tilde{k}_y = 0.4$, cylinder radius $\tilde{\rho} = 2.094$, $m_{max} = 8$. The second and third columns give the real and imaginary parts of the frequency. The fourth column gives frequencies calculated using the FDTD method [20].

Number	$\Re\tilde{\omega}_i$	$\Im\tilde{\omega}_i$	$\tilde{\omega}_i$ (FDTD)
1	0.214057	−0.000381	0.214
2	0.244246	−0.000454	0.244
3	0.257611	−0.000482	
4	0.261058	−0.000484	
5	0.262457	−0.000490	0.262
6	0.262557	−0.000489	
7	0.263206	−0.000496	
8	0.263609	−0.000497	
9	0.263828	−0.000498	
10	0.263847	−0.000497	
11	0.263902	−0.000498	
12	0.263991	−0.000499	
13	0.264076	−0.000498	
14	0.264310	−0.000495	
15	0.268499	−0.000480	0.269
16	0.277651	−0.000467	0.277

discretization, does not pick up all the modes. As we would expect from the left-hand part figure 9.9, we have two modes with lower frequency (modes 1 and 2, the plasmon-polariton modes), with the other 14 modes clustered round $\tilde{\omega}_p/\sqrt{2} = 0.264$.

Broader resonances with frequencies above the light-line are a more interesting and convincing example of complex frequencies. Taking $\tilde{\rho} = 2.8$ as an example, we see from the right-hand band structure (figure 9.9) that modes with small wave-vector at around $\tilde{\omega} \approx 0.2$ are very broad. The complex frequencies of the first three modes above the light-line are given in table 9.5 for $\tilde{k}_y = 0.1$, $m_{\mathrm{max}} = 14$, and we see that $\mathfrak{I}\omega_i$ is much greater than in table 9.4 by an order of magnitude for modes 1 and 2. Evaluating the amplitude of the Lorentzian (equation (9.88)) is not straightforward or obvious—it is not just a matter of using the renormalization formula (2.31), because the derivative $\partial\Sigma^m/\partial\omega_0^2$ is complex (section 11.2). As a second-best method, we fit the amplitudes of the three resonances to the spectral density, and figure 9.10 shows the quality of the fit. We see that the position of the

Table 9.5. Frequencies of the first three modes above the light-line at $\tilde{k}_y = 0.1$, cylinder radius $\tilde{\rho} = 2.8$, $m_{\mathrm{max}} = 14$. The second and third columns give the real and imaginary parts of the frquency.

Number	$\mathfrak{R}\tilde{\omega}_i$	$\mathfrak{I}\tilde{\omega}_i$
1	0.203854	−0.004416
2	0.229559	−0.005655
3	0.245560	−0.001697

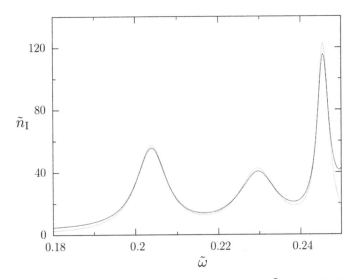

Figure 9.10. Electric spectral density integrated through region I, $\tilde{n}_{\mathrm{I}}(\tilde{\omega})$, at $\tilde{k}_y = 0.1$, cylinder radius $\tilde{\rho} = 2.8$, $m_{\mathrm{max}} = 14$. Red curve, calculation from the Green function (9.58); green curve, sum of three Lorentzians with complex frequencies given in table 9.5 and fitted amplitudes.

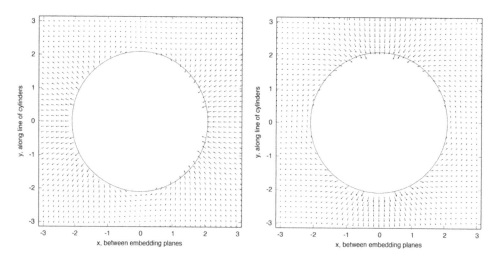

Figure 9.11. Real part of the electric field in region I for polariton modes. at $\tilde{k}_y = 0.25$ for cylinders with $\tilde{\rho} = 2.094$, $m_{max} = 8$. The left-hand figure shows $\mathbf{E}(x, y)$ for the first polariton mode, $\tilde{\omega}_1 = 0.2014$, and the right-hand figure the second polariton mode, $\tilde{\omega}_2 = 0.2247$. (From Giannakis *et al* [11].)

resonances in $\tilde{n}_1(\tilde{\omega})$ and their width is given very well by the complex eigenvalues, the small discrepancies coming from the tail of the dominant central peak at $\tilde{\omega}_p/\sqrt{2}$.

From the eigenvectors of the eigenvalue equation we can find the field distributions of the resonances—these are *resonant states* [21, 22], as discussed in section 2.6.1. For the TE modes we solve the eigenvalue problem for H_z, from which we can calculate $\mathbf{E}(x, y)$ using Maxwell's equations. Figure 9.11 shows the electric fields (arbitrarily normalized) for the two polariton modes at $\tilde{k}_y = 0.25$, for cylinders with radius $\tilde{\rho} = 2.094$, and they show a simple dipolar distribution of the field. On the other hand, the plasmon modes in the central peak of the spectral density close to $\tilde{\omega}_p/\sqrt{2}$, have a complicated field distribution, especially at large m_{max}. Figure 9.12 presents $\mathbf{E}(x, y)$ for the mode with $\tilde{\omega}_i = 0.2644$, $\tilde{k}_y = 0.5$ for cylinders with $\tilde{\rho} = 2.8$, and taking $m_{max} = 14$. The modes closest in frequency to $\tilde{\omega}_p/\sqrt{2}$ have the highest multipole moments, and this is apparent in the figure. It is also clear that the electric field is localized near the surface of the cylinder, with little interaction between the cylinders—this is why the plasmon modes are non-dispersing.

9.4 Good conductors

In chapter 5 we saw that the embedding potential for a confined system is local and energy-independent (equation (5.4)). Here, we shall show that the embedding tensor replacing a good conductor has a similar local form, though it depends on frequency, and we shall relate this to well-known results for boundary conditions in electromagnetism.

We can find $\mathbf{\Sigma}$ from equation (9.15), using the well-known solutions of Maxwell's equations near the surface of a good conductor [7]. Taking the z-axis directed into

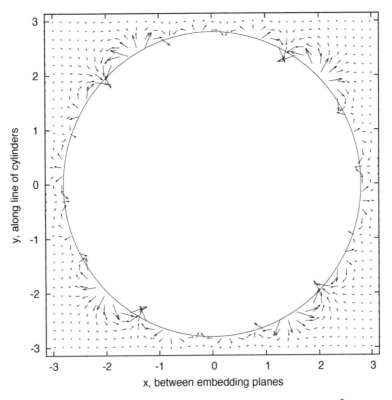

Figure 9.12. Real part of the electric field in region I for the mode with $\tilde{\omega}_i = 0.2644$, $\tilde{k}_y = 0.5$, for cylinders with $\tilde{\rho} = 2.8$, $m_{max} = 14$. (From Giannakis *et al* [11].)

the conductor with the x- and y-axes parallel (locally) to the surface, the fields inside the conductor are given by

$$\mathbf{H} = \hat{\mathbf{y}} H_0 \exp[z(i - 1)/\delta]$$
$$\mathbf{E} = \hat{\mathbf{x}} \sqrt{\frac{\omega}{8\pi\sigma}} (1 - i) H_0 \exp[(z(i - 1)/\delta], \tag{9.91}$$

where σ is the conductivity, and the skin depth δ is given by

$$\delta = \frac{c}{\sqrt{2\pi\omega\sigma}}. \tag{9.92}$$

The electric field and its curl just inside the surface of the conductor are then given by

$$\mathbf{E} = \hat{\mathbf{x}} \sqrt{\frac{\omega}{8\pi\sigma}} (1 - i) H_0, \tag{9.93}$$

$$\nabla \times \mathbf{E} = \hat{\mathbf{y}} \frac{i\omega}{c} H_0, \tag{9.94}$$

9-29

so from equation (9.15) the embedding tensor operator, which replaces the conductor, is given by [4]

$$\mathbf{\Sigma}(\mathbf{r}_S, \mathbf{r}'_S) = (i - 1)\frac{\sqrt{2\pi\omega\sigma}}{c}\delta(\mathbf{r}_S - \mathbf{r}'_S)(\hat{\mathbf{x}}\hat{\mathbf{x}} + \hat{\mathbf{y}}\hat{\mathbf{y}}). \tag{9.95}$$

This is a *local* operator, because of the short skin depth in a good conductor, so it can be used at a surface of arbitrary shape. The imaginary part of $\mathbf{\Sigma}$ is due to loss in the conductor.

In general we can regard $\mathbf{\Sigma}$ as providing a boundary condition on solutions of the electric field in region I: from this point of view, we could solve for $\mathbf{E}$ in region I using techniques such as finite element analysis rather than the variational methods which we have used in the embedding method. As the surface-parallel components of $\mathbf{E}$ and $\nabla \times \mathbf{E}$ are continuous across the surface of the conductor (assuming that $\mu = 1$), equation (9.94) gives the following boundary condition for the field in I on S,

$$\nabla \times \mathbf{E}|_y = (i - 1)\frac{\sqrt{2\pi\omega\sigma}}{c}E_x. \tag{9.96}$$

This is the Leontovich boundary condition [9, 23], well known in electromagnetic theory, with the factor multiplying E_x the inverse of the boundary impedance [5]. The subject of local boundary conditions in electromagnetism, as opposed to the more general non-local boundary condition (equation (9.15)), is the topic of entire books [9], and it would be worthwhile to explore the links with embedding further.

As a simple example, we consider electromagnetic modes confined inside a cubic metallic box, with sides d. The interior of the box constitutes region I and we replace the enclosing metal walls by the embedding tensor given by equation (9.95). We then calculate the Green function for the electric field in I using the methods described in section 9.1.3, with plane-wave basis functions for the $\mathbf{E}$-field given by

$$\mathbf{F}^\nu_{ijk} = \hat{\mathbf{e}}_\nu \exp\left(i\mathbf{g}_{ijk} \cdot \mathbf{r}\right), \qquad \mathbf{g}_{ijk} = \frac{2\pi}{D}(i, j, k). \tag{9.97}$$

The triplet of integers (i, j, k) gives the wave-vector $\mathbf{g}_{ijk}$, and ν identifies the polarization $\hat{\mathbf{e}}_\nu$. In this case we find that taking D somewhat larger than $2d$, with only the two transverse polarizations gives the 'best' basis set (we shall explore this later). Presumably as a consequence of the metallic boundary condition, spurious solutions of Laplace's equations, as in section 9.1.2, do not arise in this case [4].

Results for the electric spectral density integrated through the cube, as a function of reduced frequency $\tilde{\omega}$ are shown in figure 9.13. As conductivity σ has the same units as frequency in Gaussian units [7], we define the reduced conductivity $\tilde{\sigma}$ as

$$\tilde{\sigma} = \frac{\sigma d}{2\pi c}, \tag{9.98}$$

and the red curve in figure 9.13 was calculated with $\tilde{\sigma} = 1000$, the green curve with $\tilde{\sigma} = 100$. In both cases 342 plane-waves were used for each transverse polarization,

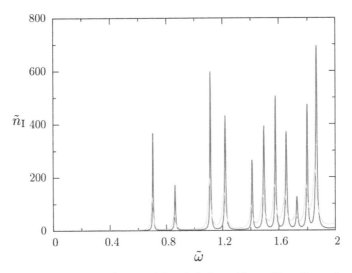

Figure 9.13. $\tilde{n}_{\rm I}$, electric spectral density integrated though the box with metallic walls as a function of $\tilde{\omega}$. Red curve, conductivity of the walls $\tilde{\sigma} = 1000$ green curve, $\tilde{\sigma} = 100$. These are calculated with 342 plane-waves for each transverse polarization, with $D/d = 2.07$.

Table 9.6. Squared frequencies of modes in the box with metallic walls, with conductivity $\tilde{\sigma} = 1000$. The second and third columns give the real and imaginary parts of $\tilde{\omega}_i^2$, the fourth column gives the degeneracy, and the fifth column gives m, n, p. The eigenvalues are calculated with 342 plane-waves for each transverse polarization, with $D/d = 2.07$.

Number	$\Re\tilde{\omega}_i^2$	$\Im\tilde{\omega}_i^2$	Degen.	mnp
1	0.4989	−0.0011	3	110
2	0.7480	−0.0020	2	111
3	1.2478	−0.0023	6	210
4	1.4966	−0.0034	6	211
5	1.9968	−0.0032	3	220
6	2.2453	−0.0047	6	221
7	2.4962	−0.0038	6	310
8	2.7446	−0.0054	3	311
9	2.7455	−0.0055	3	311
10	2.9942	−0.0058	2	222
11	3.2454	−0.0046	6	320
12	3.4935	−0.0065	6	321
13	3.4943	−0.0065	6	321

with $D/d = 2.07$. The structure consists of resonances, with a width which decreases as the skin depth decreases with increasing conductivity.

Table 9.6 shows the squared frequencies $\tilde{\omega}_i^2$ of the resonances, calculated from the embedded eigenvalue equation (9.20), together with their degeneracies. These are very close to the eigenvalues of the modes in a box with perfectly conducting walls, which are given by

$$\omega_i^2 = \frac{\pi^2}{d^2}\left(m^2 + n^2 + p^2\right), \qquad \tilde{\omega}_i^2 = \frac{1}{4}\left(m^2 + n^2 + p^2\right), \qquad (9.99)$$

where m, n, p are integers. The electric fields of these modes, which are just the modes of a square waveguide with capped ends [5], are linear combinations of the following functions,

$$\begin{aligned}
\mathbf{U} &= \hat{\mathbf{x}}\cos(\pi mx/d)\sin(\pi ny/d)\sin(\pi pz/d),\\
\mathbf{V} &= \hat{\mathbf{y}}\sin(\pi mx/d)\cos(\pi ny/d)\sin(\pi pz/d),\\
\mathbf{W} &= \hat{\mathbf{z}}\sin(\pi mx/d)\sin(\pi ny/d)\cos(\pi pz/d),
\end{aligned} \qquad (9.100)$$

where the origin is taken at the corner of the box. Now each of these waves is a mixture of longitudinal and transverse waves, and the required transverse waves, satisfying the boundary condition that the surface-parallel component of the electric field is zero at each wall, are given by [24]

$$\begin{aligned}
\mathbf{M} &= -p\mathbf{V} + n\mathbf{W},\\
\mathbf{N} &= -\left(n^2 + p^2\right)\mathbf{U} - mn\mathbf{V} - mp\mathbf{W}.
\end{aligned} \qquad (9.101)$$

The allowed values of (m, n, p) and the degeneracies which we find by solving the eigenvalue equation agree with those given by equation (9.101). As we should expect, the frequencies given in table 9.6 are slightly less than for a perfect conductor (equation (9.99)) because the finite skin depth with finite conductivity effectively increases d. One interesting point which we note in table 9.6 is that there is a small splitting in frequency of the (311) and (321) modes (and also several other modes, where the splitting is less than 10^{-4}). I don't know whether this is a real breaking of

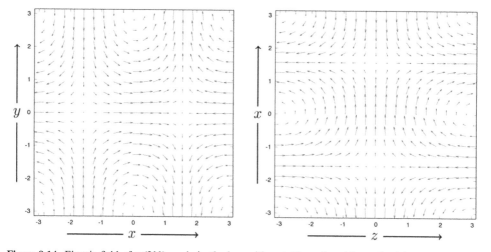

Figure 9.14. Electric field of a (211) mode in the box with metallic walls, with conductivity $\tilde{\sigma} = 1000$. The coordinate origin is taken at the centre of the box. Left-hand figure, $z = 0$ plane, right-hand figure, $y = 0$ plane. Calculated with 342 plane-waves for each transverse polarization, with $D/d = 2.07$.

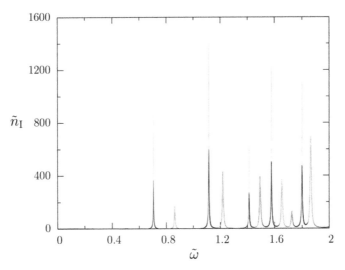

Figure 9.15. $\tilde{n}_I$, electric spectral density integrated though the box with metallic walls, as a function of $\tilde{\omega}$; conductivity of the walls $\tilde{\sigma} = 1000$. Black curve, calculated with 342 plane-waves for each transverse polarization, with $D/d = 2.07$; cyan curve, calculated with 729 plane-waves for transverse and longitudinal polarizations, with $D/d = 1.43$.

the degeneracy when the conductivity is finite, or whether it is due to numerical inaccuracy—a question of group theory which I shall leave for the time being.

As an example of an eigenstate of equation (9.20), figure 9.14 shows the electric field of a (211) mode (number 4 in table 9.6), calculated for the box with $\tilde{\sigma} = 1000$. With this value of conductivity, the electric field is pretty well perfectly normal to the confining walls.

What happens when we include longitudinal waves in the basis set given by equation (9.97)? Taking a basis with one longitudinal as well as two transverse modes for each plane-wave, we obtain the results given by the cyan curve in figure 9.15 compared with the results with just the transverse modes given by the black curve. Including the longitudinal waves doubles the degeneracy for certain modes, for example (110) and (210), as the figure clearly shows. The extra states are certainly spurious, but the other states are described accurately, and this behaviour only occurs for particular values of (m, n, p). It would be interesting to use a basis set of transverse waves plus a larger number of longitudinal waves, as we did in section 9.1.2, to see if this removes the spurious states (though these are not Laplace-type solutions, as their frequency is identical to the real modes). These extra modes are perhaps best described as ghost states.

9.5 Conclusions

We have seen in this chapter that embedding the vector fields of Maxwell's equations is much less straightforward than the scalar embedding of Schrödinger's equation, with the complications of Laplace-type solutions creeping in, as in section 9.1.2, or ghost-states as in the last section. Whether there is a single approach which can remove all the difficulties remains to be seen, though augmenting transverse basis functions with a large number of longitudinal waves works well for cylindrical geometry as well as for the cases studied in this chapter. For the moment it seems to be a matter of trial-and-error. However, in many cases, as in section 9.3, we can work with just one component of the electromagnetic field and the problem reduces to straightforward scalar embedding.

With the vast availability of commercial software for solving Maxwell's equations, using finite element methods for example, is there any reason to consider embedding? The most important aspect of embedding in general is that it properly treats the coupling of the system of interest to the rest of the world, with its continuum of states. This coupling broadens discrete states into resonances, and we have seen how the properties of these resonances can be readily calculated.

References

[1] Joannopoulos J D, Johnson S G, Winn J N and Meade R D 2008 *Photonic Crystals* 2nd edn (Princeton: Princeton University Press)

[2] Maradudin A A, Sambles J R and Barnes W L (ed) 2014 *Handbook of Surface Science volume 4, Modern Plasmonics* (Amsterdam: Elsevier)

[3] Pendry J B, Schurig D and Smith D R 2006 Controlling electromagnetic fields *Science* **312** 1780–2

[4] Inglesfield J E 1998 The embedding method for electromagnetics *J. Phys. A: Math. Gen.* **31** 8495–510

[5] Zangwill A 2013 *Modern Electrodynamics* (Cambridge: Cambridge University Press)

[6] Kemp R and Inglesfield J E 2002 Embedding approach for rapid convergence of plane waves in photonic calculations *Phys. Rev.* B **65** 115103

[7] Jackson J D 1962 *Classical Electrodynamics* 1st edn (New York: John Wiley)

[8] Tai C-T 1994 *Dyadic Green Functions in Electromagnetic Theory* (New York: IEEE Press)

[9] Senior T B A and Volakis J L 1995 *Approximate Boundary Conditions in Electromagnetism* (London: The Institution of Electrical Engineers)

[10] Pitarke J M, Inglesfield J E and Giannakis N 2007 Surface-plasmon polaritons in a lattice of metal cylinders *Phys. Rev.* B **75** 165415

[11] Giannakis N A, Inglesfield J E, Jastrzebski A K and Young P R 2013 Photonic modes of a chain of nanocylinders by the embedding method *J. Opt. Soc. Am.* B **30** 1755–64

[12] Stratton J A 1941 *Electromagnetic Theory* (New York: McGraw-Hill)

[13] Stefanou N, Yannapapas V and Modinos A 1998 Heterostructures of photonic crystals: frequency bands and transmission coefficients *Comput. Phys. Commun.* **113** 49–77

[14] Stefanou N, Yannapapas V and Modinos A 2000 MULTEM 2: a new version of the program for transmission and band-structure calculations of photonic crystals *Comput. Phys. Commun.* **132** 189–96

[15] Sözüer H S, Haus J W and Inguva R 1992 Photonic bands: Convergence problems with the plane-wave method *Phys. Rev.* B **45** 13962–72

[16] Moroz A and Sommers C 1999 Photonic band gaps of three-dimensional face-centred cubic lattices *J. Phys.: Condens. Matter* **11** 997–1008

[17] Ashcroft N W and Mermin N D 1976 *Solid State Physics* (Philadelphia: Saunders)

[18] Inglesfield J E, Pitarke J M and Kemp R 2004 Plasmon bands in metallic nanostructures *Phys. Rev.* B **69** 233103

[19] Newton R G 1982 *Scattering Theory of Waves and Particles* 2nd edn (New York: Springer-Verlag)

[20] Zhao Y and Hao Y 2007 Finite-difference time-domain study of guided modes in nano-plasmonic waveguides *IEEE Trans. Antennas Propag.* **55** 3070–7

[21] Muljarov E A, Langbein W and Zimmermann R 2010 Brillouin-Wigner perturbation theory in open electromagnetic systems *Europhys. Lett.* **92** 50050

[22] Doost M B, Langbein W and Muljarov E A 2012 Resonant-state expansion applied to planar open optical systems *Phys. Rev.* A **85** 023835

[23] Landau L D and Lifschitz E M 1960 *Electrodynamics of Continuous Media* (Oxford: Pergamon Press)

[24] Morse P M and Feshbach H 1953 *Methods of Theoretical Physics* (New York: McGraw-Hill)

Chapter 10

Time-dependent embedding

The embedding method can be generalized to solving the time-dependent Schrödinger equation [1], and although this work is still at an early stage it is worth discussing, as a new aspect of embedding, and one which will hopefully prove to be very useful.

There is growing interest in measuring time-dependent processes, for example, the use of ultra-short laser pulses in pump–probe experiments on electron emission from surfaces [2–4]. The question of time scales is a general problem in many areas of physics—how long it takes for an electron to tunnel through a potential barrier, how quickly conduction electrons relax after the creation of a core hole in photoemission, and so on. Time scales are also important in understanding the topic of chapter 7, electron transport through molecules, especially the role of electron–electron and electron–phonon interactions (coherent versus incoherent scattering, Coulomb blockade ...). To understand these processes we need to solve the time-dependent Schrödinger equation one way or another, and contributing to this is the development of time-dependent density functional theory (TDDFT), which enables the time-dependent many-electron problem to be solved within a one-electron framework—allowing, incidentally, excited states of the system to be calculated rigorously.

The reason for developing time-dependent embedding is the same as for the time-independent Schrödinger equation, to replace a substrate. As an example, time-dependent calculations have been carried out to simulate pump–probe experiments on surface emission [3]. In these calculations the surface and substrate are replaced by a slab of finite thickness with a finite region of vacuum outside. Sooner or later, the excited electron reaches the edge of the computational region, and it is important to eliminate back reflections. There has been much work in atomic physics in developing suitable boundary potentials—an absorbing potential is sometimes used at the end of the range of study [5, 6], though this inevitably gives rise to some reflection. The idea of embedding is that the embedding potential provides the

doi:10.1088/978-0-7503-1042-0ch10   10-1

correct boundary conditions on the wave-function at the boundary of the computational region, without reflection, so that time-dependent embedding can provide a solution to the boundary problem in calculations of this sort.

We might expect that the time-dependent embedding potential is just the Fourier transform of $\Sigma(E)$, given by equation (2.45), for example, for a simple one-dimensional system. However it is not quite as simple as that, because Σ varies as $E^{1/2}$ at large E, and the Fourier transform does not converge. Fortunately several authors have solved this problem, in finding an exact termination for the time-dependent Schrödinger equation [7, 8]. In this chapter I build on this work on the exact termination problem to derive time-dependent embedding, and I shall give several simple one-dimensional examples showing how it can be used.

A direct Fourier transform *does* in fact work for the embedding self-energy in a tight-binding or spatially discretized representation, and this is used in time-dependent calculations on electron transport through molecules [9]. This approach dates back many years [10, 11], preceding my work on time-dependent embedding for continuum systems. I shall discuss the time-dependent self-energy in section 10.5. (A note on units: the atomic unit of time = $2.418\,884 \times 10^{-17}$ s, so that 1 fs $\approx$ 41 a.u.)

10.1 Time-dependent embedding formalism

Our problem is to solve the time-dependent Schrödinger equation,

$$H\tilde{\Psi} = i\frac{\partial\tilde{\Psi}}{\partial t}, \tag{10.1}$$

in region I joined over surface S on to region II, solving the equation explicitly only in region I with an embedding potential on S. In general H involves a time-dependent potential $V(\mathbf{r}, t)$, but for our derivation of the embedding potential we assume that V is time-independent in region II.

We first derive the time-dependent relationships between the surface amplitude and normal derivative of the solution of the Schrödinger equation in region II, analogous to equations (2.13) and (2.15). In these equations, ψ, $\partial\psi/\partial n_S$ and G_0 are all functions of the continuous energy variable E—*not* an eigenvalue. As such, we can Fourier transform them to functions of time, the definitions of the Fourier transforms being

$$\tilde{\psi}(\mathbf{r}, t) = \frac{1}{2\pi}\int_{-\infty}^{+\infty} dE \, \exp(-iEt)\psi(\mathbf{r}, E),$$

$$\tilde{G}_0(\mathbf{r}, \mathbf{r}'; t) = \frac{1}{2\pi}\int_{-\infty}^{+\infty} dE \, \exp(-iEt)G_0(\mathbf{r}, \mathbf{r}'; E), \tag{10.2}$$

where the tilde indicates a function of time. $\tilde{\psi}$ satisfies the time-dependent Schrödinger equation in region II,

$$\left(-\frac{1}{2}\nabla^2 + V(\mathbf{r}) - i\frac{\partial}{\partial t}\right)\tilde{\psi}(\mathbf{r}, t) = 0, \tag{10.3}$$

with the inhomogeneous boundary condition that $\tilde{\psi}(\mathbf{r}_S, t)$ is specified on S. Similarly, $\tilde{G}_0$ satisfies the corresponding inhomogeneous equation, the time-dependent version of equation (2.9),

$$\left(-\frac{1}{2}\nabla^2 + V(\mathbf{r}) - i\frac{\partial}{\partial t}\right)\tilde{G}_0(\mathbf{r}, \mathbf{r}'; t) = -\delta(\mathbf{r} - \mathbf{r}')\delta(t), \qquad (10.4)$$

with the zero-derivative boundary condition on S. The Fourier transform of equation (2.13) then gives a convolution in time,

$$\tilde{\psi}(\mathbf{r}_S, t) = \frac{1}{2}\int_S d\mathbf{r}'_S \int_{-\infty}^{+\infty} dt' \ \tilde{G}_0(\mathbf{r}_S, \mathbf{r}'_S; t - t')\frac{\partial\tilde{\psi}(\mathbf{r}'_S, t')}{\partial n_S}, \qquad (10.5)$$

but because we use the retarded Green function, with

$$\tilde{G}_0(\mathbf{r}, \mathbf{r}'; t) = 0, \quad t < 0, \qquad (10.6)$$

this becomes

$$\tilde{\psi}(\mathbf{r}_S, t) = \frac{1}{2}\int_S d\mathbf{r}'_S \int_{-\infty}^{t} dt' \ \tilde{G}_0(\mathbf{r}_S, \mathbf{r}'_S; t - t')\frac{\partial\tilde{\psi}(\mathbf{r}'_S, t')}{\partial n_S}. \qquad (10.7)$$

This can be derived directly by using Green's theorem with equations (10.3) and (10.4) [7].

For embedding we need the inverse of this relation, giving the surface derivative in terms of the amplitude on S. To obtain a convergent Fourier transform, we rewrite equation (2.15) as

$$\frac{\partial\psi(\mathbf{r}_S, E)}{\partial n_S} = 2\int_S d\mathbf{r}'_S \frac{G_0^{-1}(\mathbf{r}_S, \mathbf{r}'_S; E)}{-iE}\left[-iE \ \psi(\mathbf{r}'_S, E)\right]. \qquad (10.8)$$

At large E, $G_0^{-1}(E)/E \sim E^{-1/2}$ for which the Fourier transform converges, and the Fourier transform of $-iE\psi(E)$ is $\partial\psi(t)/\partial t$. Defining the time-dependent embedding potential $\hat{\Sigma}(t)$ as the modified Fourier transform of the energy-dependent embedding potential $\Sigma(E)$,

$$\hat{\Sigma}(\mathbf{r}_S, \mathbf{r}'_S; t) = \frac{1}{2\pi}\int_{-\infty}^{+\infty} dE \ \exp(-iEt)\frac{\Sigma(\mathbf{r}_S, \mathbf{r}'_S; E)}{-iE}, \qquad (10.9)$$

the time-dependent analogue of equation (2.15) then becomes

$$\frac{\partial\tilde{\psi}(\mathbf{r}_S, t)}{\partial n_S} = -2\int_S d\mathbf{r}'_S \int_{-\infty}^{t} dt' \ \hat{\Sigma}(\mathbf{r}_S, \mathbf{r}'_S; t - t')\frac{\partial\tilde{\psi}(\mathbf{r}'_S, t')}{\partial t'}. \qquad (10.10)$$

As in equation (10.6), we have used the retarded property of $\hat{\Sigma}$ to give the upper limit of t in the integral over t'. This result has been given previously by Boucke *et al* [7]. With the relationship (10.10) we can now embed the time-dependent Schrödinger equation, and for this we will use the Dirac–Frenkel variational principle [12–14].

This not very well-known principle states that the solution $\tilde{\Psi}(\mathbf{r}, t)$ of the time-dependent Schrödinger equation satisfies

$$\delta I(t) = \int d\mathbf{r}\, \delta\tilde{\Psi}^*(\mathbf{r}, t)\left(H - i\frac{\partial}{\partial t}\right)\tilde{\Psi}(\mathbf{r}, t) = 0, \tag{10.11}$$

where the integral is over the whole of space, regions $\mathrm{I} + \mathrm{II}$.

Let us take a trial function $\tilde{\Psi}$ with the form similar to equation (2.4),

$$\tilde{\Psi}(\mathbf{r}, t) = \begin{cases} \tilde{\phi}(\mathbf{r}, t), & \mathbf{r} \in \text{region I} \\ \tilde{\psi}(\mathbf{r}, t), & \mathbf{r} \in \text{region II}, \end{cases} \tag{10.12}$$

where $\tilde{\phi}(\mathbf{r}, t)$ is a trial function in region I, and $\tilde{\psi}(\mathbf{r}, t)$ is the exact solution of the time-dependent Schrödinger equation in region II, which matches in amplitude (but not necessarily derivative) on to $\tilde{\phi}$ over S. Substituting into equation (10.11) we obtain

$$\delta I(t) = \int_{\mathrm{I}} d\mathbf{r}\, \delta\tilde{\phi}^*(\mathbf{r}, t)\left(H\tilde{\phi}(\mathbf{r}, t) - i\frac{\partial\tilde{\phi}(\mathbf{r}, t)}{\partial t}\right)$$

$$+ \frac{1}{2}\int_S d\mathbf{r}_S\, \delta\tilde{\phi}^*(\mathbf{r}_S, t)\left(\frac{\partial\tilde{\phi}(\mathbf{r}_S, t)}{\partial n_S} - \frac{\partial\tilde{\psi}(\mathbf{r}_S, t)}{\partial n_S}\right), \tag{10.13}$$

where the surface integral over S comes from $-\frac{1}{2}\nabla^2$ acting on the discontinuity in derivative of the trial function (equation (2.8)). Using equation (10.10) and the assumption that $\tilde{\psi}(\mathbf{r}_S, t) = \tilde{\phi}(\mathbf{r}_S, t)$ this becomes

$$\delta I(t) = \int_{\mathrm{I}} d\mathbf{r}\, \delta\tilde{\phi}^*(\mathbf{r}, t)\left(H\tilde{\phi}(\mathbf{r}, t) - i\frac{\partial\tilde{\phi}(\mathbf{r}, t)}{\partial t}\right) + \frac{1}{2}\int_S d\mathbf{r}_S\, \delta\tilde{\phi}^*(\mathbf{r}_S, t)\frac{\partial\tilde{\phi}(\mathbf{r}_S, t)}{\partial n_S}$$

$$+ \int_S d\mathbf{r}_S \int_S d\mathbf{r}_S' \int_{-\infty}^{t} dt'\, \delta\tilde{\phi}^*(\mathbf{r}_S, t)\hat{\Sigma}(\mathbf{r}_S, \mathbf{r}_S'; t - t')\frac{\partial\tilde{\phi}(\mathbf{r}_S', t')}{\partial t'}. \tag{10.14}$$

Setting $\delta I(t) = 0$ gives the time-dependent embedding variational principle, involving the trial function $\tilde{\phi}(\mathbf{r}, t)$ only in region I and on its boundary S. We can only apply equation (10.14) as it stands to wave-functions which are initially confined to region I, and spread out into region II in the course of time. This is to provide a lower limit to the time-integral of the embedding term, which we normally take as $t = 0$.

To find $\tilde{\phi}(\mathbf{r}, t)$ such that $\delta I(t) = 0$ for all variations, we use a basis set expansion with time-dependent coefficients,

$$\tilde{\phi}(\mathbf{r}, t) = \sum_n a_n(t)\chi_n(\mathbf{r}). \tag{10.15}$$

It simplifies the formalism and calculations if the χ_n are orthonormal over region I, and we can construct these from a starting set of functions using the procedure described in section 5.6: an orthogonal basis set is given by the eigenfunctions of the overlap matrix, and these can be normalized by the eigenvalues. As always with embedding, we start off with a basis satisfying an homogeneous boundary condition on a surface beyond S. Substituting equation (10.15) into equation (10.14) gives

$$\delta I(t) = \sum_{mn}\delta a_m^*(t)\left(H_{mn}(t)a_n(t) + \int_0^t dt'\,\hat{\Sigma}_{mn}(t-t')\frac{da_n(t')}{dt'}\right) - i\sum_m \delta a_m^*(t)\frac{da_m(t)}{dt},$$

(10.16)

where

$$H_{mn}(t) = \frac{1}{2}\int_I dr[\nabla\chi_m(\mathbf{r})\cdot\nabla\chi_n(\mathbf{r}) + \chi_m(\mathbf{r})V(\mathbf{r},t)\chi_n(\mathbf{r})],$$

$$\hat{\Sigma}_{mn}(t) = \int_S d\mathbf{r}_S \int_S d\mathbf{r}_S'\,\chi_m(\mathbf{r}_S)\hat{\Sigma}(\mathbf{r}_S,\mathbf{r}_S';t)\chi_n(\mathbf{r}_S').$$

(10.17)

The variational principle $\delta I(t) = 0$ for any variation $\delta a_m^*(t)$ then implies

$$\sum_n\left(H_{mn}(t)a_n(t) + \int_0^t dt'\,\hat{\Sigma}_{mn}(t-t')\frac{da_n(t')}{dt'}\right) = i\frac{da_m(t)}{dt},$$

(10.18)

a time-dependent matrix equation which we solve for $t > 0$ with the $a_m(0)$, determined by the initial wave-function $\tilde{\phi}(\mathbf{r},0)$ inside region I [1]. The integral over time implies that the time-development of the wave-function is non-Markovian, depending as it does on the previous history [11].

10.2 Model atomic problem

We now test the formalism developed in the last section, applying it to a one-dimensional model atomic problem [1]. The model we use, which is one considered by several previous authors [7, 15], is to calculate the time evolution of an initial wave-function,

$$\tilde{\phi}(z,0) = \frac{1}{\sqrt{2}\cosh(z)},$$

(10.19)

the bound state of the potential $V(z) = -1/\cosh^2(z)$, in a time-dependent electric field $\mathcal{E} = \mathcal{E}_0\sin(\omega t)$. This seems an unlikely choice for our test of embedding, as the electric field extends through all space and our formalism is specifically designed for a time-dependent potential restricted to region I. However, this system can be converted via the Kramers–Henneberger transformation[1] into the problem of the wave-function evolving in the oscillatory potential given by

$$V(z,t) = -\frac{1}{\cosh^2[z + \xi_0\sin(\omega t)]},$$

(10.20)

---

[1] This unitary transformation into an accelerated frame of reference is used to study the interaction of intense laser beams with atomic electrons [16].

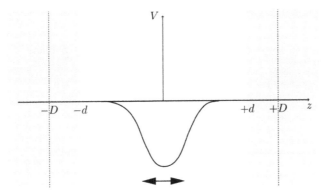

Figure 10.1. Oscillating atom model potential. Region I, treated explicitly, lies between $z = \pm d$, and regions II (shaded green) are replaced by time-dependent embedding potentials. The potential in region I is $V(z, t) = -1/\cosh^2[z + \xi_0 \sin(\omega t)]$, oscillating with frequency ω and amplitude ξ_0. D defines the basis functions in region I.

which tends rapidly to zero at large $|z|$. In equation (10.20) ξ_0 is the amplitude of an electron oscillating classically in field $\mathcal{E}_0 \sin(\omega t)$, with $\xi_0 = \mathcal{E}_0/\omega^2$. To treat this problem by embedding, we divide the system into region I lying between $z = \pm d$, enclosing the oscillating atom, and regions II consisting of the semi-infinite regions with $V = 0$ on either side (figure 10.1). The first stage in the calculation is to find the free-electron embedding potential to replace these regions II.

10.2.1 Time-dependent embedding potential

To calculate time-dependent embedding potentials, we evaluate the integral in equation (10.9) by taking E just above the real axis, as we do when we evaluate the density of states (equation (2.35)). We can complete the contour of integration by an appropriate semicircle, in the upper half-plane for $t < 0$, and in the lower half-plane for $t > 0$ (figure 10.2). As $\Sigma(E)$ is analytic in the upper half E-plane (it is, after all, a Green function, see equation (6.61)), the contour for $t < 0$ contains no poles and $\hat{\Sigma}(t < 0) = 0$. On the other hand, the contour for $t > 0$ encloses singularities, and $\hat{\Sigma}(t > 0)$ is finite.

With constant potential $V = 0$ in region II, $\Sigma(E)$ is given by

$$\Sigma(E) = \begin{cases} \sqrt{-E/2}, & E < 0 \\ -i\sqrt{E/2}, & E > 0, \end{cases} \tag{10.21}$$

corresponding to taking E just above the real axis (equation (2.29)). Substituting into equation (10.9) we obtain

$$\hat{\Sigma}(t) = \frac{1}{2\pi} \int_{-\infty}^{+\infty} dE \, \exp(-iEt) \begin{cases} -i/\sqrt{-2E}, & E < 0 \\ 1/\sqrt{2E}, & E > 0. \end{cases} \tag{10.22}$$

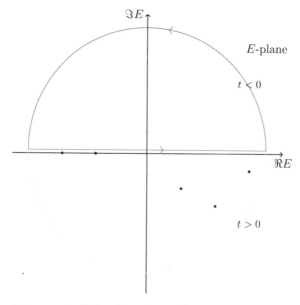

Figure 10.2. Contours in the complex E-plane for evaluating $\hat{\Sigma}(t)$ (equation (10.9)). Singularities in $\Sigma(E)$ are shown by dots lying on the $\Re E$-axis or in the lower half-plane. The contour is taken just above the $\Re E$-axis (blue line), closed by the red semicircle for $t < 0$ or the green semicircle for $t > 0$.

This integral can be performed analytically using contour integration [7], and the result for the time-dependent free-electron embedding potential is

$$\hat{\Sigma}(t) = \begin{cases} 0, & t < 0 \\ (1 - \mathrm{i})/2\sqrt{\pi t}, & t > 0. \end{cases} \tag{10.23}$$

10.2.2 Time-evolution in region I

With this free-electron embedding potential, we can now calculate the time evolution of the wave-function given by equation (10.19) when the time-dependent potential (equation (10.20)) is switched on at $t = 0$. As $\tilde{\phi}(z, 0)$ is the bound state of $V(z, 0)$ we can think of this as starting the oscillations of the potential at $t = 0$, or in terms of the Kramers–Henneberger transformation, switching on the perturbing electric field. As basis functions to expand $\tilde{\phi}(z, t)$ (equation (10.15)) we use

$$\chi_n(z) = \begin{cases} \cos n\pi z/2D, & n \text{ even} \\ \sin n\pi z/2D, & n \text{ odd}, \end{cases} \tag{10.24}$$

orthogonalized and normalized within region I, $|z| < d$ (figure 10.1); as usual, D is chosen somewhat greater than d to give flexibility in amplitude and derivative at the boundaries.

We have tried several integration schemes to calculate the time development of the expansion coefficients $a_n(t)$. In abstract form, equation (10.18) can be written as

$$\frac{\mathrm{d}a}{\mathrm{d}t} + \mathrm{i}Ha = -\mathrm{i}\Gamma, \tag{10.25}$$

where a is the vector a_m, H is the matrix H_{mn}, and Γ is the vector representing the embedding terms in equation (10.18),

$$\Gamma_m(t) = \chi_m(-d)\int_0^t \mathrm{d}t'\, \hat{\Sigma}(t - t')\frac{\partial \tilde{\phi}(-d, t')}{\partial t'} + \chi_m(+d)\int_0^t \mathrm{d}t'\, \hat{\Sigma}(t - t')\frac{\partial \tilde{\phi}(+d, t')}{\partial t'}. \tag{10.26}$$

A first-order time-integration scheme is then given by

$$a(t + \delta t) = [1 + \mathrm{i}\delta tH(t)]^{-1}[a(t) - \mathrm{i}\delta t\Gamma(t)], \tag{10.27}$$

but it is more accurate (and remains stable) to expand the time-evolution operator to second order in δt, giving

$$a(t + \delta t) = [1 + \mathrm{i}\delta tH(t) - \delta t^2\, H(t)^2/2]^{-1}[a(t) - \mathrm{i}\delta t\Gamma(t)]. \tag{10.28}$$

This is the scheme we use for this example.

The parameters we use in $V(z, t)$ (equation (10.20)) are the same as those used by Boucke *et al* [7], with an amplitude of oscillation $\xi_0 = 2.5$ a.u. and a frequency $\omega = 0.2$ a.u. Region I is taken with $d = 10$ a.u., and the basis functions are defined with $D = 13$ a.u.; we present results for basis sets of 25 and 40 basis functions. An interval $\delta t = 0.01$ a.u. is used in the time-integration (equation (10.28)). As a benchmark we compare the embedding calculation with the full wave-function $\tilde{\Psi}(z, t)$ calculated over an extended range, $|z| < 400$ a.u., with spatial as well as temporal finite differences.

In the upper graphs in figure 10.3 and figure 10.4 we compare the embedded and finite-difference wave-functions in region I for $t = 80$ and 400 a.u. (the period of the oscillating potential is 31.42 a.u.): we can be encouraged with the agreement. At the scale of the upper part of figure 10.3, $|\tilde{\phi}(z, t = 80)|$ calculated from embedding with 40 basis functions is indistinguishable from the extended range finite-difference wave-function $|\tilde{\Psi}(z, t = 80)|$. The lower part of this figure, plotted on a larger scale, shows how close the two are at the edge of region I. At $t = 400$ a.u. we see from the upper part of figure 10.4 that the difference between $|\tilde{\phi}|$ and $|\tilde{\Psi}|$ is a bit bigger: this is partly due to the accumulation of errors in the time integration (equation (10.28)), but the oscillations we see in $|\tilde{\Psi}(z, t = 400)|$ are due to the wave-function in the finite-difference calculation being reflected from the ends of the range at $z = \pm 400$ a.u., giving rise to interference. This can be seen very clearly in the lower part of figure 10.4, which shows $|\tilde{\Psi}(z, t = 400)|$ over the whole range, $|z| < 400$ a.u. As much as anything in this model calculation, this shows the value of embedding— removing spurious effects which arise sooner or later when an infinite region is replaced by a finite (though very large) range.

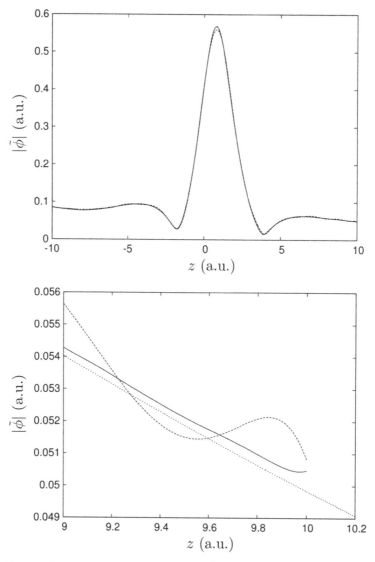

Figure 10.3. Time-development in the oscillating potential: $|\tilde{\phi}(z)|$ at $t = 80$ a.u., 2.55 periods. The upper figure is plotted in region I between $z = \pm 10$ a.u., and the lower figure over a short range around the embedding point at $z = +10$ a.u. Solid line, calculated with embedding using 40 basis functions; dashed line, calculated with embedding using 25 basis functions; the short-dashed line shows $|\tilde{\Psi}(z, t = 80)|$ from the extended range finite-difference calculation. (From Inglesfield [1].)

10.3 Time-evolution of extended states

The formalism we have developed up to now assumes that the wave-function $\tilde{\phi}(\mathbf{r}, t)$, whose evolution we calculate in the embedded region I, has zero amplitude at the embedding surface and in region II for $t < 0$. This is all well and good for the time-development of atomic wave-functions, but in condensed matter we are also

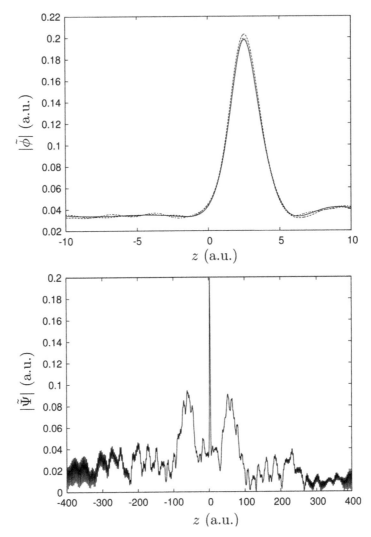

Figure 10.4. Time-development in the oscillating potential. Upper figure: $|\tilde{\phi}(z)|$ at $t = 400$ a.u., 12.73 periods, plotted in region I between $z = \pm 10$ a.u. Solid line, calculated with embedding using 40 basis functions; dashed line, calculated with embedding using 25 basis functions; the short-dashed line shows $|\tilde{\Psi}(z, t = 400)|$ from the extended range finite-difference calculation. Lower figure: $|\tilde{\Psi}(z, t = 400)|$ over the extended range between $z = \pm 400$ a.u. (From Inglesfield [1].)

interested in the evolution of states which extend beyond region I, for example, bulk states at a surface subject to an external time-dependent perturbation. We must extend the formalism to cover this type of problem [1, 14].

Before the perturbation is switched on at $t = 0$, the unperturbed wave-function is $\tilde{\Xi}(\mathbf{r}, t)$, a stationary state of H_0,

$$\tilde{\Xi}(\mathbf{r}, t) = \xi(\mathbf{r})\exp(-iEt), \quad \text{with} \quad H_0\xi = E\xi, \tag{10.29}$$

$\xi(\mathbf{r})$ can extend into region II as well as region I. We now switch on the perturbation $\delta V(\mathbf{r}, t)$ in region I, and see how $\tilde{\Xi}$ develops. Writing the full time-dependent Hamiltonian as H and the evolving wave-function as $\tilde{\Psi}$, we have for $t \leqslant 0$,

$$H = H_0, \qquad \tilde{\Psi}(\mathbf{r}, t) = \tilde{\Xi}(\mathbf{r}, t), \qquad \mathbf{r} \in \text{I} + \text{II}, \qquad (10.30)$$

and for $t > 0$,

$$H = H_0 + \delta V(\mathbf{r}, t), \qquad \tilde{\Psi}(\mathbf{r}, t) = \tilde{\Xi}(\mathbf{r}, t) + \tilde{\phi}(\mathbf{r}, t), \qquad \mathbf{r} \in \text{I},$$
$$H = H_0, \qquad \tilde{\Psi}(\mathbf{r}, t) = \tilde{\Xi}(\mathbf{r}, t) + \tilde{\psi}(\mathbf{r}, t), \qquad \mathbf{r} \in \text{II}. \qquad (10.31)$$

As before, $\tilde{\phi}(\mathbf{r}, t)$ is the trial function in region I, and $\tilde{\psi}(\mathbf{r}, t)$ is the exact solution of the time-dependent Schrödinger equation in region II, which matches in amplitude (but not necessarily derivative) on to $\tilde{\phi}$ over S.

Substituting $\tilde{\Psi}$ into equation (10.11) we obtain

$$\delta I = \int_{\text{I}} \mathrm{d}\mathbf{r}\, \delta\tilde{\phi}^*(\mathbf{r}, t)\left(H\tilde{\phi}(\mathbf{r}, t) + \delta V(\mathbf{r}, t)\tilde{\Xi}(\mathbf{r}, t) - \mathrm{i}\frac{\partial\tilde{\phi}(\mathbf{r}, t)}{\partial t} \right)$$
$$+ \frac{1}{2}\int_S \mathrm{d}\mathbf{r}_S\, \delta\tilde{\phi}^*(\mathbf{r}_S, t)\left(\frac{\partial\tilde{\phi}(\mathbf{r}_S, t)}{\partial n_S} - \frac{\partial\tilde{\psi}(\mathbf{r}_S, t)}{\partial n_S} \right), \qquad t > 0, \qquad (10.32)$$

the same as equation (10.13) with the addition of the extra inhomogeneous term $\delta V(\mathbf{r}, t)\,\tilde{\Xi}(\mathbf{r}, t)$ in the integral over region I. The variation of the functional then becomes

$$\delta I = \int_{\text{I}} \mathrm{d}\mathbf{r}\, \delta\tilde{\phi}^*(\mathbf{r}, t)\left(H\tilde{\phi}(\mathbf{r}, t) + \delta V(\mathbf{r}, t)\tilde{\Xi}(\mathbf{r}, t) - \mathrm{i}\frac{\partial\tilde{\phi}(\mathbf{r}, t)}{\partial t} \right)$$
$$+ \frac{1}{2}\int_S \mathrm{d}\mathbf{r}_S\, \delta\tilde{\phi}^*(\mathbf{r}_S, t)\frac{\partial\tilde{\phi}(\mathbf{r}_S, t)}{\partial n_S}$$
$$+ \int_S \mathrm{d}\mathbf{r}_S \int_S \mathrm{d}\mathbf{r}_S' \int_0^t \mathrm{d}t'\, \delta\tilde{\phi}^*(\mathbf{r}_S, t)\hat{\Sigma}(\mathbf{r}_S, \mathbf{r}_S'; t - t')\frac{\partial\tilde{\phi}(\mathbf{r}_S', t')}{\partial t}, \qquad t > 0,$$
$$\qquad (10.33)$$

where the lower limit of $t = 0$ in the integral over the embedding potential reflects the fact that $\tilde{\phi} = 0$ for $t < 0$. We use the same basis set expansion as before (equation (10.15)) for $\tilde{\phi}(\mathbf{r}, t)$, and the requirement that $\delta I = 0$ then gives us the generalization of equation (10.18) [14],

$$\sum_n \left(H_{mn}(t)a_n(t) + \int_0^t \mathrm{d}t'\, \hat{\Sigma}_{mn}(t - t')\frac{\mathrm{d}a_n(t')}{\mathrm{d}t'} \right) + e_m(t) = \mathrm{i}\frac{\mathrm{d}a_m(t)}{\mathrm{d}t}, \qquad t > 0,$$
$$\qquad (10.34)$$

where the Hamiltonian matrix element is now given by

$$H_{mn}(t) = \frac{1}{2} \int_I dr\{\nabla\chi_m(\mathbf{r}) \cdot \nabla\chi_n(\mathbf{r}) + \chi_m(\mathbf{r})[V(\mathbf{r}) + \delta V(\mathbf{r}, t)]\chi_n(\mathbf{r})\}, \quad (10.35)$$

$\hat{\Sigma}_{mn}$ is the same as in equation (10.17) and vector e_m is given by

$$e_m(t) = \int_I dr\, \chi_m(\mathbf{r})\delta V(\mathbf{r}, t)\tilde{\Xi}(\mathbf{r}, t). \quad (10.36)$$

Equation (10.34) is an inhomogeneous equation which we can solve for $a_m(t)$, with the initial condition that $a_m(t = 0) = 0$.

10.4 Excitation of electrons at the Cu(111) surface

As an example of time-dependent embedding of extended states, we shall consider the excitation of electrons at a model metal surface, in bulk and surface states—we recall from section 3.3.2 that Shockley surface states usually decay quite slowly into the bulk, and extend beyond region I in embedded surface calculations. The surface we shall study is Cu(111), using the one-dimensional Chulkov model potential [17]. This has the form

$$V(z) = \begin{cases} A_1 \cos(2\pi z/a), & z < 0 \\ -A_{10} - A_{20} + A_2 \cos(\beta z), & 0 < z < z_1 \\ -A_{10} + A_3 \exp[-\alpha(z - z_1)], & z_1 < z < z_{im} \\ -A_{10} + \left(\exp[-\lambda(z - z_{im})] - 1\right)/4(z - z_{im}), & z > z_{im}, \end{cases} \quad (10.37)$$

where a is the bulk interlayer spacing in the $\langle 111 \rangle$ direction, and z_{im} is the image plane from which the asymptotic image potential is measured. On the one side $V(z)$ is sinusoidal, and on the other side it goes asymptotically to the image potential. The parameters in equation (10.37) are fitted to measured surface state energies, with the requirement that $V(z)$ and its derivative are continuous across the potential boundaries at $z = 0$, z_1, and z_{im}: figure 10.5 shows the resulting Cu(111) surface potential.

10.4.1 Cu(111) surface electronic structure

Before we calculate the effects of a time-dependent perturbation on electrons at Cu(111), we need to know the unperturbed, static electronic structure in the surface region, embedded on to the bulk at z_b and on to the Coulomb tail of the image potential at z_C (figure 10.5).

To find the bulk embedding potential $\Sigma_b(E)$—at this stage a function of complex energy E—we could use the transfer matrix method described in section 3.6.1. However, in this one-dimensional problem there is a particularly neat method involving integrating the Schrödinger equation through a unit cell in each direction to give two independent solutions, ϕ_1 and ϕ_2. The starting point is the Green

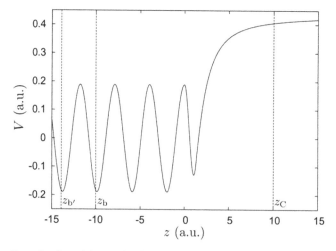

Figure 10.5. One-dimensional model potential for the Cu(111) surface. The top layer of atoms is at $z = 0$, and the average bulk potential is $V = 0$. Region I is taken between z_b, where it is embedded on to the bulk crystal, and z_C, where it is embedded on to the Coulomb tail of the image potential. The Schrödinger equation is integrated through the unit cell between $z_{b'}$ and z_b to evaluate the bulk embedding potential. (From Inglesfield [14].)

function formula, which follows from equation (2.11), for the wave-function $\psi(z)$ in some interval in terms of its derivatives at the end of the range[2],

$$\psi(z) = \frac{1}{2}[G_0(z, 0)\psi'(0) - G_0(z, a)\psi'(a)], \qquad (10.38)$$

where $\psi'(z) = d\psi/dz$, and we take the range to be the unit cell of the infinite crystal between $z = 0$ and $z = a$. $G_0(z)$ is the Green function with the zero-derivative boundary condition at the ends of the range. Using the Bloch property of $\psi(z)$, and evaluating $\psi(z)$ at $z = 0$ and a we obtain

$$\psi(0) = \frac{1}{2}\psi'(0)[G_0(0, 0) - G_0(0, a)\exp(ika)],$$

$$\psi(a) = \frac{1}{2}\psi'(a)[G_0(a, 0)\exp(-ika) - G_0(a, a)], \qquad (10.39)$$

where k is the Bloch wave-vector. Now, the logarithmic derivative ψ'/ψ is invariant to a lattice displacement, so comparing these two equations we obtain

$$\cos(ka) = \frac{G_0(0, 0) + G_0(a, a)}{2G_0(0, a)}. \qquad (10.40)$$

The Green function can be written in terms of the wave-functions $\phi_1(z)$ and $\phi_2(z)$ which satisfy the Schrödinger equation with the boundary conditions of the Green

---

[2] This formula is analogous to the equation in electrostatics giving the potential inside some region of space in terms of the boundary values of the electric field, and can be derived in the same way.

function at each end of the unit cell; ϕ_1 is integrated from $z = 0$ to $z = a$, with $\phi_1(0) = 1$, $\phi_1'(0) = 0$, and ϕ_2 is integrated from $z = a$ to $z = 0$, with $\phi_2(a) = 1$, $\phi_2'(a) = 0$, giving

$$G_0(z, z') = 2\frac{\phi_1(z_<)\phi_2(z_>)}{W(\phi_1, \phi_2)}, \tag{10.41}$$

where W is the Wronskian,

$$W(\phi_1, \phi_2) = \phi_1\phi_2' - \phi_2\phi_1'. \tag{10.42}$$

Substituting equation (10.41) into equation (10.40) gives the remarkably simple expression for k,

$$\cos(ka) = \frac{\phi_1(a) + \phi_2(0)}{2}. \tag{10.43}$$

A related expression has been given by Kohn [18] and Butti [19] (see also section 28.2 (iii) of the NIST Handbook in the chapter on Mathieu functions [20]).

We can now evaluate the embedding potential for embedding on to a semi-infinite crystal to the left of $z = a$. Remembering that dz is directed in this case *into* region I, we have

$$\Sigma_b = \frac{1}{2}\frac{\psi'(a)}{\psi(a)}, \tag{10.44}$$

which from equations (10.39) and (10.41) becomes

$$\Sigma_b(E) = \frac{W(\phi_1, \phi_2)}{2[\exp(-ika) - \phi_1(a)]}. \tag{10.45}$$

Here, E has a small positive imaginary part (at some point on the blue line in figure 10.2) and k is the solution of equation (10.43) with a negative imaginary part corresponding to a wave decaying to the left into the semi-infinite bulk. To obtain the embedding potential at z_b (figure 10.5) we take the unit cell between $z = z_{b'}$ and z_b in evaluating ϕ_1 and ϕ_2.

In the vacuum region outside the surface, the electron asymptotically feels the Coulomb tail of the image potential,

$$V(z) = -A_{10} - \frac{1}{4(z - z_{\text{im}})}. \tag{10.46}$$

The embedding potential $\Sigma_C(E)$ at $z = z_C$ (figure 10.5) is then given in terms of the Coulomb function H_0^- [21], as we described in section 3.1.4.

We now use Σ_b and Σ_C to find the Cu(111) surface density of states integrated through region I between $z_b = -10$ a.u. and $z_C = +10$ a.u. using the usual expansion methods for calculating the Green function (section 2.5). Figure 10.6 shows the surface density of states $n_s(E)$ at $\mathbf{k}_\parallel = 0$, where $\omega = E + i\varepsilon$ with an energy

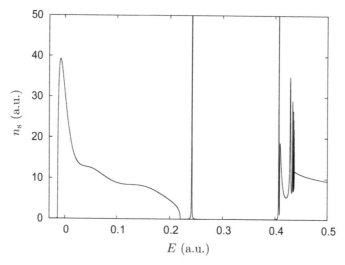

Figure 10.6. Cu(111) surface density of states $n_s(E)$ at $\mathbf{k}_{\parallel} = 0$ evaluated with the Chulkov potential. The density of states is integrated through region I between $z_b = -10$ a.u. and $z_C = +10$ a.u. (figure 10.5), embedded at z_b on to the bulk embedding potential Σ_b and at z_C a.u. on to the embedding potential for the Coulomb tail Σ_C. 80 basis functions are used and the imaginary part of the energy is $\eta = 1 \times 10^{-5}$ a.u. (From Inglesfield [14].)

broadening $\eta = 1 \times 10^{-5}$ a.u. We see the Shockley surface state at $E = 0.2415$ a.u., close to the bottom of the band gap at 0.2201 a.u., with the first image state (section 4.2) at $E = 0.4072$ a.u., immediately below the top of the gap at 0.4087 a.u. The higher members of the image Rydberg series overlap the continuum, and become image resonances. For comparison, the Fermi energy is at 0.2556 a.u. and the vacuum zero is at 0.4371 a.u. [17]. (We should note that the Chulkov potential is designed to describe the states close to the band gap, and it does not contain the Cu d-bands.)

10.4.2 Time-dependent embedding potentials for Cu(111)

With the energy-dependent embedding potentials, we can now carry out the (modified) Fourier transform (equation (10.9)) to calculate time-dependent embedding potentials $\hat{\Sigma}_b(t)$ and $\hat{\Sigma}_C(t)$, for embedding on to the bulk crystal and Coulomb tail at the Cu(111) surface. This involves a numerical integration along the blue contour shown in figure 10.2, which for the bulk embedding potential, say, becomes

$$\hat{\Sigma}_b(t) = \frac{\mathrm{i}}{2\pi} \int_{-\infty+\mathrm{i}\varepsilon}^{+\infty+\mathrm{i}\varepsilon} \mathrm{d}E \, \exp(-\mathrm{i}Et) \frac{\Sigma_b(E)}{E}, \qquad (10.47)$$

with a similar expression for the Coulomb embedding potential. At large $|E|$ the energy-dependent embedding potentials have free-electron behaviour, proportional

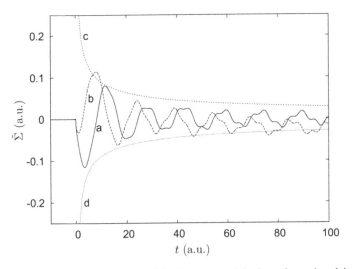

Figure 10.7. Time-dependent embedding potentials. Curves a and b show the real and imaginary parts of $[\hat{\Sigma}_b(t) - \hat{\Sigma}_f(t)]$, where $\hat{\Sigma}_b(t)$ is the embedding potential for matching on to Cu(111) at $z_b = -10$ a.u., and $\hat{\Sigma}_f(t)$ is the free-electron embedding potential with real and imaginary parts shown in curves c and d. (From Inglesfield [14].)

to $|E|^{1/2}$, and to speed up the convergence of the integral we can subtract this in the integrand and add on the corresponding time-dependent potential,

$$\hat{\Sigma}_b(t) = \frac{\mathrm{i}}{2\pi} \int_{-\infty+\mathrm{i}\varepsilon}^{+\infty+\mathrm{i}\varepsilon} \mathrm{d}E \frac{\exp(-\mathrm{i}Et)}{E}[\Sigma_b(E) - \Sigma_f(E)] + \hat{\Sigma}_f(t), \qquad (10.48)$$

where the free-electron $\Sigma_f(E)$ is given by equation (10.21) and $\hat{\Sigma}_f(t)$ by equation (10.23). We evaluate this integral by discretization within finite limits (the convergence trick keeps these reasonable), smearing out singularities in $\Sigma_b(E)$ with a small but finite value of η. However, we use only the real part of E in the $\exp(-\mathrm{i}Et)$ factor.

The time-dependent embedding potential for matching on to bulk Cu(111) is shown in figure 10.7. Curves a and b give the real and imaginary parts of $[\hat{\Sigma}_b(t) - \hat{\Sigma}_f(t)]$, bulk embedding with the free-electron embedding potential subtracted, while curves c and d show the real and imaginary parts of $\hat{\Sigma}_f(t)$. In evaluating the integral in equation (10.48), the energy is discretized with $\delta E = 1.25 \times 10^{-4}$ a.u. between energy limits of ± 50 a.u., and the energy broadening is $\eta = 2.5 \times 10^{-4}$ a.u. The accuracy of this evaluation is demonstrated by the fact that $\hat{\Sigma}_b(t)$ is almost exactly equal to zero for $t < 0$, with only very tiny Gibbs oscillations barely visible just below $t = 0$.[3] The structure of $\hat{\Sigma}_b(t)$ for $t > 0$ is interesting, with several periodicities visible in figure 10.7. The fundamental

---

[3] This shows the amazing power of complex variable theory, with causality, demonstrated in figure 10.2, emerging from the numerical evaluation of $\hat{\Sigma}_b(t)$.

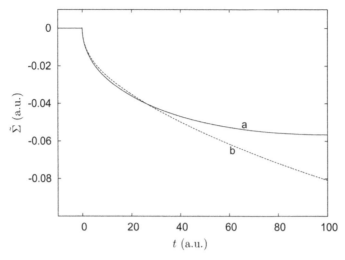

Figure 10.8. Time-dependent embedding potential for matching on to the Cu(111) image potential. Curves a and b show the real and imaginary parts of $[\hat{\Sigma}_C(t) - \hat{\Sigma}_f(t)]$, where $\hat{\Sigma}_C(t)$ is the embedding potential for matching on to the Coulomb tail at $z_C = +10$ a.u., and $\hat{\Sigma}_f(t)$ is the free-electron embedding potential. (From Inglesfield [14].)

periodicity is $\Delta t \approx 15.2$ a.u., which in Fourier transform corresponds to $\Delta E \approx 0.41$ a.u. This is very close to the band gap of 0.42 a.u. in the Cu(111) band structure.

It is more demanding to evaluate $\hat{\Sigma}_C(t)$, the time-dependent embedding potential for matching on to the Coulomb tail of the image potential. For this we take a smaller energy discretization, $\delta E = 1 \times 10^{-5}$ a.u., but even with this tiny value of δE, $\hat{\Sigma}_C(t)$ turns out to be finite—though small—for $t < 0$. This error, which is in fact constant for all t, is proportional to δE, so we simply subtract it off, in this way going to the limit $\delta E \to 0$. Figure 10.8 shows $[\hat{\Sigma}_C(t) - \hat{\Sigma}_f(t)]$ calculated in this way, with $\hat{\Sigma}_C(t)$ very close to zero for $t < 0$, with even smaller Gibbs oscillations than in the case of $\hat{\Sigma}_b(t)$.

10.4.3 Excitation of the Cu(111) surface state

With these embedding potentials, we are now ready to calculate the time-development of Cu(111) wave-functions under the influence of a time-dependent perturbation, working entirely in region I between the embedding planes at z_b and z_C. As a perturbing potential we take

$$\delta V(z, t) = A \exp(-z^2/\Delta)\sin(\omega t), \tag{10.49}$$

switched on at $t = 0$—this roughly models the perturbation in surface photoemission.

To solve the time-dependent Schrödinger equation we use a different scheme from equation (10.28). We write the matrix equation (10.34) in the same abstract form as equation (10.25), but with Γ containing the inhomogeneous vector e_m (equation (10.36)) as well as the embedding terms,

$$\Gamma_m(t) = e_m(t) + \chi_m(z_b) \int_0^t dt'\, \hat{\Sigma}_b(t - t') \frac{\partial \tilde{\phi}(z_b, t')}{\partial t'} + \chi_m(z_C) \int_0^t dt'\, \hat{\Sigma}_C(t - t') \frac{\partial \tilde{\phi}(z_C, t')}{\partial t'}.$$

(10.50)

Using central differences, equation (10.25) gives to first order in δt,

$$a(t + \delta t) = a(t) - i\delta t[H(t + \delta t/2)a(t) + \Gamma(t + \delta t/2)].$$

(10.51)

But an alternative first order equation for $a(t + \delta t)$ is

$$a(t + \delta t) = a(t) - i\delta t[H(t + \delta t/2)a(t + \delta t) + \Gamma(t + \delta t/2)].$$

(10.52)

Adding these two equations gives us

$$a(t + \delta t) = \left[1 + i\frac{\delta t}{2} H(t + \delta t/2) \right]^{-1} \left\{ \left[1 - i\frac{\delta t}{2} H(t + \delta t/2) \right] a(t) - i\delta t \Gamma(t + \delta t/2) \right\},$$

(10.53)

which is the equation we use to advance the wave-function forward in time. Without Γ, this is the Crank–Nicolson method [22], a unitary Cayley expansion of the time evolution operator $\exp(-i\delta t H)$ [23]; in our case, there is no conservation of probability within region I by itself, as we shall see shortly.

Although equation (10.53) is very accurate, much more so than equation (10.28), it turns out to be stable only for small values of δt. This is connected with the choice of basis set, and the closer the basis is to linear dependence or over-completeness, the greater the tendency to be unstable. In this work we use equation (10.15) as the basis set with the mid-point between z_b and z_C as the origin, and with a length of region I $(z_C - z_b) = 20$ a.u. and $2D = 24$ a.u., $\delta t = 0.002$ a.u. gives stability up to 40 basis functions. With $(z_C - z_b) = 40$ a.u. and $2D = 44$ a.u., this value of δt gives stability up to 75 basis functions. We can always obtain stability by using a smaller δt, or by bringing $2D$ closer to $(z_C - z_b)$, but this stability issue is one which needs further work.

We shall now calculate the excitation of the Shockley surface state at $E = 0.2415$ a.u., $\mathbf{k}_\parallel = 0$ (figure 10.6). First we study the accuracy of the embedding method, by calculating the electron density with different choices of region I; in this work, we take a perturbing potential (equation (10.49)) with a width parameter $\Delta = 2$ a.u., a frequency $\omega = 0.5$ a.u., and an amplitude $A = 0.2$ a.u. The initial surface state wave-function $\xi(z)$ in equation (10.29) is found by very accurate integration of the time-independent Schrödinger equation, and is normalized to unity over the whole of space (remember, it extends beyond region I into the crystal as well as into the vacuum); the corresponding electron density is given by the long-dashed line in figure 10.9. We calculate the time-development of this state using equation (10.53), and the solid line and the short-dashed line in figure 10.9 show the electron density $|\tilde{\Psi}(z)|^2$ at $t = 200$ a.u. for a short and long region I, respectively: the solid line is calculated with region I taken between $z_b = -10$ a.u. and $z_C = +10$ a.u., and the short-dashed line with region I between $z_b = -20$ a.u. and $z_C = +20$ a.u.

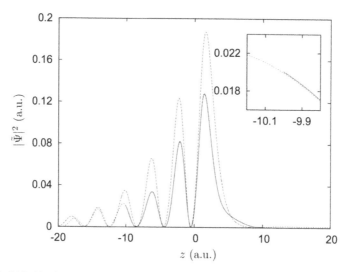

Figure 10.9. Cu(111) Shockley surface state electron density at $t = 200$ a.u. after switching on the time-dependent surface perturbation (equation (10.49)) with $\Delta = 2$ a.u., $\omega = 0.5$ a.u., $A = 0.2$ a.u. Solid line: calculated with region I between ± 10 a.u., using 40 basis functions with $2D = 24$ a.u.; short-dashed line: calculated with region I between ± 20 a.u., using 70 basis functions with $2D = 44$ a.u. A time-step $\delta t = 0.002$ a.u. is used in both calculations. The long-dashed curve shows the initial surface state density, normalized to unity over all space. The inset compares the densities around $z = -10$ a.u. with the two choices of region. (From Inglesfield [14].)

The difference between the two curves is invisible on the scale of the graph, and is barely visible on the scale of the enlarged inset showing the density around $z = -10$ a.u. This shows just how well time-dependent embedding works.

It's obvious from figure 10.9 that the surface state density integrated through region I has deceased with time—the electrons have been emitted into region II, into the crystal and into the vacuum. To study this in more detail, we calculate $Q(t)$, the number of electrons in region I in the surface state wave-function,

$$Q(t) = \int_{z_b}^{z_C} dz \, |\tilde{\Psi}(z, t)|^2. \tag{10.54}$$

Now continuity of probability density requires that

$$Q(t) + J_b(t) + J_C(t) = \text{constant}, \tag{10.55}$$

where J_b and J_C are the time-integrated probability currents leaving region I across the left- and right-hand embedding planes,

$$J_b(t) = -\int_0^t dt' \Im\left(\tilde{\Psi}^*(z_b, t')\frac{\partial \tilde{\Psi}(z_b, t')}{\partial z}\right),$$

$$J_C(t) = \int_0^t dt' \Im\left(\tilde{\Psi}^*(z_C, t')\frac{\partial \tilde{\Psi}(z_C, t')}{\partial z}\right). \tag{10.56}$$

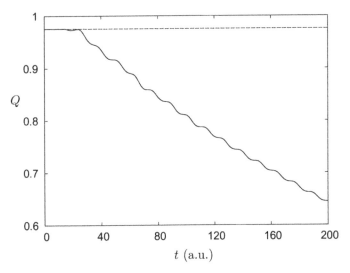

Figure 10.10. The solid line shows $Q(t)$, the surface state electron number in region I as a function of time, after switching on the time-dependent surface perturbation (equation (10.49)) with $\Delta = 2$ a.u., $\omega = 0.5$ a.u., $A = 0.2$ a.u. The dashed line shows $(Q + J_b + J_C)$, where J_b and J_C are the time-integrated currents across the embedding planes. Calculated with region I between ± 20 a.u., using 70 basis functions with $2D = 44$ a.u. A time-step $\delta t = 0.002$ a.u. is used to calculate the time evolution of the wave-function. (From Inglesfield [14].)

The expressions for the current densities are the same as equation (7.3), with a minus sign in the case of the current at z_b as we require the current *leaving* region I. We evaluate the derivatives $\partial \tilde{\Psi}/\partial z$ using the embedding expression (10.10), which amounts to using the time-embedding analogue of equation (7.6). Taking region I between $z_b = -20$ a.u. and $z_C = +20$ a.u., 70 basis functions with $2D = 44$ a.u., and the same perturbation as in figure 10.9 we obtain results for $Q(t)$ and $(Q(t) + J_b(t) + J_C(t))$ shown by the solid line and dashed line in figure 10.10. We see that $Q(t)$ decreases steadily, apart from small oscillations, as time progresses, but $(Q + J_b + J_C)$ stays almost perfectly constant—the maximum variation is less than 10^{-4} across the whole time range.

We study the currents into the crystal and into the vacuum separately in figure 10.11, using a perturbation with a smaller amplitude than in figure 10.10, $A = 0.1$ a.u., and a frequency $\omega = 0.6585$ a.u. (this apparently curious choice of frequency gives a final state energy of 0.9 a.u., the same as we use for the continuum state calculation which we shall discuss in the next section). $Q(t)$ decreases steadily, at a slower rate than in figure 10.10, because the amplitude of the perturbation is half the size. We also see that $J_b(t)$ has similar small oscillations to $Q(t)$, showing that this feature is due to the crystal potential. But the really significant feature of figure 10.11 is the fact that the slopes of J_b and J_C are quite close, so that the currents into the crystal and into the vacuum are roughly the same. This means that momentum is almost conserved in the emission process, though the surface step in the potential (figure 10.5) provides a source of momentum [24].

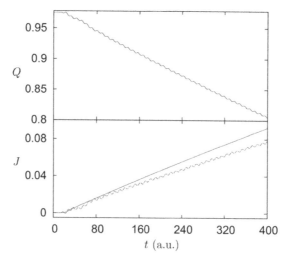

Figure 10.11. Upper figure: $Q(t)$, the surface state electron number in region I as a function of time, after switching on the time-dependent surface perturbation (equation (10.49)) with $\Delta = 2$ a.u., $\omega = 0.6585$ a.u., $A = 0.1$ a.u. Lower figure: time-integrated currents across the embedding planes; the solid line shows $J_C(t)$, the current into vacuum at $z_C = +20$ a.u., and the dashed line $J_b(t)$, the current into the crystal at $z_b = -20$ a.u. (From Inglesfield [14].)

10.4.4 Excitation of a Cu(111) bulk state

The behaviour of $Q(t)$ and the currents is quite different when we excite a bulk state. We take the bulk state with energy $E = 0.1$ a.u., $\mathbf{k}_\parallel = 0$, and start off with the energy-normalized wave-function, applying a perturbing potential with $\Delta = 2$ a.u., $A = 0.1$ a.u., as for figure 10.11, and a frequency $\omega = 0.8$ a.u. to give the same final state.

Both $Q(t)$ and $J_b(t)$ show persistent short-period oscillations due to the crystal potential, just as we saw in figure 10.11, but averaging these by eye $Q(t)$ settles down to an almost constant value, after initial transients have died down. This is because there is a current *entering* region I from the bulk which cancels the current emitted into vacuum—we see that $J_b(t)$ is negative, with a slope equal in magnitude but opposite in sign to that of $J_C(t)$. This remarkable result shows that in emission from a bulk state, far from this state being depleted, it is constantly replenished from the bulk. An interesting question is to what extent this picture is modified by the quantum mechanics of detecting an emitted electron—it's a single electron which is detected, with a certain probability; presumably it is at this stage that the hole may be filled from the bulk, again with a certain probability. This process also competes with many-body processes which can also fill the hole left after emission.

There is other physics in these results, such as the time-lag in figures 10.11 and 10.12 before emission becomes appreciable. What does this mean? It certainly seems relevant to the sort of laser physics experiments currently being performed on emission from electron states at surfaces [2, 4].

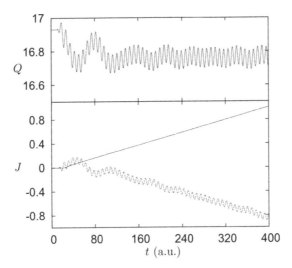

Figure 10.12. Upper figure: $Q(t)$, electron number in region I of the energy-normalized bulk state with $E = 0.1$ a.u., as a function of time, after switching on the time-dependent surface perturbation (equation (10.49)) with $\Delta = 2$ a.u., $\omega = 0.8$ a.u., $A = 0.1$ a.u. Lower figure: time-integrated currents across the embedding planes; the solid line shows $J_C(t)$, the current into vacuum at $z_C = +20$ a.u., and the dashed line $J_b(t)$, the current into the crystal at $z_b = -20$ a.u. (From Inglesfield [14].)

10.4.5 Outlook

The results presented in section 10.4 show the strengths and limitations of this approach to time-dependent embedding. Not only does it work, but as figure 10.10 shows, it is also very accurate. However, there is still a long way to go before it can be applied to more realistic systems than the one-dimensional model surface, over longer time scales. The most urgent problem to tackle is improving the time-evolution algorithm—equation (10.53) is accurate, but the whole stability issue needs sorting out so that the time interval δt can be increased for arbitrary size of basis. Can the time-integral over the embedding potential (equation (10.10)) be simplified? The fact that this integral goes from the starting time at $t = 0$ all the way up to time t means that the computing time increases exponentially with t, and with typical desktop computing power, it's impractical to go much beyond $t = 400$ a.u., or 10 fs. An answer to both these questions has been partly provided by the self-energy calculations which we discuss in section 10.5.

What I envisage is the development of the method so that time-dependent processes for realistic systems can be carried out—within the framework of time-dependent density functional theory, so that time-dependent screening effects can be included. But at least there is such a thing as time-dependent embedding!

10.5 Time-dependent embedding in a localized basis

A time-dependent self-energy can be used to embed the time-dependent Schrödinger equation expressed in a discrete basis, either tight-binding/linear combination of atomic orbitals (LCAO) (section 6.1) or with spatial discretization of the continuum

Hamiltonian (section 6.3) [9]. In fact this is the original approach to time-dependent embedding [11], and although the formalism is similar in many ways to the methods we have described in sections 10.1 and 10.3, there are as we shall see some notable differences. As continuum and discrete embedding can be related, at least for the time-independent Schrödinger equation (section 6.3.1), no doubt these differences can be resolved. The time-dependent embedding self-energy has proved particularly useful in molecular transport studies (mostly for model systems), especially within the framework of TDDFT to include many-body effects [25].

The sort of systems we are considering are shown schematically in figures 6.1 and 6.6, with a time-dependent Schrödinger equation given by

$$\begin{pmatrix} H_{11} & H_{12} \\ H_{21} & H_{22} \end{pmatrix} \begin{pmatrix} \tilde{\phi}_1 \\ \tilde{\psi}_2 \end{pmatrix} = i\frac{\partial}{\partial t} \begin{pmatrix} \tilde{\phi}_1 \\ \tilde{\psi}_2 \end{pmatrix}. \tag{10.57}$$

We are using the same notation as in section 6.1.1, with vector $\tilde{\phi}_1(t)$ representing the time-dependent wave-function in region I, which we want to find explicitly, and $\tilde{\psi}_2(t)$ the wave-function in II, which we want to eliminate. We neglect differential overlap (corresponding to $O_{ij} = \delta_{ij}$ in equation (6.4)). The Hamiltonian matrix H is in general time-dependent, and unlike the assumption we made in sections 10.1 and 10.3, we can have time dependence in region II with $H_{22}(t)$ [9].

As in equation (6.5), let's assume that we know $\tilde{\phi}_1(t)$. Then the second line of equation (10.57) gives

$$H_{22}\tilde{\psi}_2 - i\frac{\partial\tilde{\psi}_2}{\partial t} = -H_{21}\tilde{\phi}_1, \tag{10.58}$$

an inhomogeneous equation which we can solve using the time-dependent Green function $\tilde{\mathcal{G}}_{22}(t, t')$ for region II. This satisfies the matrix equation

$$H_{22}\tilde{\mathcal{G}}_{22}(t, t') - i\frac{\partial\tilde{\mathcal{G}}_{22}(t, t')}{\partial t} = -I\delta(t - t'), \tag{10.59}$$

within region II decoupled from region I. Here, I is the unit matrix. The *formal* solution of this equation is given by

$$\tilde{\mathcal{G}}_{22}(t, t') = \begin{cases} -i\exp[-iH_{22}(t - t')], & t > t' \\ 0, & t < t'. \end{cases} \tag{10.60}$$

Formal, as H_{22} in the exponential is a matrix; if H_{22} is time-independent $\tilde{\mathcal{G}}_{22}(t, t')$ is given by the Fourier transform (equation (10.2)) of $\mathcal{G}_{22}(E)$ in equation (6.6). The solution of equation (10.58) is then given by

$$\tilde{\psi}_2(t) = \int^t dt' \, \tilde{\mathcal{G}}_{22}(t, t')H_{21}\tilde{\phi}_1(t'), \tag{10.61}$$

where we have deliberately left the lower limit of the integral as undetermined. Now we can add a solution of the homogeneous version of equation (10.58) on to equation (10.61), in order to satisfy the boundary condition that $\tilde{\psi}_2(t)$ has a particular value at some starting time, which we shall take as $t = 0$. This solution is

$$\tilde{\psi}_2(t) = i\tilde{\mathcal{G}}_{22}(t, 0)\tilde{\psi}_2(0), \tag{10.62}$$

and putting equations (10.61) and (10.62) together gives us

$$\tilde{\psi}_2(t) = i\tilde{\mathcal{G}}_{22}(t, 0)\tilde{\psi}_2(0) + \int_0^t dt'\ \tilde{\mathcal{G}}_{22}(t, t')H_{21}\tilde{\phi}_1(t'), \quad t > 0, \tag{10.63}$$

the solution of equation (10.58) which has the given value of $\tilde{\psi}_2(0)$ at $t = 0$. We can now substitute this into the first line of equation (10.57) to obtain an embedded equation for $\tilde{\phi}_1$ [9, 11],

$$H_{11}\tilde{\phi}_1(t) + \int_0^t dt'\ \tilde{\Sigma}_{11}(t, t')\tilde{\phi}_1(t') + iH_{12}\tilde{\mathcal{G}}_{22}(t, 0)\tilde{\psi}_2(0) = i\frac{\partial \tilde{\phi}_1}{\partial t}, \quad t > 0, \tag{10.64}$$

where the embedding self-energy in time is given by

$$\tilde{\Sigma}_{11}(t, t') = H_{12}\tilde{\mathcal{G}}_{22}(t, t')H_{21}. \tag{10.65}$$

The time-dependent self-energy has the same structure as equation (6.10).

Comparing equation (10.64) with equation (10.34) for the continuum, we see that they have similar structure, a time-dependent Schrödinger equation in region I embedded into region II by an integral over the embedding potential up to time t: this is the *memory* term. But both equations contain an inhomogeneous third term on the left-hand side: $e_m(t)$ in equation (10.34) and $iH_{12}\tilde{\mathcal{G}}_{22}(t, 0)\tilde{\psi}_2(0)$ in equation (10.64). This is sometimes called the *source* term. This term arises in both cases because at $t = 0$, when the solution of the time-dependent Schrödinger equation starts, there is already a wave in region II. The difference between the two cases is that in equation (10.34) $\tilde{\phi}(\mathbf{r}, t)$ is the change in wave-function, whereas in equation (10.64) $\tilde{\phi}_1$ describes the full state. (This difference has nothing to do with the discrete basis, and we could have let $\tilde{\phi}_1$ describe the change in wave-function, exactly as in section 10.3.) In fact, $e_m(t)$ is much easier to evaluate than the source term in equation (10.64), because it only involves knowing the initial wave-function in region I, whereas to evaluate $H_{12}\tilde{\mathcal{G}}_{22}(t, 0)\tilde{\psi}_2(0)$ we need (in principle) to know the initial state throughout region II. On the other hand, the formalism provided by equation (10.64) is flexible enough to describe a time-dependent $H_{22}(t)$ [9], whereas equation (10.34) is restricted to time-dependence only in region I.

10.5.1 Time-dependent self-energy

Perhaps the most significant difference between the discrete and continuum forms of time-dependent embedding is that in equation (10.64) the discrete embedding term is

$$\int_0^t dt'\ \tilde{\Sigma}_{11}(t, t')\tilde{\phi}_1(t'), \tag{10.66}$$

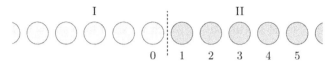

Figure 10.13. One-dimensional tight-binding chain. Green circles represent orbitals in region I, and brown circles orbitals in region II. Orbital 0 is the orbital in region I on which the embedding self-energy is to be added, replacing region II.

whereas in equation (10.26), for example, the continuum embedding contribution looks like

$$\int_0^t dt'\, \hat{\Sigma}(t - t')\frac{\partial \tilde{\phi}(t')}{\partial t}. \tag{10.67}$$

The time-derivative of $\tilde{\phi}$ in equation (10.67) is of course due to the fact that $\hat{\Sigma}(t)$ is the modified Fourier transform of $\Sigma(\omega)$ (equation (10.9)), with the extra factor of ω^{-1} to ensure convergence. If H_{22} is time-independent, $\tilde{\Sigma}_{11}(t)$ is just the straightforward Fourier transform of $\Sigma_{11}(\omega)$ (equation (6.10))—this converges, because in tight-binding/LCAO Σ_{11} is bounded as a function of ω.

The time-dependent self-energy can be found analytically for the one-dimensional chain shown in figure 10.13 [11]. We write the nearest-neighbour tight-binding Hamiltonian in region II as

$$H_{ij} = \begin{cases} \epsilon_0, & i = j \\ -v, & i = j \pm 1 \end{cases}, \tag{10.68}$$

where i, j label the individual orbitals rather than a group of orbitals as in equations (10.57)–(10.65). (Note the minus in front of v; with the discretized version of the free-electron Hamiltonian, v is positive.) Then the Green function for region II, decoupled from region I by breaking the link between orbitals 0 and 1 (figure 10.13), is given by [11]

$$\tilde{\mathcal{G}}_{ij}(t) = \begin{cases} i \exp(-i\epsilon_0 t)\left[i^{i+j}J_{i+j}(2vt) - i^{i-j}J_{i-j}(2vt)\right], & t > 0 \\ 0, & t < 0 \end{cases}, \quad i, j \geqslant 1, \tag{10.69}$$

where J_l is the cylindrical Bessel function. Using the sum rules for Bessel functions [20], we can show by direct substitution that this satisfies the matrix Schrödinger equation in region II,

$$\sum_k H_{ik}\tilde{\mathcal{G}}_{kj}(t) = i\frac{\partial \tilde{\mathcal{G}}_{ij}(t)}{\partial t}, \quad t > 0, \tag{10.70}$$

and it is clear immediately that

$$\tilde{\mathcal{G}}_{ij}(0+) - \tilde{\mathcal{G}}_{ij}(0-) = -i\delta_{ij}. \tag{10.71}$$

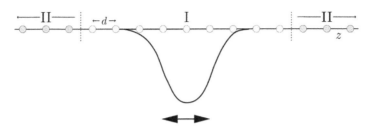

Figure 10.14. Discretization of the oscillating atom model. The green circles show the discretization points in region I, and the brown circles points in region II. The separation of the points is d. The oscillating potential is shown schematically, as in figure 10.1.

Hence $\tilde{\mathscr{G}}_{ij}(t)$ satisfies the inhomogeneous Schrödinger equation in region II

$$\sum_k H_{ik}\tilde{\mathscr{G}}_{kj}(t) - i\frac{\partial \tilde{\mathscr{G}}_{ij}(t)}{\partial t} = -\delta_{ij}\delta(t), \tag{10.72}$$

the same as equation (10.59), and is the Green function which we require to construct the self-energy. The Green function on orbital 1 of region II (figure 10.13) is then given by

$$\tilde{\mathscr{G}}_{11}(t) = -i\exp(-i\epsilon_0 t)\frac{J_1(2vt)}{vt}, \tag{10.73}$$

so from equation (10.65) the time-dependent embedding self-energy on orbital 0 in region I is given by

$$\tilde{\Sigma}_{00}(t, t') = \begin{cases} -iv\exp[-i\epsilon_0(t - t')]\dfrac{J_1[2v(t - t')]}{t - t'}, & t > t' \\ 0, & t < t' \end{cases}. \tag{10.74}$$

10.5.2 Discretized model atomic problem

As a test of this formalism, we shall apply it to the time-dependent model atomic problem which we discussed in section 10.2, discretizing the whole problem including embedding on to the semi-infinite free-space regions on either side of the oscillating 'atom'. We discretize the wave-functions on a one-dimensional grid with interval d (figure 10.14), and the time-dependent Schrödinger equation then becomes (compare with equation (6.54))

$$\sum_j H_{ij}(t)\tilde{\Psi}(z_j, t) = i\frac{\partial \tilde{\Psi}(z_i, t)}{\partial t}, \tag{10.75}$$

where z_i is the coordinate of the ith lattice point; we are using the same notation as in equation (10.12) and elsewhere in this chapter, that $\tilde{\Psi}(z_i, t)$ is the wave-function in the whole system, and $\tilde{\phi}(z_i, t)$ is the wave-function in embedded region I. The discretized Hamiltonian matrix is given by

$$
H_{ij}(t) = \begin{cases} \dfrac{1}{d^2} + V(z_i, t), & i = j \\[2mm] -\dfrac{1}{2d^2}, & i = j \pm 1 \end{cases}, \tag{10.76}
$$

where $1/d^2$ on the diagonal and $-1/2d^2$ in the off-diagonal elements represent the discretization of the kinetic energy $-\frac{1}{2}\frac{d^2}{dz^2}$, and $V(z_i, t)$ is given by equation (10.20).

To solve the time-dependent Schrödinger equation for the whole system, treating only region I explicitly, we break the links across the dashed lines between regions I and II (figure 10.14), and add the embedding self-energy given by equation (10.66) to the diagonal matrix elements on the left- and right-hand orbitals of region I. The embedding self-energy in equation (10.66) is given by equation (10.74) with $\epsilon_0 = 1/d^2$ and $v = 1/2d^2$,

$$
\tilde{\Sigma}_{00}(t, t') = \begin{cases} -\dfrac{i}{2d^2} \exp[-i(t - t')/d^2]\dfrac{J_1[(t - t')/d^2]}{t - t'}, & t > t' \\[2mm] 0, & t < t' \end{cases}, \tag{10.77}
$$

representing and replacing the semi-infinite free-electron regions II on either side of region I. The subscript 0 here corresponds to the right-hand orbital of region I, but of course the embedding self-energy is the same at the left-hand end. Taking $d = 0.15$ a.u. as a 'typical' grid spacing, we then obtain the time-dependent self-energy shown in figure 10.15. The most striking features of $\tilde{\Sigma}(t)$ are the rapid oscillations, and from equation (10.77) we see that the shorter the grid interval, the more rapid the oscillations of $\tilde{\Sigma}$ as a function of time. This obviously has implications for evaluating the self-energy integral (equation (10.66)), making it more difficult to calculate this accurately. On the other hand, $\tilde{\Sigma}(t)$ decays to zero very rapidly, so that the memory of the system is relatively short-lived; this overcomes the problem discussed in section 10.4.5, of the computer time growing exponentially with t.

We use the same starting wave-function (10.19) and time-dependent perturbing potential (equation (10.20)) as in section 10.2, discretized on the grid z_i. To step the embedded Schrödinger equation forward in time we use the adapted Crank–Nicolson method of equation (10.53), with a time-step $\delta t = 0.0002$ a.u. In evaluating the embedding contributions to the Hamiltonian we make use of the short memory of $\tilde{\Sigma}_{00}(t)$, replacing equation (10.66) with

$$
\int_{t_0}^{t} dt' \, \tilde{\Sigma}_{00}(t, t')\tilde{\phi}_0(t'). \tag{10.78}
$$

We take $t - t_0 = 20$ a.u., and evaluate the integral using the trapezoidal rule[4].

---

[4] These calculations are meant to be illustrative—preliminary—and I've made little effort to optimize the parameters or the method of evaluating the embedding integrals.

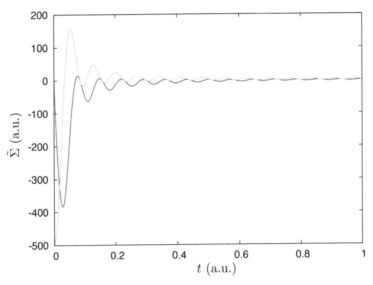

Figure 10.15. Time-dependent embedding self-energy for embedding on to a discretized free-electron half-space. The red and green curves show the real and imaginary parts of $\tilde{\Sigma}(t)$, calculated with a grid spacing $d = 0.15$ a.u.

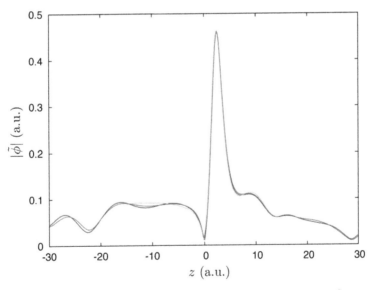

Figure 10.16. Time-development in the oscillating potential with spatial discretization: $|\tilde{\phi}(z)|$ at $t = 160$ a.u. The red curve shows results for embedded region I with $|z| < 30$ a.u., and the green curve with $|z| < 15$ a.u. The blue curve shows $|\tilde{\Psi}(z, t = 160)|$ from the extended range calculation without embedding.

Figure 10.16 shows the time-evolved wave-function at $t = 160$ a.u., calculated using the embedded Hamiltonian for region I with $|z| < 30$ a.u. (red curve) and a shorter region I with $|z| < 15$ a.u. (green curve). The blue curve shows the 'exact' result, calculated using discretization points with $|z| < 2250$ a.u. (essentially infinity at this relatively short value of *t*) without any embedding. It is clear that this discretized embedding procedure is working—the embedded wave-functions are fairly close to the exact result. However, it is less accurate than the basis set method for the same problem, as we see by comparing figure 10.16 (section 10.2) with figure 10.3. The reasons for this loss of accuracy are not yet clear.

10.5.3 Time-dependent calculations for molecular transport

The most productive applications of the time-dependent self-energy formalism are to time-dependent LCAO calculations of molecular transport, with the self-energy describing the leads, as in time-independent calculations (section 7.2). TDDFT then allows many-body effects in the molecule, such as Coulomb blockade [26], to be treated within a single-particle picture [25]. (In TDDFT the one-electron time-dependent Schrödinger equation contains a time-dependent exchange-correlation potential $V_{xc}[\rho](\mathbf{r}, t)$, local in space and time, which is a non-local functional of the time-dependent charge density [27]; a useful summary of TDDFT in this context is given in [28].) It also provides a framework for studying the effects of electron–phonon interactions, in which vibrations of the conducting molecule can scatter the electrons [29].

It is the paper by Kurth *et al* [9] which really provides the basis of a practical scheme for time-dependent calculations of transport through molecules, using discrete basis functions. This paper builds on the idea of Cini [30] that we should start off with the molecule in equilibrium with the leads, without any bias voltage, and then the bias is switched on—this is what happens in real experiments and can be contrasted with the approach of Caroli *et al* [31] in which a bias is established in disconnected electrodes in the distant past, and the contact then switched on adiabatically. The time-dependent embedded equation (10.64) is derived in this paper, but just as significant from our perspective is that these authors derive a Crank–Nicolson algorithm for time-evolution appropriate to the embedded discretized system. Judging from the results, this is much more efficient and accurate than our rather ad hoc formula (10.53).

As a first example of this method we consider the paper by Verdozzi *et al* [29], on the effect of nuclear motion on electron transport through a model molecule, the so-called Holstein wire. This consists of a row of atomic orbitals, with electron hopping *V* between nearest neighbours, and the interaction between the electrons and nuclear motion given by the following term in the Hamiltonian [32],

$$H_{e-ph} = -g \sum_{i=-M}^{M} x_i \hat{n}_i. \tag{10.79}$$

The sum is over the atoms labelled from $i = -M$ to M, x_i is the displacement of the *i*th atom, and $\hat{n}_i$ is the operator giving the number of electrons on atom *i* (this is the same second quantization notation for the number operator as in equation (6.37)). This adds a term $-gx_i$ to the *i*th diagonal element of the Hamiltonian matrix in the electron equation of motion (equation (10.64)); in Ehrenfest dynamics [33], the motion of atom i with mass m_i is given by the semi-classical equation of motion,

$$\ddot{x}_i = -\omega_0^2 - \frac{1}{m_i} \frac{\partial}{\partial x_i} \langle \tilde{\phi}(t) | H_{\text{e-ph}} | \tilde{\phi}(t) \rangle,$$
(10.80)

where ω_0 is the bare frequency of vibration.

The results from this calculation are shown in figure 10.17, for a molecule consisting of 15 atoms, connected to leads with half-filled bands. The coupling parameter λ, measuring the strength of the electron–phonon interaction, is defined as $\lambda = g^2/(2V\omega_0)$ and the vibration frequency is fixed at $\omega_0 = 0.1V$. The left-hand panels in figure 10.17 show the atomic displacements and electronic densities of states (insets) before the potential bias is switched on at $t = 0$. In the weak-coupling case ($\lambda = 0.1$, upper panel), the atoms show a roughly periodic displacement, and three bound states are pulled off the bottom of the band; in the strong-coupling case, on the other hand ($\lambda = 1$, lower panel), there is rather uniform displacement of

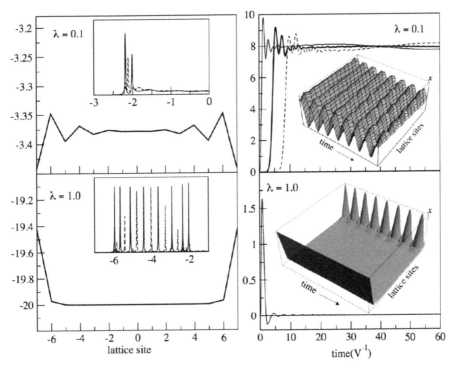

Figure 10.17. The left panels show the ground-state atomic displacements in the embedded 15-atom molecule before the bias is applied to the leads; the insets show the electron local densities of states on the left-hand atom (dashed line) and the middle atom (solid line). The right panels show the current $I(t)$ after switching on a bias $U/V = 0.5$: the solid line gives the current at the left-hand end of the molecule, and the dashed line the current in the middle; the insets show $x_i(t)$. The upper panels correspond to weak coupling, $\lambda = 0.1$, and the lower panels strong coupling, $\lambda = 1$. Units are described in the text. (From Verdozzi, Stefanucci and Almbladh [29].)

atoms and 15 bound states appear in the density of states. The right-hand panels show $I(t)$, the current as a function of time after applying a potential bias—a potential of $0.5V$ is applied to the left-hand electrode. We see that for $\lambda = 0.1$, after initial transients the current settles down quickly to a constant value, on which is superimposed a weak oscillation with frequency ω_0; the atoms oscillate with the current, as shown in the inset. With $\lambda = 1$ there is an initial transient at the left-hand end of the molecule, but the current quickly drops to zero as we would expect with the bound states.

To show how the method can be used within the framework of TDDFT we turn to the 2010 paper of Kurth *et al* [25], in which the effects of Coulomb blockade on transport are studied, using a single-level 'quantum dot' (in fact a single atomic orbital) embedded as before on to semi-infinite leads. The Hamiltonian for the isolated quantum dot is given by

$$H_{QD} = v_{ext} \sum_\sigma \hat{n}_{0\sigma} + U\hat{n}_{0\uparrow}\hat{n}_{0\downarrow}, \qquad (10.81)$$

where the sum is over spin, $\hat{n}_{0\sigma}$ is the number operator for electrons with spin σ on the quantum dot, v_{ext} is an external potential, and in the second term, U is the interaction between spin-up and spin-down electrons—the same as in equation (6.36). The quantum dot is coupled to the leads on either side by the nearest-neighbour hopping matrix element V_{link}, and within the leads (replaced by the embedding self-energy) the hopping is V. In TDDFT, the interacting Hamiltonian is replaced by a Kohn–Sham Hamiltonian $H_{KS}(t)$ acting on non-interacting electrons, and in an adiabatic approximation the Kohn–Sham potential $v_{KS}(t)$ is a functional of $n_0(t)$, the simultaneous number of electrons on the quantum dot,

$$H_{KS}(t) = \sum_\sigma v_{KS}[n_0(t)]\hat{n}_{0\sigma}. \qquad (10.82)$$

In this work, Kurth *et al* use a form of $v_{KS}[n_0(t)]$ which reproduces the exact exchange-correlation potential for an isolated quantum dot, and which is also in good agreement with calculations on the Anderson model.

For calculating the current through the quantum dot in the Coulomb blockade regime, the parameters were chosen as follows (in units of V): $V_{link} = 0.3$, $v_{ext} = 2$, $U = 2$, and the Fermi energy was set at 1.5. With the system in equilibrium for $t < 0$, a bias W_L applied to the left-hand electrode was switched on at $t = 0$ and the system allowed to develop in time. The results for $n_0(t)$ and $I_0(t)$, the time-dependent current through the quantum dot, are shown in figure 10.18 for three values of bias, $W_L = 1.3$ (black continuous curve), 1.6 (red dot-dashed curve) and 1.9 (green dashed curve). After a transient period lasting about $t = 10$ (time is measured in units of V^{-1}), both $n_0(t)$ and $I_0(t)$ show continuous self-sustaining oscillations, rather small in the case of the number (the inset in the top panel), but much more apparent in the current. These are associated with jumps in $v_{KS}(t)$ (the thin lines in the three middle panels) as $n_0(t)$ passes through 1—the exchange-correlation potential is discontinuous at half-filling of the quantum dot level when $n_0 = 1$ [25, 34]. The authors conclude that this is a real physical phenomenon, and that Coulomb blockade

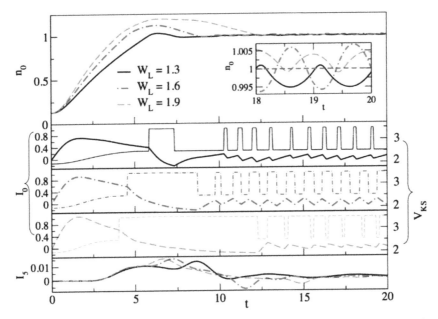

Figure 10.18. Transport through a quantum dot for different bias voltages applied to the left-hand electrode. The top panel shows $n_0(t)$, the number of electrons on the quantum dot, as a function of time after switching on the bias, for $W_L = 1.3$ (black continuous curve), 1.6 (red dot-dashed curve) and 1.9 (green dashed curve). The inset shows $n_0(t)$ on a larger scale. The middle three panels show $I_0(t)$, the current through the quantum dot, for the three bias voltages with the same colour code; the thin lines show $v_{KS}(t)$. The bottom panel shows $I_5(t)$ on the fifth site away from the quantum dot. Time is measured in units of V^{-1}. (From Kurth et al [25].)

manifests itself in transport by oscillatory behaviour as the quantum dot charges and discharges.

We mentioned in section 10.5 that the methods used here have the flexibility to include time dependence in the potential in the leads. This was used in the Coulomb blockade study to see the effect of switching on the bias voltage gradually—the oscillations in the charge and current decrease as the switching becomes more gradual. A time-dependent bias—an AC voltage—was also used in the work on transport through a vibrating molecule [29] to study the way that this affected the dissociation of the molecule.

This time-dependent embedding self-energy approach to transport through molecules is obviously very powerful, especially in the way that it can incorporate many-body effects via TDDFT (though there are questions to be answered over the time-dependent exchange-correlation functional itself [35]). Although the applications up to now have been to model systems, this method can certainly be extended to more realistic systems. Already there are powerful TDDFT programs such as OCTOPUS [36] for solving the time-dependent Schrödinger equation in molecular systems, which uses a real-space grid discretization and includes electron–phonon coupling.

10.6 Conclusions

The time-dependent embedding methods described in this chapter show their potential and demonstrate that the termination problem—how to treat the boundaries of the region treated explicitly in the time-dependent Schrödinger equation—can be solved. All the applications have been to more-or-less model systems, and there is a need to tackle *real* problems. One way forward may be to combine the discrete self-energy described in section 10.5, with its finite memory, with the basis set methods for region I described in section 10.1. Perhaps we can learn a thing or two from atomic and molecular physics, where successful ways of using the R-matrix method have been developed to solve time-dependent problems, such as the excitation of atomic electrons in multiphoton processes using ultra-short laser pulses [23, 37]. This is just the sort of problem which we would like to solve for surfaces [4].

References

[1] Inglesfield J E 2008 Time-dependent embedding *J. Phys.: Condens. Matter* **20** 095215
[2] Cavalieri A L *et al* 2007 Attosecond spectroscopy in condensed matter *Nature* **449** 1029–32
[3] Kazansky A K and Echenique P M 2009 One-electron model for the electronic response of metal surfaces to subfemtosecond photoexcitation *Phys. Rev. Lett.* **102** 177401
[4] Neppl S, Ernstorfer R, Bothschafter E M, Cavalieri A L, Menzel D, Barth J V, Krausz F, Kienberger R and Feulner P 2012 Attosecond time-resolved photoemission from core and valence states of magnesium *Phys. Rev. Lett.* **109** 087401
[5] Manolopoulos D E 2002 Derivation and reflection properties of a transmission-free absorbing potential *J. Chem. Phys.* **117** 9552–9
[6] Varga K 2011 Time-dependent density functional study of transport in molecular junctions *Phys. Rev.* B **83** 195130
[7] Boucke K, Schmitz H and Kull H-J 1997 Radiation conditions for the time-dependent Schrödinger equation: Application to strong-field photoionization *Phys. Rev.* A **56** 763–71
[8] Ehrhardt M 1999 Discrete transparent boundary conditions for general Schrödinger-type equations *VLSI Des.* **9** 325–38
[9] Kurth S, Stefanucci G, Almbladh C-O, Rubio A and Gross E K U 2005 Time-dependent quantum transport: A practical scheme using density functional theory *Phys. Rev.* B **72** 035308
[10] Wingreen N S, Jauho A-P and Meir Y 1993 Time-dependent transport through a mesoscopic structure *Phys. Rev.* B **48** 8487–90
[11] Hellums J R and Frensley W R 1994 Non-Markovian open-system boundary conditions for the time-dependent Schrödinger equation *Phys. Rev.* B **49** 2904–6
[12] Dirac P A M 1930 Note on exchange phenomena in the Thomas atom *Proc. Camb. Phil. Soc.* **26** 376–85
[13] Tannor D J 2007 *Introduction to Quantum Mechanics* (Sausalito: University Science Books)
[14] Inglesfield J E 2011 A time-dependent embedding calculation of surface electron emission *J. Phys.: Condens. Matter* **23** 305004
[15] Ermolaev A M, Puzynin I V, Selin A V and Vinitsky S I 1999 Integral boundary conditions for the time-dependent Schrödinger equation: Atom in a laser field *Phys. Rev.* A **60** 4831–45
[16] Walter C 1968 Henneberger. Perturbation method for atoms in intense light beams *Phys. Rev. Lett.* **21** 838–41

[17] Chulkov E V, Silkin V M and Echenique P M 1999 Image potential states on metal surfaces: binding energies and wave functions *Surf. Sci.* **437** 330–52

[18] Kohn W 1959 Analytic properties of Bloch waves and Wannier functions *Phys. Rev.* **115** 809–21

[19] Butti G 2005 *Surface Specific Electronic Structure on Extended Substrates* PhD thesis (Università degli Studi di Milano-Bicocca)

[20] Olver F W J, Lozier D W, Boisvert R F and Clark C W 2010 *NIST Handbook of Mathematical Functions* (Cambridge: Cambridge University Press)

[21] Thompson I J and Barnett A R 1986 Coulomb and Bessel functions of complex arguments and order *J. Comput. Phys.* **64** 490–509

[22] Crank J and Nicolson P 1947 A practical method for numerical evaluation of solutions of partial differential equations of the heat-conduction type *Proc. Cambridge Philos. Soc.* **43** 50–67

[23] Burke P G and Burke V M 1997 Time-dependent R-matrix theory of multiphoton processes *J. Phys. B: At. Mol. Opt. Phys.* **30** 383–91

[24] Feibelman P J and Eastman D E 1974 Photoemission spectroscopy-correspondence between quantum theory and experimental phenomenology *Phys. Rev.* B **10** 4932–47

[25] Kurth S, Stefanucci G, Khosravi E, Verdozzi C and Gross E K U 2010 Dynamical Coulomb blockade and the derivative discontinuity of time-dependent density functional theory *Phys. Rev. Lett.* **104** 236801

[26] Datta S 2005 *Quantum Transport: Atom to Transistor* (Cambridge: Cambridge University Press)

[27] Runge E and Gross E K U 1984 Density-functional theory for time-dependent systems *Phys. Rev. Lett.* **52** 997–1000

[28] Ventra M D 2008 *Electrical Transport in Nanoscale Systems* (Cambridge: Cambridge University Press)

[29] Verdozzi C, Stefanucci G and Almbladh C-O 2006 Classical nuclear motion in quantum transport *Phys. Rev. Lett.* **97** 046603

[30] Cini M 1980 Time-dependent approach to electron transport through junctions: General theory and simple applications *Phys. Rev.* B **22** 5887–99

[31] Caroli C, Combescot R, Nozières P and Saint-James D 1971 Direct calculation of the tunneling current *J. Phys. C: Solid St. Phys.* **4** 916–29

[32] Kopidakis G, Soukoulis C M and Economou E N 1995 Electron-phonon interaction, localization, and polaron formation in one-dimensional systems *Phys. Rev.* B **51** 15038–52

[33] Horsfield A P, Bowler D R, Fisher A J, Todorov T N and Sánchez C 2004 Beyond Ehrenfest: correlated non-adiabatic molecular dynamics *J. Phys.: Condens. Matter* **16** 8251–66

[34] Lima N A, Oliveira L N and Capelle K 2002 Density-functional study of the Mott gap in the Hubbard model *Europhys. Lett.* **60** 601–7

[35] Ramsden J D and Godby R W 2012 Exact density-functional potentials for time-dependent quasiparticles *Phys. Rev. Lett.* **109** 036402

[36] Andrade X *et al* 2012 Time-dependent density-functional theory in massively parallel computer architectures: the OCTOPUS project *J. Phys.: Condens. Matter* **24** 233202

[37] Lysaght M A, van der Hart H W and Burke P G 2009 Time-dependent R-matrix theory for ultrafast atomic processes *Phys. Rev.* A **79** 053411

IOP Publishing

The Embedding Method for Electronic Structure

John E Inglesfield

Chapter 11

Connections

This chapter is devoted to connections between the embedding method, and two other techniques of theoretical physics. First I shall describe the links to the R-matrix method in atomic and molecular physics [1], whose origins go back to nuclear physics papers by Wigner and Bloch [2, 3] in the 1940s and 50s. The connections between the R-matrix method and embedding were first discussed by Zou [4]. In this method, typically used to describe electron scattering by an N-electron atom, space is divided into two (or more) regions, with different representations of the wave-function in each; the R-matrix describes the $N + 1$ electron system in the inner region (the atom), and enables the scattering of the electron in the outer region to be calculated. It is in some sense many-body embedding.

Next I shall explore the properties of resonances and their associated resonant states [5, 6], which we introduced in section 2.6 and applied to plasmon resonances in section 9.3.4. In section 11.2 we shall use embedding to calculate the poles of the Green function as a function of complex wave-vector rather than energy: in simple cases this allows the Green function to be found exactly in terms of these poles alone, eliminating the awkward cut in the complex-energy plane which we found in section 2.6.1. This enables us to make contact with the modern theory of resonant-state expansions [7–9].

The link between the two parts of this chapter is that they describe techniques originally discussed in the context of nuclear physics. The link with embedding is that they are all ways of dealing with open systems—systems in which a finite region of interest can interact with the rest of the world, infinitely extended and with a continuum of states.

11.1 Embedding and R-matrix theory

The R-matrix method was introduced by Wigner and Eisenbud [2] as a way of studying the scattering of nucleons by nuclei, and subsequently used to study electron–atom scattering, following the work of Burke, Hibbert and Robb [10].

doi:10.1088/978-0-7503-1042-0ch11

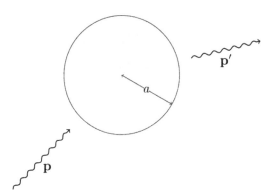

Figure 11.1. Electron–atom scattering in the R-matrix method: space is divided into an internal region, the R-matrix sphere containing the atom and all electron–electron interactions, and an external region. The electron is scattered from an initial state with momentum $\mathbf{p}$, and the atom in state i, to a final state with electron momentum $\mathbf{p}'$, and the atom in state i'. Radius a is chosen sufficiently large so that the wave-functions of the atomic target are completely contained in the R-matrix sphere.

In such scattering experiments, the electron (or nucleon) can be scattered either elastically, or inelastically, in which case the state of the target atom/nucleus changes. The different possible states of the target after scattering correspond to different *channels* of the scattered electron or nucleon. The aim of R-matrix theory (any theory of scattering in fact) is to calculate the transition probability between the incident electron in an initial state $|\mathbf{k}\rangle$ and the target atom in some state i (to be specific we shall concentrate on electron–atom scattering), and the scattered electron in a final state $|\mathbf{k}'\rangle$ and the target in state j. There is a very extensive literature on R-matrix theory and its applications—useful sources are [1, 11, 12]—and here I shall give a minimalist version [13].

We first divide the system into two regions, an internal region inside the so-called R-matrix sphere, containing the target atom, and an external region in which the scattering experiment is carried out (figure 11.1). The radius a of the R-matrix sphere is chosen to be large enough so that the N-electron target wave-functions have more-or-less zero amplitude at the sphere boundary—these include not only the initial state i of the atom, but also the final states j. The full solutions of the scattering problem satisfy the Schrödinger equation with the $(N + 1)$-electron Hamiltonian H_{N+1},

$$H_{N+1}\Psi(\mathbf{x}_1, \mathbf{x}_2,\dots \mathbf{x}_N, \mathbf{x}_{N+1}) = E\Psi(\mathbf{x}_1, \mathbf{x}_2,\dots \mathbf{x}_N, \mathbf{x}_{N+1}). \qquad (11.1)$$

Here, Ψ is an $(N + 1)$-electron wave-function with energy E, describing both the atom and the scattering electron, and because of the spherical symmetry we can specify the total angular momentum $\mathbf{L}$, total spin $\mathbf{S}$, and their z-components. The $\mathbf{x}_i$ electron coordinates include spin σ_i as well as the spatial part $\mathbf{r}_i$; the $\mathbf{r}_i$ coordinates vary over the whole of space, both the internal and external regions. We now construct $(N + 1)$-electron wave-functions ψ_k for which the spatial coordinates are confined to the internal region, satisfying a zero-derivative boundary condition on the sphere boundary. We saw how to do this in section 5.5 for a one-electron system—the generalization of equation (2.23) to the $(N + 1)$-electron case is

$$\left(H_{N+1} + L_{N+1}\right)\psi_k(\mathbf{x}_1, \mathbf{x}_2, \ldots \mathbf{x}_N, \mathbf{x}_{N+1}) = E_k\psi_k(\mathbf{x}_1, \mathbf{x}_2, \ldots \mathbf{x}_N, \mathbf{x}_{N+1}), \qquad (11.2)$$

where L_{N+1} is given by the sum over electron coordinates,

$$L_{N+1} = \frac{1}{2}\sum_i^{N+1}\delta(r_i - a)\frac{\partial}{\partial r_i}. \qquad (11.3)$$

This term ensures Hermiticity; it was first introduced by Bloch [3], and in the literature is called the Bloch operator. In abstract notation, the Green function operator corresponding to the Hamiltonian in equation (11.2) is then given by

$$(E - H_{N+1} - L_{N+1})^{-1} = \sum_k \frac{|\psi_k\rangle\langle\psi_k|}{E - E_k}. \qquad (11.4)$$

This is the zero-derivative Green function inside the R-matrix sphere, equivalent to G_0 in equations (2.9) and (2.10). We now write equation (11.1) as

$$\left(H_{N+1} + L_{N+1} - E\right)|\Psi\rangle = L_{N+1}|\Psi\rangle. \qquad (11.5)$$

Then substituting from equation (11.4) we obtain

$$|\Psi\rangle = \sum_k |\psi_k\rangle\frac{1}{E_k - E}\langle\psi_k|L_{N+1}|\Psi\rangle, \qquad (11.6)$$

which (in the coordinate representation) relates the wave-function $\Psi(\mathbf{x}_1, \mathbf{x}_2, \ldots \mathbf{x}_N, \mathbf{x}_{N+1})$ inside the R-matrix sphere to its derivative over the surface— the $(N + 1)$-electron generalization of equation (2.13). This result could have been just as well derived using Green's theorem [14], but the advantage of this operator derivation is that we don't get bogged down in numerous volume integrals over the R-matrix sphere.

The next stage is to expand the ψ_ks in terms of N-electron states Φ_i of the atom itself, together with a complete set of one-electron wave-functions $u_j(r)$, which describe the motion of the scattering electron. The target wave-functions satisfy

$$H_N\Phi_i(\mathbf{x}_1, \mathbf{x}_2, \ldots \mathbf{x}_N) = \epsilon_i\Phi_i(\mathbf{x}_1, \mathbf{x}_2, \ldots \mathbf{x}_N). \qquad (11.7)$$

We remember that the R-matrix sphere is chosen large enough so that the Φ_i's have negligible amplitude at $r = a$. These are then coupled to angular and spin functions of the scattering electron to form *channel functions* $\bar{\Phi}_t(\mathbf{x}_1, \mathbf{x}_2, \ldots \mathbf{x}_N; \hat{\mathbf{r}}_{N+1}, \sigma_{N+1})$ with the required total angular momentum $\mathbf{L}$ and spin $\mathbf{S}$; $\hat{\mathbf{r}}_{N+1}$ and σ_{N+1} are the angular and spin coordinates of the $(N + 1)$ th scattering electron. The one-electron functions $u_j(r)$ are eigenstates of the one-electron radial Schrödinger equation,

$$-\frac{1}{2r^2}\frac{\mathrm{d}}{\mathrm{d}r}\left(r^2\frac{\mathrm{d}u_j}{\mathrm{d}r}\right) + \frac{l(l + 1)}{2r^2}u_j(r) + U_0(r)u_j(r) = E_j^0 u_j(r), \qquad (11.8)$$

satisfying the zero-derivative boundary condition on the R-matrix sphere boundary,

$$\left.\frac{du_j}{dr}\right|_{r=a} = 0. \tag{11.9}$$

$U_0(r)$ is an effective potential representing the interaction of the scattering electron with the other N electrons—we might use Hartree or LDA. The ψ_k's can then be written as

$$\psi_k(\mathbf{x}_1, \mathbf{x}_2, \ldots \mathbf{x}_N, \mathbf{x}_{N+1}) = \mathcal{A} \sum_{ij} a_{kij} \bar{\Phi}_i(\mathbf{x}_1, \mathbf{x}_2, \ldots \mathbf{x}_N; \hat{\mathbf{r}}_{N+1}, \sigma_{N+1}) u_j(r_{N+1}), \tag{11.10}$$

where $\mathcal{A}$ is the antisymmetrization operator. This expression is then substituted into the matrix form of the Schrödinger equation (11.2),

$$\langle \psi_k | H_{N+1} + L_{N+1} | \psi_l \rangle = E_k \delta_{kl}, \tag{11.11}$$

which we can diagonalize to find the coefficients a_{kij} and the eigenvalues E_k.

We can now simplify equation (11.6), to obtain a separation of the motion of the scattering electron. First we multiply the Green's theorem equation (11.6) by the channel state $\langle \bar{\Phi}_i |$ on the left-hand side to give

$$F_i(a) = \sum_k w_{ik} \frac{1}{E_k - E} \langle \psi_k | L_{N+1} | \Psi \rangle, \tag{11.12}$$

where F_i and w_{ik} are the projection integrals,

$$F_i(r) = \langle \bar{\Phi}_i | \Psi \rangle', \qquad w_{ik} = \langle \bar{\Phi}_i | \psi_k \rangle'_{r=a}. \tag{11.13}$$

The prime indicates that the integration is carried out over all electron space and spin coordinates, except the radial coordinate of the scattered electron, which we call r. Next we insert the unit operator in the space of N-electrons, given by $\sum_\kappa |\bar{\Phi}_\kappa\rangle\langle\bar{\Phi}_\kappa|$, in the middle of the matrix element in equation (11.12) to give

$$\langle \psi_k | L_{N+1} | \Psi \rangle = \frac{a^2}{2} \sum_\kappa w_{\kappa k} \left.\frac{dF_\kappa}{dr}\right|_{r=a}. \tag{11.14}$$

The factor of a^2 comes from the integration of the spatial coordinate of the scattered electron over the surface of the R-matrix sphere, implicit in the matrix element. Putting this together we finally obtain

$$F_i(a) = \sum_\kappa R_{i\kappa}(E) \left.\frac{dF_\kappa}{dr}\right|_{r=a}, \tag{11.15}$$

where the R-matrix is defined as

$$R_{i\kappa}(E) = \frac{a^2}{2} \sum_k \frac{w_{ik} w_{\kappa k}}{E_k - E}. \tag{11.16}$$

How do we interpret these equations? Firstly, what *is* $F_i(r)$ in equation (11.15)? From equation (11.13), it's just the one-electron wave-function of the scattered electron when the atom is in the target state corresponding to $\bar{\Phi}_i$, and as such it satisfies a one-electron Schrödinger equation in the external region at an energy $E - \epsilon_i$. Equation (11.15) gives the boundary conditions on the one-electron solutions in the external region (equivalent to region I in embedding) so that they match on to the $(N + 1)$-electron problem inside the R-matrix sphere (or region II). The R-matrix is an inverse logarithmic derivative—a many-electron version of G_0 in equation (2.13).

11.1.1 Electron energy-loss spectroscopy from NiO

The R-matrix provides a multi-channel boundary condition for one-electron wave-functions in the external region—in a very real sense, embedding the one-electron wave-functions on to a many-body system. However, it is a theory in which there is only one electron outside the interacting region; for this reason it can only have limited application to the many-body problems of condensed matter physics, where typically a strongly interacting region is immersed in a Fermi sea of electrons. There are, of course, experiments in which electrons are scattered off solids, and the R-matrix method has been applied to the interpretation of low-energy electron energy-loss spectroscopy (EELS) experiments from NiO [13, 17]. In these experiments, electrons with an energy typically of 20–100 eV are scattered off the NiO surface in the same geometry as in low-energy electron diffraction, with energy losses coming from 3d–3d excitations on the Ni ions [15, 16].

Using R-matrix theory, we calculate inelastic electron scattering from a Ni^{2+} ion, with a 3d^8 electron configuration, in the crystal field of the surrounding ions [17]. This means that we have to adapt the usual R-matrix methods, which usually deal with atoms in a spherically symmetric potential, to handle the cubic symmetry felt by the Ni^{2+} ion in NiO. The crystal field potential has the form [18]

$$V_c(\mathbf{r}) = \beta r^4 \left\{ Y_{40}(\hat{\mathbf{r}}) + \left(\frac{5}{14} \right)^{1/2} \left[Y_{44}(\hat{\mathbf{r}}) + Y_{4-4}(\hat{\mathbf{r}}) \right] \right\}, \qquad (11.17)$$

where β is a parameter fitted to measured excitation energies. The target states of the Ni^{2+} ion are calculated in the Hartree–Fock approximation, with V_c added on to the potential due to the ion core, and the electron–electron Coulomb interaction scaled by 0.7 to simulate the effects of hybridization with the ligands [19]. Table 11.1 shows the target state energies[1] (ϵ_i in equation (11.7)) calculated in this way [17], compared with the excitation energies from the EELS experiments [15, 16], and we see that agreement is satisfactory.

From these target states we next construct the channel functions. As these must be symmetry-adapted to the octahedral environment, the channel functions acquire rather a lot of subscripts and superscripts, $\bar{\Phi}^{\Gamma}_{i(p_1)p_2 hl}$. Suffice it to say that the

---

[1] A_2, T_1, etc, label the irreducible representations of the octahedral group to which the different many-electron wave-functions belong, with subscript g indicating *gerade*, or even under inversion. The superscripts give $(2S + 1)$ (either 1 or 3), where S is the total spin.

Table 11.1. Ground state and excited states of the $3d^8$ electron configuration in NiO. The middle column shows excitation energies measured by EELS [15, 16], and the right-hand column energies calculated using the crystal field model [17]. Energies are in eV.

Symmetry	EELS	Theory
$^3A_{2g}$	0.00	0.00
$^3T_{2g}$	1.10	1.05
^1E_g	1.60	1.70
$^3T_{1g}$	1.70	1.75
$^1T_{2g}$	2.75	2.70
$^1A_{1g}$	2.81	2.80
$^3T_{1g}$	3.00	3.13
$^1T_{1g}$	3.55	3.28
^1E_g	...	4.06
$^1T_{2g}$	...	4.12
$^1A_{1g}$	...	7.04

superscript Γ gives the representation (P), spin (S) and parity (π) of the channel function as a whole, and the subscripts give the symmetries of the target state and the scattering electron. Further details of the group theoretical aspects of the channel functions are given in the paper by Jones *et al* [13]. The one-electron functions $u_j(r)$ are calculated with an effective potential, as in equation (11.8), and from these and the channel functions we calculate the $(N + 1)$-electron functions ψ_k^Γ—note that the superscript Γ is the same as that on the channel functions in terms of which ψ_k^Γ is expanded (equation (11.10)). From the $\bar{\Phi}_{i(p_1)p_2hl}^\Gamma$ and ψ_k^Γ, with their energy eigenvalues E_k, we can finally calculate the R-matrix (equation (11.16)).

From the wave-function outside the R-matrix sphere, where the electron satisfies the free-electron Schrödinger equation with the boundary condition given by equation (11.15), we can find the t-matrix; this gives the amplitude f for scattering between an incident electron with momentum $\mathbf{p}$ and the target in state i, and a scattered electron with momentum $\mathbf{p}'$ and target in state i' [13],

$$f(\mathbf{p}', i' \leftarrow \mathbf{p}, i) \propto \langle \mathbf{p}', i'|t|\mathbf{p}, i\rangle. \tag{11.18}$$

We can then find the various scattering cross-sections, and we find for example that the total cross-section averaged over incident electron angles and target spins for scattering from target state i to i' is given by

$$\bar{\sigma}(i' \leftarrow i) = \frac{4\pi^3}{p_i^2} \sum_{PS} \sum_{p_2hl;p_2'h'l'} \frac{N_P(2S+1)}{2N_i(2S_i+1)} \left| \left\langle \left\langle \bar{\Phi}_{i'(p_1')p_2'h'l'}^\Gamma |t| \bar{\Phi}_{i(p_1)p_2hl}^\Gamma \right\rangle \right\rangle \right|^2. \tag{11.19}$$

Here p_i is the momentum of the incident electron, and the summations are over the quantum numbers P and S of the compound channel functions, and over the quantum numbers of the scattering electron before and after scattering. N_i and N_P

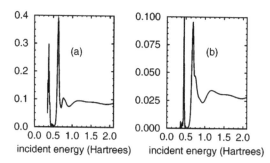

Figure 11.2. Averaged total cross sections in inelastic electron scattering from Ni^{2+} in NiO, as a function of incident electron energy. (a) $^3A_{2g} \to {}^3T_{2g}$, energy loss = 1.05 eV; (b) $^3A_{2g} \to {}^1T_{2g}$, energy loss = 2.70 eV. Atomic units are used so that the cross-section on the y-axis is measured in (Bohr radius)^2. (From Michiels *et al* [17].)

are the degeneracies of the target state and the channel state in their representations p_1 and P, and $(2S_i + 1)$ and $(2S + 1)$ are the corresponding spin degeneracies. The differential cross sections can also be found, but as the expressions are even more complicated than equation (11.19) I shall not give them.

Results for the total cross section as a function of the energy of the incident electron energy are shown in figure 11.2 for two typical 3d–3d excitations [17]. The inelastic cross-sections are several orders of magnitude smaller than the cross section for elastic scattering, which is what is found in the EELS experiments. There is some interesting physics in these results: below an incident electron energy of ≈ 0.7 a.u. there is resonant structure, which is due to the coupling of the 3d target with $l = 1$ and 2 angular momentum components of the scattering electron. Above this energy, in the energy range where the experiments are carried out, the cross sections are relatively featureless. In general the multiplicity-changing transitions (that is, spin-triplet to spin-singlet) are weaker than the triplet-triplet transitions; this is because triplet-singlet transitions only involve the exchange scattering, whereas triplet-triplet involve both direct and exchange scattering mechanisms [17].

The differential cross-sections are very different for the different transitions [13]. Figure 11.3 shows polar plots of the differential cross-sections with an incident electron energy of 20 eV, for elastic scattering (left-hand plot) and inelastic scattering for the $^3A_{2g} \to {}^1E_g$ transition at an energy loss of 1.70 eV (right-hand plot). We see that elastic scattering is mainly in the forward direction, whereas the inelastic scattering—much weaker than elastic—looks much more like a d orbital in this case. It's these differential cross sections which really show what the R-matrix calculations are capable of, and show that they are in a real sense equivalent to many-electron embedding.

When we compare calculation with experiment we must remember that these results show the scattering of a single ion in the NiO environment; in reality, multiple scattering occurs with a single inelastic scattering event sandwiched between many elastic scattering events. Moreover, this calculation only describes intra-atomic excitations, whereas inter-atomic excitations can occur, with charge transfer excitations across the band gap. Nevertheless, agreement with experiment is quite good, as discussed in the papers by Michiels *et al* [17] and Jones *et al* [13].

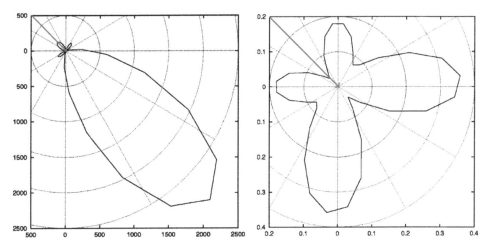

Figure 11.3. Polar plot of spin-averaged differential cross-sections in scattering from Ni^{2+} in NiO, at an incident electron energy of 20 eV; the red arrows indicate the direction of the incident electron, with the arrow tip at the atom. Left-hand plot, elastic scattering, $^3A_{2g} \rightarrow {}^3A_{2g}$; right-hand plot, inelastic scattering, $^3A_{2g} \rightarrow {}^1E_g$, energy loss = 1.70 eV. (From Jones *et al* [13].)

11.2 Resonant states

In our earlier discussion of resonances in section 2.6 we considered the analytic properties of $G(E)$, the Green function as a function of complex energy E, but it can be more useful to consider $G(k)$, with the wave-vector k as the complex variable. In some geometries $G(k)$ is a meromorphic function with simple poles [6], and $G(k)$ can be found using the Mittag-Leffler theorem [20, 21] in terms of the residues at the poles, without any cut contribution as we found to be necessary when we considered $G(E)$ (section 2.6.1).

The pole structure of $G(k)$ for the the one-dimensional square well, for example, which has been known for a long time [1, 7], is shown schematically in figure 11.4. The first quadrant of the complex-k plane corresponds to outgoing and decaying waves, and with $E = k^2/2$ this maps to the upper half of the complex E-plane, on the physical sheet (figure 2.8). The poles on the positive $\Im k$-axis in figure 11.4 (black circles) correspond to bound states of the system, with negative energy, and there are no other poles in this quadrant. However, there are poles in the fourth quadrant (positive $\Re k$, negative $\Im k$) corresponding to resonances (red circles), and which map on to the resonance poles in the complex E-plane (figure 2.10). Each pole with positive $\Re k$ (outgoing resonances in the nomenclature of Lind [7]) is matched by one at negative $\Re k$ (green triangles, incoming resonances), so that if there is a pole at $k = \kappa - i\gamma$ there is another at $k = -\kappa - i\gamma$. Finally, there are poles down the negative $\Im k$-axis (blue square), corresponding to *antibound* or *virtual* states as they are called.

To show how we can calculate the poles and residues of $G(k)$ using embedding, we consider the s-wave Green function of the spherically symmetric square well, using the same parameters as in section 2.6.1 with $v = 1$ a.u., $r_s = 2$ a.u. We solve the same embedded matrix eigenvalue problem (equation (2.25)) with the basis set given by equation (2.52), but now we consider the two possible values of $k = \sqrt{2E}$ when we

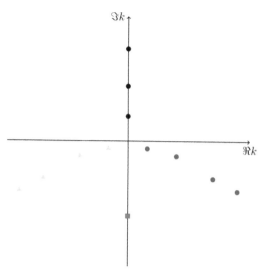

Figure 11.4. Poles of $G(k)$ (schematic): there are simple poles at the black circles (bound states), red circles (outgoing resonances), green triangles (incoming resonances) and the blue square (an antibound or virtual state).

evaluate Σ and $\partial\Sigma/\partial E$. Taking the root with positive $\Re k$, or positive $\Im k$ when $\Re k = 0$ (this is the Fortran convention [22]) gives us the bound states and the outgoing resonances; taking the other root automatically gives us any antibound or virtual states and the incoming resonances. We use exactly the same procedure that we used in section 2.6, starting off with a trial value of E and iterating on the eigenvalue. This works just as well with the antibound states and incoming resonances as it does with the more 'physical' states. (Of course these states can be found by a variety of methods [23], including elementary wave-function matching for the spherically symmetric square well [24].) The Green function can then be written using the Mittag-Leffler theorem as a sum over all the poles–incoming as well as outgoing resonances, antibound as well as bound states,

$$G(\mathbf{r}, \mathbf{r}'; k) = \sum_i \frac{R(\mathbf{r}, \mathbf{r}'; k_i)}{k - k_i}. \tag{11.20}$$

The residue $R(\mathbf{r}, \mathbf{r}'; k_i)$ has the same form as equation (2.58), with an additional factor of k_i in the denominator coming from differentiating $E = k^2/2$,

$$R(\mathbf{r}, \mathbf{r}'; k_i) = \frac{\phi_i(\mathbf{r})\phi_i(\mathbf{r}')}{k_i\left[\int_I \mathrm{d}\mathbf{r}\,\phi_i(\mathbf{r})^2 - \int_S \mathrm{d}\mathbf{r}_S \int_S \mathrm{d}\mathbf{r}'_S\,\phi_i(\mathbf{r}_S)\frac{\partial\Sigma}{\partial E}\Big|_{k=k_i}\phi_i(\mathbf{r}'_S)\right]}, \tag{11.21}$$

where $\phi_i(\mathbf{r})$ is the eigenvector of the matrix equation corresponding to the pole at k_i.

We now calculate the Green function in the square well using the Mittag-Leffler expression (11.20). To find the poles and residues we solve the embedded matrix eigenvalue equation with 30 basis functions defined with $D = 2.5$ a.u. Figure 11.5

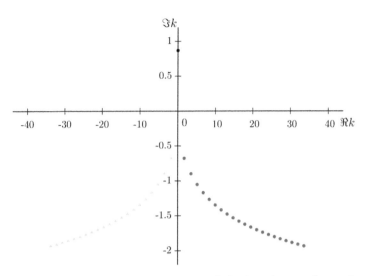

Figure 11.5. Poles of $G(k)$ for s-waves in a spherical square well, depth $v = 1$ a.u., radius $r_s = 2$ a.u., calculated with 30 basis functions defined with $D = 2.5$ a.u. The black circle at $\Im k = 0.86856$ a.u. corresponds to the bound state, the red circles to outgoing resonances, and the green triangles to incoming resonances.

shows the resulting poles of $G(k)$ plotted in the complex k-plane, with the one bound state and outgoing and incoming resonances (for a square well with $r_s = 2$ a.u. there are no antibound states for $V < 2.65$ a.u.). We have plotted the first 21 resonances; these were all very simple to locate to a high degree of accuracy, using simple iteration, but going to higher resonances has turned out to be more difficult and will require further numerical work. As in section 2.6.1 we evaluate $g_\rho(E)$, the Green function integrated over a sphere of radius ρ (equation (2.59)), changing from the variable k in equation (11.20) to real energy E.

The results for $-\Im g_\rho(E)$ for sphere radii $\rho = 1.8$ a.u. and $\rho = 1.95$ a.u. are shown in figure 11.6, and we see that the agreement between the Mittag-Leffler results and the exact results is remarkable, especially for $\rho = 1.8$ a.u. We see that for the Green function integrated through the smaller radius, with 39 poles in total—the bound state + 19 resonances of each type—the results are extremely close. With the larger radius, almost the radius of the square well, there is some divergence at larger E with 39 poles, but by increasing the number of poles to 41 we again obtain excellent agreement with the exact result (not quite as good as with $\rho = 1.8$ a.u.). More work needs to be done to explore the behaviour of the Mittag-Leffler expansion as $\rho \to r_s$, or rather, the embedding surface on which the embedding potential determines the boundary conditions.

The functions $\phi_i(\mathbf{r})$ which appear in equation (11.21), the outgoing and incoming resonance wave-functions, with the wave-functions of any bound and antibound states, form a complete (actually over-complete) basis [7], which can be used as an expansion set to solve a different Schrödinger equation. The advantage of such a *resonant-state expansion* is that it provides a discrete basis in an open system, able to represent the continuum of states above the zero of energy. This has been

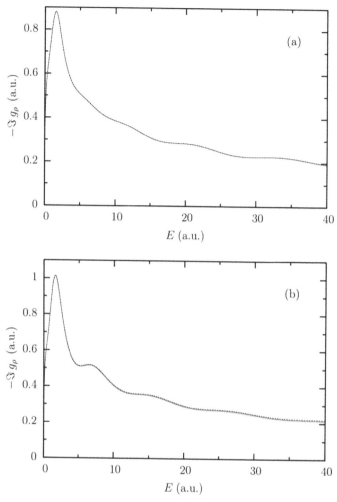

Figure 11.6. $-\Im g_\rho(E)$ for the spherical square well, $v = 1$ a.u., $r_s = 2$ a.u. (a) $\rho = 1.8$ a.u.: red curve, calculated using Mittag-Leffler with the bound state pole and 2×19 resonance poles; dashed black line, exact result. (b) $\rho = 1.95$ a.u.: green curve, calculated with the bound state pole and 2×19 resonance poles; red curve, calculated with the bound state pole and 2×21 resonance poles; dashed black line, exact result. The poles and residues are calculated with the embedded matrix eigenvalue equation using 30 basis functions, defined by $D = 2.5$ a.u. Note that here E is real energy.

applied to Maxwell's equations by Muljarov *et al* [8], who developed Brillouin–Wigner perturbation theory for open electromagnetic systems: the resonant states for a relatively simple dielectric object are first calculated, and then used as a basis to find the states for a perturbed system. The advantage of the method is that the basis of resonant states can be calculated once and for all, and then the states for the perturbed system can be found by solving a linear eigenvalue problem. There has been a series of impressive applications of the method, from planar dielectrics [9] to systems with cylindrical symmetry [25], and three-dimensional systems in which transverse magnetic and transverse electric modes have to be treated together [26]. To take the one-dimensional case, the resonant states of a single planar dielectric can be used to expand the states in a Bragg reflector consisting of dielectric layers, and from these states the Green function, hence the transmission properties can be found [9]. This work involves calculating hundreds of resonant states, and Doost [27] has made a very careful study of convergence with the number of states in the basis set.

My aim in this section has been to show the connection between embedding and resonances in open systems, in particular to show how resonant states can be found from the embedded matrix eigenvalue equation, and how the continuum Green function can be found from these states using the Mittag-Leffler theorem. In fact, for many open systems, such as electron states at a surface, we wouldn't attempt to use Mittag-Leffler with resonant states to find G because of the complicated analytic structure of Σ (though the resonant states and analytic structure might be interesting in themselves—as in [28]). In such cases inverting the embedded Schrödinger equation (section 2.5) is the only viable method of finding the Green function. But even in such cases we can study individual resonances in terms of the complex eigenvalues of the embedded matrix eigenvalue equation, either by analytic continuation of $\Sigma(E)$ in the complex energy plane, or simply by evaluating Σ on the real energy (or frequency) axis (c.f. Tables 9.4 and 9.6).

References

[1] Burke P G 2011 *R-Matrix Theory of Atomic Collisions* (Berlin: Springer-Verlag)

[2] Wigner E P and Eisenbud L 1947 Higher angular momenta and long range interaction in resonance reactions *Phys. Rev.* **72** 29–41

[3] Bloch C 1957 Une formulation unifiée de la théorie des réactions nucléaires *Nucl. Phys.* **4** 503–28

[4] Zou P F 1992 A regional embedding method *Int. J. Quantum Chem.* **44** 997–1013

[5] Siegert A J F 1939 On the derivation of the dispersion formula for nuclear reactions *Phys. Rev.* **56** 750–2

[6] More R M and Gerjuoy E 1973 Properties of resonance wave functions *Phys. Rev. A* **7** 1288–303

[7] Lind P 1993 Completeness relations and resonant state expansions *Phys. Rev. C* **47** 1903–20

[8] Muljarov E A, Langbein W and Zimmermann R 2010 Brillouin-Wigner perturbation theory in open electromagnetic systems *Europhys. Lett.* **92** 50050

[9] Doost M B, Langbein W and Muljarov E A 2012 Resonant-state expansion applied to planar open optical systems *Phys. Rev. A* **85** 023835

[10] Burke P G, Hibbert A and Robb W D 1971 Electron scattering by complex atoms *J. Phys. B: At. Mol. Phys.* **4** 153–61

[11] Burke P G and Berrington K A (ed) 1993 *Atomic and Molecular Processes: An R-matrix Approach* (Bristol: Institute of Physics Publishing)

[12] Bell K L, Berrington K A, Crothers D S F, Hibbert A and Taylor K F (ed) 1999 *Supercomputing, Collision Processes, and Applications* (New York: Kluwer Academic/Plenum Publishers)

[13] Jones P, Inglesfield J E, Michiels J J M, Noble C J, Burke V M and Burke P G 2000 *R*-matrix approach to low-energy electron energy-loss spectroscopy from NiO *Phys. Rev. B* **62** 13508–21

[14] Inglesfield J E 1999 Embedding and *R*-matrix methods at surfaces *Supercomputing, Collision Processes, and Applications* ed K L Bell, K A Berrington, D S F Crothers, A Hibbert and K F Taylor (Kluwer Academic) pp 183–95

[15] Gorschlüter A and Merz H 1994 Localized d - d excitations in NiO(100) and CoO(100) *Phys. Rev.* B **49** 17293–302

[16] Fromme B, Möller M, Anschütz Th, Bethke C and Kisker E 1996 Electron-exchange processes in the excitations of NiO(100) surface d states *Phys. Rev. Lett.* **77** 1548–51

[17] Michiels J J M, Inglesfield J E, Noble C J, Burke V M and Burke P G 1997 Atomic theory of electron energy loss from transition metal oxides *Phys. Rev. Lett.* **78** 2851–4

[18] Sugano S, Tanabe Y and Kamimura H 1970 *Multiplets of Transition-metal Ions in Crystals* (New York: Academic Press)

[19] Lynch D W and Cowan R D 1987 Effect of hybridization on $4d \rightarrow 4f$ spectra in light lanthanides *Phys. Rev.* B **36** 9228–33

[20] Jeffreys S H and Swirles B (Lady Jeffreys) 1972 *Methods of Mathematical Physics* 3rd edn (Cambridge: Cambridge University Press)

[21] Zeidler E 2004 *Oxford Users' Guide to Mathematics* (Oxford: Oxford University Press)

[22] Metcalf M, Reid J and Cohen M 2004 *Fortran 95/2003 Explained* (Oxford: Oxford University Press)

[23] Tolstikhin O I, Ostrovsky V N and Nakamura H 1997 Siegert pseudo-states as a universal tool: resonances, S matrix, Green function *Phys. Rev. Lett.* **79** 2026–9

[24] Nussenzveig H M 1959 The poles of the S-matrix of a rectangular potential well or barrier *Nucl. Phys.* **11** 499–521

[25] Doost M B, Langbein W and Muljarov E A 2013 Resonant state expansion applied to two-dimensional open optical systems *Phys. Rev.* A **87** 043827

[26] Doost M B, Langbein W and Muljarov E A 2014 Resonant-state expansion applied to three-dimensional open optical systems *Phys. Rev.* A **90** 013834

[27] Doost M B 2014 *Resonant State Expansion Applied to Open Optical Systems* PhD thesis (Cardiff University)

[28] Eden R J and Taylor J R 1964 Poles and shadow poles in the many-channel S matrix *Phys. Rev.* **133** 1575–80

Lightning Source UK Ltd.
Milton Keynes UK
UKHW05n0614180618
324301UK00003B/62/P